LAND USE - ASSESSING THE PAST, ENVISIONING THE FUTURE

Edited by **Luís Carlos Loures**

Land Use - Assessing the Past, Envisioning the Future
http://dx.doi.org/10.5772/intechopen.75839
Edited by Luís Carlos Loures

Contributors

Rui Alexandre Castanho, Sérgio Antonio Neves Lousada, José Manuel Naranjo Gómez, Patrícia Carlota Escórcio, José Cabezas, Luis Fernández-Pozo, Luís Carlos Loures, Jon Burley, Rüya Yilmaz, Chung Qing Liu, Raed Najjar, Zhen Wu, Zhi Yue, Chun-Hua Guo, Yiwen Xu, Na Li, Stefan Hörtenhuber, Michaela C. Theurl, Gerhard Piringer, Werner Zollitsch, Nararuk Boonyanam, Isaac Oluwatayo, Daniela Manuschevich, Patchareeya Chaikaew, Huan Vu Duc, Chalermphol Chaikaew, Meg Sherval, Carlos Nobre, Ismael Nobre, Maria Ilhéu, Paula Matono, Elsa Paula Morgado Sampaio, Teresa Batista

First published in London, United Kingdom, 2019 by IntechOpen
IntechOpen is the global imprint of INTECHOPEN LIMITED, registered in England and Wales, registration number: 11086078, The Shard, 25th floor, 32 London Bridge Street
London, SE19SG – United Kingdom
Printed in Croatia

British Library Cataloguing-in-Publication Data
A catalogue record for this book is available from the British Library

Additional hard copies can be obtained from orders@intechopen.com

Land Use - Assessing the Past, Envisioning the Future, Edited by Luís Carlos Loures
p. cm.
Print ISBN 978-1-78985-703-0
Online ISBN 978-1-78985-704-7

For EU product safety concerns:
IN TECH d.o.o., Prolaz Marije Krucifikse Kozulić 3, 51000 Rijeka, Croatia,
info@intechopen.com or visit our website at intechopen.com.

Meet the editor

Luís Loures is a Landscape Architect and Agronomic Engineer, who holds a PhD in Planning and a Post-Doc in Agronomy. Since graduating, he has published several peer-reviewed papers and been a guest researcher and lecturer both at Michigan State University (USA) and the University of Toronto (Canada) where he has developed part of his Ph.D. research with the financial support from the Portuguese Foundation for Science and Technology (Ph.D. grant). During his academic career, he has taught different courses at several Universities, mainly regarding the fields of landscape architecture, urban and environmental planning, and sustainability. Currently, he is a researcher at VALORIZA, Research Center for Endogenous Resource Valorization - Polytechnic Institute of Portalegre.

Contents

Preface

Landscape planning is generally described as a multi-disciplinary field that incorporates several branches of knowledge, considering at the same time science, technology, and art. In this scenario, each and every single landscape and planning specialist should be perceived as someone who is able to promote the definition of land use, thus enabling landscape alterations that should ensure sustainable development, while protecting the environment, preserving natural and cultural assets, and improving people's quality of life. Unfortunately, sometimes this is not true.

In this regard, the transformation of landscapes worldwide has raised global concerns, increasing the need to rethink landscape planning and protect the environment. This is especially true for previously developed areas that are now abandoned or underused. Instead of consuming green lands, the brown and grey lands need to be redeveloped and given new life, achieving a more sustainable urban setting. In this sense, although landscape reclamation plays a very important role for societal development, the continuous use of green fields continues to have deep economic, social, and ecological impacts that require special attention. The new environmental paradigms associated with globalization, progressive climate change, and increasing food production needs will intensify the entropy and the instability in most of the existing natural land. This reality creates the perfect momentum to assess these issues.

The present book aims to highlight the opportunities and challenges associated with the development of new sustainable landscapes, considering current and future challenges related to land use change, planning, and development, considering not only the different sustainability pillars, but also the impacts these changes might have in each one of them.

This book covers a wide range of research domains and issues associated with land use change and redevelopment such as public involvement, landscape quality assessment, land use resilience, land policies, urban planning, and landscape reclamation, among others. The book covers a wide range of topics related to land use change and planning, assessing the impact of contemporary needs and constraints and landscape management strategies both on planning, ecosystem, and landscape design.

As a landscape architect and agronomic engineer with research interests deeply related to the field of sustainability, landscape planning and design, my main investigation goals are directly connected to fitting design to the needs and desires of contemporary life, addressing in equal measure society, the natural landscape, heritage and culture, and economic issues. I believe landscape planning is increasingly missing a vision towards future planning processes that differ from many contemporaries in its philosophical grounding in the social as well as creative matrices, calling for a comprehensive view of the different components of landscape design, acknowledging the need for an interrelated analysis of the ecological, cultural and socioeconomic issues in planning and design processes.

Luís Carlos Loures
Polytechnic Institute of Portalegre, Portugal

Section 1

Introduction

Chapter 1

Introductory Chapter: Land-Use Planning and Land-Use Change as Catalysts of Sustainable Development

Luis Carlos Loures

Additional information is available at the end of the chapter

http://dx.doi.org/10.5772/intechopen.84520

1. Land-use planning: paradigm changes and future perspectives—a brief overview

Land-use change has often been one of the main drivers of economic growth, social change and innovations of the government. For this reason, as mentioned by Magalhães [1], the analysis and comprehension of the processes, which throughout time, influenced landscape form and patterns (and thus land use), constitute an essential feature for those aiming to work in and with it. Thus, this subject has been widely addressed considering not only the historical role of cities but also the problem that land-use change had caused throughout time ([2–6]). Still, the analysis of land-use change is generally associated to the impacts of growth, and to the implications it had on environmental, economic and social development dimensions ([7–12]). In fact, the environmental movement marked somehow by the publication of the book *Silent Spring* by Rachel Carson in 1962 may be considered a good example of this association, not only because Carson's book exposed the negative environmental impacts of land-use change considering the unchecked impact of industrial development both on natural ecosystems and human health, but also because the conversion of natural land into urbanized one started to be viewed as a possible threat to future of the planet.

This movement, which had a great impact in terms of land-use planning, gained a special momentum in 1969, the year of the first 'Earth Day', which revealed the environment to be a powerful political issue. It was the year of the formation of the U.S. Environmental Protection Agency (EPA), which enabled a wide range of laws to control existing and potential threats to the environment, thus affecting land use; and it was the publication year of the book *Design with Nature* by the landscape architect Ian McHarg, which according to Andresen [13] introduced the direct application of ecological principles in planning.

Since then, several steps were taken in order to mitigate land-use conflicts representing a response to different environmental paradigms. Still, according to Saraiva [14], since the beginning of the environmental movement, new concepts have emerged, including different variables into the existing models, enabling the creation of new environmental paradigms which may be divided in three phases (**Figure 1**): the first regarding *Environmental Protection* during the 60s and 70s—in which the imposition of limits to economic growth and to pollution were the main concern; the second regarding *Resource Management* during the 1970s and 1980s—considering the unmeasured consumption of natural resources; and the third during the 1980s and 1990s regarding *Sustainable Development*, and the need to consider social, economic and environmental aspects in development policies, taking into account environmental preservation in a way human needs can be met not only in the present but also for future generations.

These paradigms were and continue to be important steps in order to solve or reduce most of the land-use problems created during the last century.

Commoner ([15], cited by Lyle [16]) argues that the main problem lies in our means of production and that in order to solve environmental land-use problems, we need to change not only the location of certain activities but also the ways of making things. As it has been expressed, understanding this phenomenon is perhaps one of the most relevant consequences of assessing the history of land-use development (especially since industrial revolution), given that it becomes simpler not only to comprehend the current state of the art as it applies to us, but also to envision possible solutions for present and future problems ([17–21]; Loures [22]). In a period when cities have become places of diversity and contrast, of abundant wealth and abject poverty [47], of opportunity and threat [48], places where beauty and ugliness lie in close proximity and where the future collides with the past [23], it is increasingly necessary to understand its processes and the problems inherent to them, which are now substantially different from what they were in the beginning, and which are directly dependent on land-use change and evolution [24, 25], progressively moving from liner planning strategies to circular planning strategies (**Figure 2**) in which land use is defined considering not only present solutions but also landscape resilience, bearing in mind that imperatively, humanized landscapes are all transitional places.

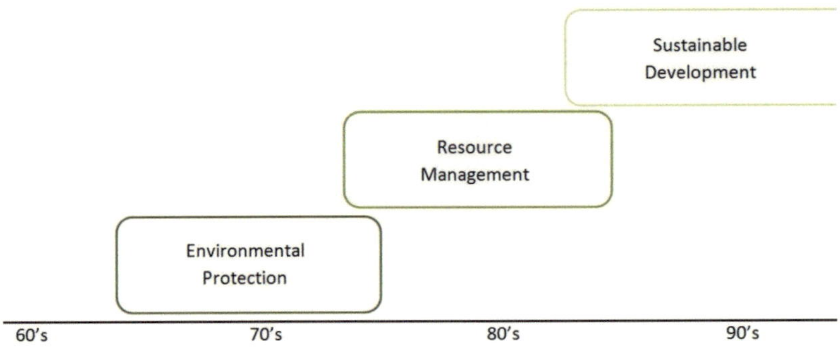

Figure 1. Evolution of the environmental paradigms. Source: Loures [10].

http://dx.doi.org/10.5772/intechopen.84520

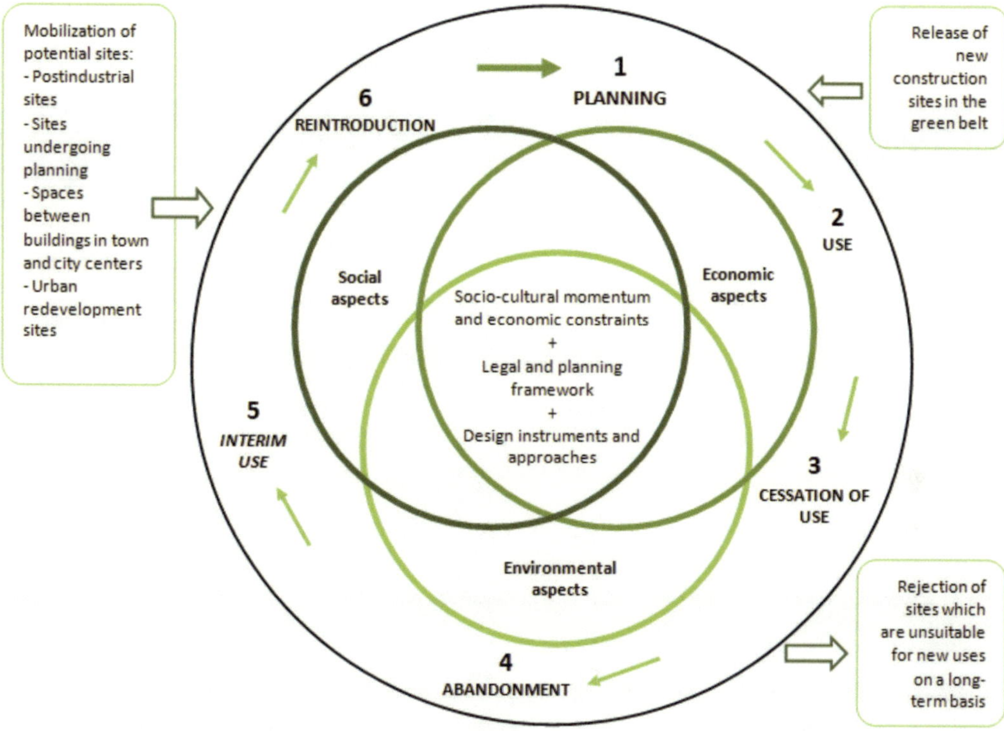

Figure 2. Circular land use management plan—Loures [10].

2. Sustainable land-use planning—from growth to development

It is a given that land-use change provides constant new opportunities for those who have the desire and the ability to seize landscape, regardless of their nature [21, 26–29]. In this regard, land-use change and planning are considered to be a significant resource for achieving sustainable development [30–33], contributing as well to improve life's quality. Nonetheless, it needs to be thought in terms of the town-nature reconnection, considering previously developed ecological and sustainability theories and principles. Indeed, future land-use planning needs to be redeveloped in an integrated multifunctional way, emphasizing the fact that envisioned planning alternatives should not only offer different multipurpose uses, in order to be more attractive and viable, but also incorporate sustainability (considering socio-cultural, economic and environmental and aesthetic dimensions) at various levels, from national and regional planning to individual construction sites.

Land use needs to be thought in terms of *sustainability* and/or *sustainable development*, terms that get used a lot these days, and which since their appearance have been faced as new development paradigms introduced in land-use matters, merging social, economic and environmental 'dimensions' [34], and putting nations to work together in the definition of new principles and frameworks towards sustainable development (**Figure 3**).

Figure 3 . Schematic overview of some conferences and publications regarding sustainable development issues. Source: Loures [10].

In this regard, the application of sustainability principles to land-use change and land-use planning, the concept of sustainable development suggests that growth must occur, but that it must be quality growth. Still, encouraging growth while reducing resource consumption, is according to North [35] in fundamental opposition. This idea, though acceptable, is arguably not totally correct, for example, under the scope of landscape redevelopment, is arguably not totally correct, once, while fostering growth, greyscape redevelopment (considered here as an alternative to greenfield development) reduces land consumption. Growth does not always have to mean 'new or more', the problem is that the relatively vague construction of the concept, depending on the context or purpose of its use, turned it into a kind of 'catch-all' term that now refers to almost anything [36].

Still, questions such as: How to achieve sustainability? How to measure it? And how does sustainable development improve current design and planning practices and *vice-versa*? continue to nourish the discussion around the concept of sustainability ([35, 37]; Manta-Conroy [38]), indicating that there is still a poor understanding of what it means [39], and of its influence in current landscape use and planning practices. This happens not only because sustainability may be interpreted in two ways: the first refers to landscape conservation, regarding the continuation of practices that maintain and organize it; the second, to the idea of perceiving sustainability as a main principle for future land use, considering the potential landscapes have to enhance sustainability [40], but also because the notion of a sustainable landscape development may involve a contradiction based on the fact that landscapes evolve somehow in a more or less chaotic way as a reflection of social and economic needs [40]. In this way, as mentioned by Potschin and Young ([41], p. 157) landscapes *may contribute to sustainability, but they are not sustainable in themselves*.

Considering this, at the planning level, the idea expressed by the Portuguese architect Nuno Portas [42] may represent in some way a vision of what planners and designers might understand by sustainable land-use planning. According to Portas, in order to be sustainable, once we are living in a period of great uncertainties, planning and design decisions should be flexible, leaving space for possible amendments. He goes further and states: *for example, I am not thankful to Le Corbusier, by determining, in the beginning of the twentieth century, how modern cities should be. Today we know enough to say that the cities He envisioned were not good*. Sustainability land-use planning means in this way, the capacity to develop resilient landscapes, that is, landscapes with the capacity to absorb disturbance and reorganize while undergoing change so as to retain essentially the same function, structure, identity and feedbacks ([43, 44]), including also the capacity to recover from management mistakes.

These ideas are in some regard connected with the vision highlighted by the Dutch architects Rem Koolhaas and Adrian Geuze according to which planning and design should be faced as an opportunity to sketch out a future development without entirely fixing it [45], creating landscapes with the potential to change and evolve in accordance to still unknown usages. In fact, the integration of sustainability in planning and design processes represents a paradigm shift to the extent that it reflects not only changes in the manner in which development is planned, but also in the organization of the socio-cultural and economic mechanisms that control and implement planning, and in the role of the community in those land-use planning processes. This new reality enabled in part, by significant economic changes, led urban planning and design to another level, in which places and people acquired an increasing significance in economic redevelopment (**Figure 4**). Though, these efforts operate in contradiction to factors such as an increasing population, and a growing use of resources, many of them non-renewable [46].

Even if throughout recent years several normative theories regarding sustainable land use, considering both design and planning principles towards sustainable communities, were created, defining not only the ways in which land-use planning should be envisioned but also the ways in which new developments should be created, the answer to this question is far from being achieved. From an overall viewpoint, sustainable land use represents a subject of real sustainable dimensions, given that it tackles environmental, social and economic issues, which are the main dimensions of sustainability.

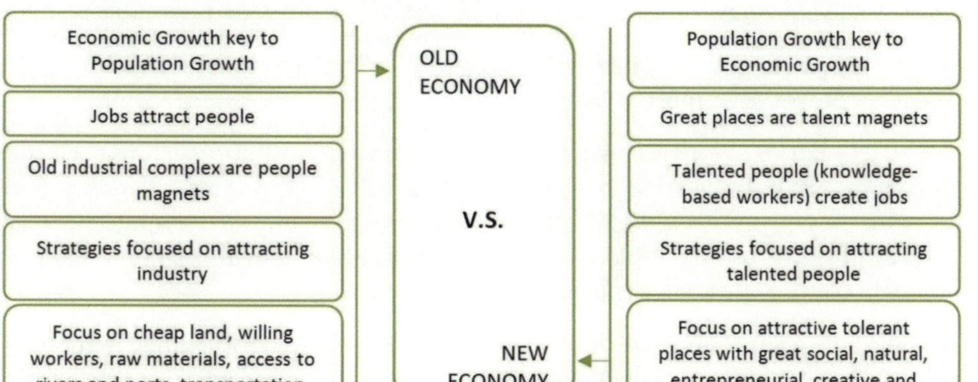

Figure 4 . Economic paradigm shift—old economy versus new economy. Source: Loures [10].

The present book considers a set of subjects which highlight the diverse nature of the scientific domains associated to land-use planning, emphasizing the need to acknowledge not only that environmental land use is not sufficient, but also that the contribution of each sustainability pillar is equally important, offering complementary development opportunities, while enabling landscapes to fulfil multiple functions in an integrated way, underlining the relevance of multifunctionality to promote sustainable land-use, planning strategies and policies.

Author details

Luis Carlos Loures

Address all correspondence to: lcloures@gmail.com

VALORIZA–Research Center for Endogenous Resource Valorization, Polytechnic Institute of Portalegre, Portugal

References

[1] Magalhães M. A Arquitectura Paisagista: Morfologia e Complexidade. Lisboa: Editorial Estampa; 2001

[2] Jacobs J. The Economy of Cities. New York: Vintage Books; 1970. pp. 3-48

[3] Loures L. Postindustrial landscapes as drivers for urban redevelopment: Public versus expert perspectives towards the benefits and barriers of the reuse of postindustrial site. Habitat International. 2015;**45**:7281

[4] Lynch K. A Theory of Good Urban Form. Cambridge: MIT Press; 1981

[5] Rae D. City: Urbanism and Its End. New Haven: Yale University Press; 2005

[6] Mumford L. The City in History: Its Origins, Its Transformations and Its Prospects. London: Secker & Warburg; 1961

[7] Loures L, Burley J. Post-industrial land transformation–An approach to sociocultural aspects as catalysts for urban redevelopment. In: Advances in Spatial Planning. Jaroslav Burian: IntechOpen. DOI: 10.5772/36380. Available from: https://www.intechopen.com/books/advances-in-spatial-planning/post-industrial-land-transformation-an-approach-to-socio-cultural-aspects-as-catalysts-for-urban-red

[8] Loures L. Industrial heritage: A gear to redevelopment. In: Proceedings of the EURAU 08—Cultural Landscape, 4th European Symposium on Research in Architecture and Urban Design; 16-19 January 2008; Madrid, Spain. pp. 1-7; 2008

[9] Loures L. Industrial heritage: The past in the future of the city. WSEAS Transactions on Environment and Development. 2008;**4**(9):784-793

[10] Loures L. Planning and Design in Post-industrial Land Transformation: East bank Arade river, Lagoa–case study. Faculdade de Ciências e Tecnologia, Universidade do Algarve, Faro, Portugal—Dissertação de Doutoramento em Planeamento Urbano; 2011

[11] Loures L, Panagopoulos T, Burley J. Assessing user preferences on post-industrial redevelopment. Environment and Planning. B, Planning & Design. 2016;**43**(5):871-892

[12] Loures L, Crawford P. Democracy in progress: Using public participation in post-industrial landscape (re)-development. WSEAS Transactions on Environment and Development. 2008;**4**(9):794-803

[13] Andresen T. Para a Crítica da Paisagem [Doctoral dissertation]. Aveiro: Universidade de Aveiro; 1992

[14] Saraiva M. O Rio Como Paisagem. Lisboa: Fundação Calouste Gulbenkian e Fundação para a Ciência e a Tecnologia; 1999

[15] Commoner B. Making Peace with the Planet. New York: Pantheon Books; 1990

[16] Lyle J. Regenerative Design for Sustainable Development. New York: John Wiley and Sons; 1994

[17] Loures L, Burley J, Panagopoulos T. Postindustrial landscape redevelopment: Addressing the past, envisioning the future. International Journal of Energy and Environment. 2011;**5**(5):714-724

[18] Loures L, Panagopoulos T. Sustainable reclamation of industrial areas in urban landscapes. In: Kungolas A, Brebbia C, Beriatos E, editors. Sustainable Development and Planning III. Southampton: WIT Press; 2007. pp. 791-800

[19] Loures L, Panagopoulos T. From derelict industrial areas towards multifunctional landscapes and urban renaissance. WSEAS Transactions on Environment and Development. 2007;**3**(10):181-188

[20] Loures L, Horta D, Santos A, Panagopoulos T. Strategies to reclaim derelict industrial areas. WSEAS Transactions on Environment and Development. 2006;**2**(5):599-604

[21] Panagopoulos T, Loures L. Reclamation of derelict industrial land in Portugal: Greening is not enough. In: Book of Abstracts of the 10th European Forum on Urban Forestry; 16-19 May 2007; Gelsenkirchen, Germany. 2007. pp. 71-727

[22] Loures L. Post-Industrial landscapes as renaissance locus - the case study research methods. In: Brebbia C, Gospodini A, Tiezzi E, editors. Southampton, Sustainable City V: WIT Press; 2008

[23] European Union. State of European Cities Report. 2007. Available from http://ec.europa.eu/regional_policy/sources/docgener/studies/pdf/urban/stateofcities_2007 [Retrieved: May 01, 2009]

[24] Loures L, Heuer T, Horta D, Silva S, Santos R. Multifunctional clusters in post-industrial landscapes: Rising from what's left. WSEAS Transactions on Environment and Development. 2008;**4**(8):619-628

[25] Loures L, Heuer T, Horta D, Silva S, Santos R. Reinventing the post-industrial landscape: A multifunctional cluster approach as redevelopment strategy. In: Panagopoulos T, Burley J, (Eds.). New Aspects of Landscape Architecture, Athens: WSEAS Press; 2008. pp.123-129

[26] Ferreira V, Panagopoulos T, Andrade R, Guerrero C, Loures L. Spatial variability of soil properties and soil erodibility in the Alqueva dam watershed, Portugal. Solid Earth Discussions. 2015;**7**:301-327

[27] Loures L, Loures A, Nunes J, Panagopoulos T. Landscape valuation of environmental amenities throughout the application of direct and indirect methods. Sustainability. 2015;**7**(1):794-810

[28] Loures L, Santos R, Panagopoulos T. Urban parks and sustainable city planning—The case of Portimão, Portugal. WSEAS Transactions on Environment and Development. 2007;**3**(10):171-180

[29] Lu D, Burley J, Crawford P, Schutzki R, Loures L. Quantitative methods in environmental and visual quality mapping and assessment: A Muskegon, Michigan watershed case study with urban planning implications. In: Urban Planning. INTech; 2011. pp.127-142

[30] Castanho R, Loures L, Cabezas L, Fernández-Pozo L. Cross-border cooperation (CBC) in Southern Europe—An iberian case study. The Eurocity Elvas-Badajoz. Sustainability. 2017;**9**:360

[31] Castanho R, Loures L, Fernandez J, Pozo L. Identifying critical factors for success in cross border cooperation (CBC) development projects. Habitat International. 2018;**72**:92-99

[32] Rato Nunes J, Ramos-Miras J, Lopez-Piñeiro A, Loures L, Gil C, Coelho J, et al. Concentrations of available heavy metals in mediterranean agricultural soils: A case study in typical mediterranean soil. Sustainability. 2014;**6**(12):9124-9138

[33] Vargues P, Loures L. Using geographic information systems in visual and aesthetic analysis: The case study of a golf course in Algarve. WSEAS Transactions on Environment and Development. 2008;**4**(9):774-783

[34] Singh P, Sharma A. Integrated approach to improve quality of life in urban distress areas sustainable urban regeneration. The International Journal of Environmental, Cultural, Economic and Social Sustainability. 2009;**5**(2):121-128

[35] North A. Sick of sustainability. In: Morris A, Yagod M, editors. Architecture, Landscape, and Design Annual. Toronto: Faculty of Architecture, Landscape and Design, and Graduate Architecture, Landscape and Design Student Union; 2007

[36] Stallworth H. The Economics of Sustainability. Washington DC: United States Environmental Protection Agency (US EPA), Office of Sustainable Ecosystems and Communities; 1997

[37] Bell S, Morse S. Measuring Sustainability: Learning by Doing. London: Earthscan; 2003

[38] Berke P. Manta-Conroy M. Are we planning for sustainable development? An evaluation of 30 comprehensive plans. Journal of the American Planning Association. 2000;**66**(1):21-33

[39] Pepperdine S. Social indicators of rural community sustainability: An example from the woady yaloak catchment. In: First National Conference on the Future of Australiahhy Towns; June 28-30 2000; Bendigo, Australia; 2000

[40] Antrop M. Sustainable landscapes: Contradiction, fiction or utopia? Landscape and Urban Planning. 2006;**75**(3-4):187-197

[41] Potschin M, Young R. Landscapes and sustainability. Landscape and Urban Planning. 2006;**75**:155-161

[42] Portas N. Nesta Época de Incertezas, Nossas Decisões Devem ser Flexíveis, Passíveis de Aperfeiçoamento. 2001. Available from http://www.arcoweb.com.br/entrevista/nuno-portas-nesta-epoca-19-04-2001.html [Retrieved: March 01, 2010]

[43] Berkes F, Colding J, Folke C. Navigating Social-Ecological Systems: Building Resilience for Complexity and Change. Cambridge: Cambridge University Press; 2003

[44] Holling C. Resilience and stability of ecological systems. Annual Review of Ecology and Systematics. 1973;**4**:1-23

[45] Borret K. The "Void" as a productive concept for urban public space. In: Ghent Urban Studies Team (GENT), editor. The Urban Condition: Space, Community and Self in the Contemporary Metropolis. Rotterdam: 010 Publisher; 1999. pp. 236-251

[46] Elkin T, McLaren D, Hillman M. Reviving the City: Towards Sustainable Urban Development. London: Friends of the Earth; 1991

[47] Loures L. Post-industrial landscapes: Dereliction or heritage? In: 1st WSEAS International Conference on Landscape Architecture and 4th IASME/WSEAS International Conference on Energy, Environment, Ecosystems and Sustainable Development; July 2008; Faro, Portugal; 2008

[48] Loures L, Panagopoulos T. Recovering derelict industrial landscapes in Portugal: Past interventions and future perspectives. In: Proceeding of the 3rd IASME/WSEAS the 3rd International Conference on Energy, Environment, Ecosystems and Sustainable Development; July 2007; Agios Nikolaos, Crete Island, Greece

Section 2

Land-Use Change Modelling and Monitoring Processes

Chapter 2

Dynamics of the Land Use Changes and the Associated Barriers and Opportunities for Sustainable Development on Peripheral and Insular Territories: The Madeira Island (Portugal)

Rui Alexandre Castanho, Sérgio Lousada,
José Manuel Naranjo Gómez, Patrícia Escórcio,
José Cabezas, Luis Fernández-Pozo and Luís Loures

Additional information is available at the end of the chapter

http://dx.doi.org/10.5772/intechopen.80827

Abstract

Considering the complex dynamics, patterns, and particularities that peripheral and insular territories/regions present—e.g., as the fragility, they show to achieve a sustainable development and growth—a study that analyzes the land uses of this territories is seen as pivotal to identify barriers and opportunities for a long-term sustained development. Contextually, a general analysis was carried out through case study research methods covering those territorial typologies of the insular territory of Madeira Island, Portugal. The study, which was carried out through GIS mapping tools, enabled us to identify the land use changes in the last decades over the territory—allowing to establish a relation and identification of the associated barriers and opportunities presented by the territories to face the emerging sustainable development challenges. The study reveals the evident limitations of "ultra-peripheral" territories not only by the physical spatial dimensions but also by the difficulty to reconvert land uses. Thus, the main actors and their policies over the territory are even more relevant and need to be conducted in a more reasonable way—considering the fragility of this regions; such actions present higher impact over the territory and over their inhabitants' life's quality standards and finally on the long-term sustainability.

Keywords: insular territories, peripheral regions, land use changes, sustainability, territorial dynamics

1. Introduction

The need for knowledge and information on the state of our planet's surface and its occupation has boosted several initiatives to study land uses and cover and their patterns and dynamics [1, 2]. Several sets of global or continental land cover data, most of them from the Earth's observation by satellite, were promoted and created, and there is a variety of different mapping standards [2].

The quantity of available products reflects the wide scope of interests. It is important to highlight the Global Land Cover (GLC2000) created for the year 2000 at a global level in Europe, the Pan-European Land Cover Monitoring (PELCOM) based in 1996 images, and the Coordination of Information on the Environment (CORINE) maps at regional and national levels [1].

In Europe, many efforts to quantify a standardized way of the land cover have been done. In this regard, the CORINE Land Cover (CLC) has been created and processed by the EEA based on the guidelines for "land and ecosystems" of the System of Environmental and Economic Accounting (SEEA), and it is used by many of the organizations [3].

Since 2006, in several countries, such as Germany, Austria, the United Kingdom, Sweden, Switzerland, and others, the map is obtained from generalization techniques from national maps with greater detail than the traditional photo satellite interpretation [4]. These different methods conduct to heterogeneity in the land cover maps, which have been a discussion topic [4]; nevertheless, the different ways of producing CORINE maps have been used to analyze soil applications [1].

The land use and land cover maps can play an important role in the balance of the socioeconomic, political, cultural, and environmental factors of a certain territory [5]. In fact, they allow analyzing significant changes in the landscape, study cycles, and trends. Several studies have been conducted in the European territories concerning land uses and their patterns and dynamics; however, in relation to the case of peripheral and insular territories/regions as is the case of the Autonomous Region of Madeira (RAM) (Portuguese Island), such typology of studies has not been carried out—increasing the relevance of the work toward a better understanding of the territorial dynamics, barriers, and opportunities for a sustainable growth and development.

Contextually, insular territories are affected by their geographic position, which gives them a high degree of isolation and their small dimensions (spatial constraints), and so they represent a specific challenge and fragility in the face of changes [6]. This typology of territory is affected directly and indirectly by the proximity to the sea and is considered a coastal territory.

In this sense, the territorial planning is a fundamental instrument to attribute conditions of prosperity to its inhabitants and consequently to future generations, promoting the mitigation of social inequalities and spatial imbalances, as well as a catalyst tool for sustainable development. In this context, the sustainable development allows not only to respond to the problems discussed above but also to create opportunities and more competitive territories.

Facing today's society and its demands, territorial planning must inevitably consider its future, and it should be constructed in an organized way to satisfy the public needs and not be dictated by a casuistic and uncontrolled evolution from the point of political and/or individual interests. Thus, sustainable development and growth are undoubtedly the main concerns and objectives of the regional territories [7–10].

With the Brundtland report [11], the sustainability has become a worldwide concern, since we are stakeholders in the process. Issues like the meaning or how to measure sustainability, which strategies should be implemented, have been studied [12, 13]. Although there is a high interrelation between the economic, social, and environmental dimensions, in practice they are considered separately, which can lead to non-sustainable trends [12]. Limited land and water resources make the insular territories a case, where the harmonization of the different dimensions of sustainability is a challenging process since it can lead, for example, to the degradation of the natural habitats [12, 14].

The overall objective of this study is to analyze and assess the land use changes in peripheral or "ultra-peripheral" and insular territories—i.e., islands, through a practical approach to a case study—the Madeira Island, Portugal. Moreover, through the understanding of the land use changes and consequently the territorial dynamics and tendencies, barriers and opportunities for a sustainable growth and development will be explored and addressed.

2. Material and methods

The present study is based essentially on CORINE Land Cover (CLC). The CLC is a vector map with a scale of 1:100000, a minimum cartographic unit (MCU) of 25 ha, and a geometric accuracy better than 100 m. It maps homogeneous landscape patterns, i.e., more than 75% of the pattern has the characteristics of a given class from the nomenclature. This nomenclature is a three-level hierarchical classification system and has 44 classes at the third and most detailed level (**Table 1**). To deal with areas smaller than 25 ha, a set of generalization rules were defined [15].

In this regard, the years of 1990, 2000, 2006, and 2012 were analyzed through direct and indirect tools and methods. Thus, exploratory tools were used as is the case of GIS tools, CLC, or the site analysis conducted by the authors. Moreover, a literature review has been performed in order to properly describe, discuss, and understand the obtained results—the land use change dynamics in Madeira Island.

Nevertheless, later in the present chapter, these methods will be exposed and further developed.

2.1. Case study: the Madeira Island

The Madeira Archipelago is located in the North Atlantic Ocean. Covering an area of 802 km^2, the Madeira Archipelago is composed of the following islands: Madeira (742 km^2) (**Figure 1**),

Level 1	Level 2	Level 3
1 Artificial surfaces	11 Urban fabric	111 Continuous urban fabric
		112 Discontinuous urban fabric
	12 Industrial, commercial, and transport units	121 Industrial or commercial units
		122 Road and rail networks and associated land
		123 Port areas
		124 Airports
	13 Mine, dump, and construction sites	131 Mineral extraction sites
		132 Dump sites
		133 Construction sites
	14 Artificially and nonagriculturally vegetated areas	141 Green urban areas
		142 Sport and leisure facilities
2 Agricultural areas	21 Arable land	211 Nonirrigated arable land
		212 Permanently irrigated land
		213 Rice fields
	22 Permanent crops	221 Vineyards
		222 Fruit trees and berry plantations
		223 Olive groves
	23 Pastures	231 Pastures
	24 Heterogeneous agricultural areas	241 Annual crops associated with permanent crops
		242 Complex cultivation patterns
		243 Land principally occupied by agriculture, with significant areas of natural vegetation
		244 Agroforestry areas
3 Forest and seminatural areas	31 Forests	311 Broad-leaved forest
		312 Coniferous forest
		313 Mixed forest
	32 Scrub and/or herbaceous vegetation associations	321 Natural grasslands
		322 Moors and heathland
		323 Sclerophyllous vegetation
		324 Transitional woodland shrub
	33 Open spaces with little or no vegetation	331 Beaches, dunes, and sands
		332 Bare rocks
		333 Sparsely vegetated areas
		334 Burned areas
		335 Glaciers and perpetual snow

Level 1	Level 2	Level 3
4 Wetlands	41 Inland wetlands	411 Inland marshes
		412 Peat bogs
	42 Maritime wetlands	421 Salt marshes
		422 Salines
		423 Intertidal flats
5 Water bodies	51 Inland waters	511 Water courses
		512 Water bodies
	52 Marine waters	521 Coastal lagoons
		522 Estuaries
		523 Sea and ocean

Source: http://www.igeo.pt/gdr/pdf/CLC2006_nomenclature_addendum.pdf

Table 1. CLC nomenclature.

Figure 1. Madeira Archipelago—Madeira Island, Portugal.

Porto Santo (43 km^2), Desertas (14 km^2), and Selvagens (3 km^2). The main features of Madeira Island will be exposed in the following (**Table 2**).

Madeira Archipelago presents particular conditions for the occurrence of potential natural disasters—i.e., wildfires; the high exposition of urban areas to natural disasters (rugged terrain promotes vertical impulsion of maritime tropical air masses coming from the southwest); free surface flow's fast convergence into the river channels and the high drainage density levels (floods); deeply changed volcanic geological substrate and consequently less permeable; and embedded V-shaped valleys, enabling greater interaction between landslides and river patterns, among many other extreme phenomena [19–21]. Also, the specific and rough geomorphology of Madeira Island occupies most of the land cover—i.e., 120 watersheds occupying almost the entire territory (741 km^2) correspond to 40% of the total island surface [21, 23].

Physical features	
Average altitude	646 m
Highest peak	Pico Ruivo (1862 m)
Average slope	56%
Perimeter	177.3 km
Area	742 km^2
Predominant soils	Andosols (42%)
Average daily temperature: Maximum (August) Minimal (February)	23 °C 5.4 °C
Prevailing winds: Direction Maximum average velocity (and direction)	N-NE 30 km/h (S-SW)
Weighted average annual precipitation	1628 mm
Population	256,424 (inhabitants)

Table 2. Main features of Madeira Island [16–22].

Moreover, the high human pressure under the territory should be also considered. In fact, in these particular territories, human activities and densities are more critical for the success or failure of a sustainable development and growth—considering the limitations presented by these "ultra-peripheral" territories.

3. Results and discussion

The results come from the analysis of the land use changes for RAM in the years 1990, 2000, 2006, and 2012. The results will be exposed through the graphs and tables. This typology of results exposed allows to extract the most relevant information and to characterize the evolution of land use based on the 44 uses of the soil determined by CLC. The information is organized as presented from **Tables 3–7**, in percentage.

At **Table 3**, it is possible to analyze the behavior of the artificial surfaces; by far the highest values have been found on the land use 112 (discontinuous urban fabric), which also increased over the years (where the tendency is located). The second most representative land use, considering artificial surfaces, is for the uses 121, 124, and 142 (industrial or commercial units, airports, and sport and leisure facilities)—presenting close values and oscillation patterns among them. The policies of urban and infrastructural expansion carried out by the successive autonomic governments for the Madeira Island territory may explain these results. For example, the case of the touristic *boom* that the region as felt in the last few decades was lead

Code		111	112	121	122	123	124	131	132	133	141	142
Year	1990	0.21	**9.16**	0.10	0.00	0.06	0.22	0.00	0.00	0.06	0.05	0.00
	2000	0.22	**12.76**	0.22	0.04	0.08	0.26	0.00	0.04	0.10	0.05	0.22
	2006	0.31	**13.16**	0.32	0.04	0.08	0.26	0.15	0.08	0.09	0.01	0.33
	2012	0.31	**13.20**	0.35	0.04	0.08	0.26	0.15	0.06	0.01	0.01	0.36

Bold identifies the higher value founded.

Table 3. Artificial surfaces.

Code		211	212	221	222	231	241	242	243	244
Year	1990	0.16	0.49	0.11	0.74	0.94	0.90	4.83	**10.68**	0.25
	2000	0.16	0.49	0.11	0.46	0.91	0.90	3.02	**9.47**	0.47
	2006	0.04	0.04	0.18	0.31	0.32	0.23	2.80	**10.05**	0.00
	2012	0.04	0.04	0.21	0.31	0.32	0.23	2.77	**10.08**	0.00

Bold identifies the higher value founded.

Table 4. Agricultural areas.

Code		311	312	313	321	322	324	331	332	333	334
Year	1990	**20.89**	5.90	14.24	8.34	10.50	6.56	0.11	0.07	1.91	0.00
	2000	**20.67**	5.74	13.99	8.22	10.41	6.33	0.11	0.07	1.91	0.06
	2006	**20.82**	5.29	13.14	9.07	10.77	6.25	0.11	1.52	1.75	0.24
	2012	**19.27**	4.72	11.98	8.70	9.06	5.58	0.11	1.52	1.75	6.23

Bold identifies the higher value founded.

Table 5. Forest and seminatural areas.

Code		523
Year	1990	**2.55**
	2000	2.52
	2006	2.25
	2012	2.25

Bold identifies the higher value founded.

Table 6. Water bodies.

Level		Artificial surfaces	Agricultural areas	Forest and seminatural areas	Water bodies
Year	1990	9.86	19.10	**68.52**	2.55
	2000	13.99	15.99	**67.51**	2.52
	2006	14.83	13.97	**68.96**	2.25
	2012	14.83	14.00	**68.92**	2.25

Bold identifies the higher value founded.

Table 7. Evolution of the occupied surfaces, according to levels (in the period 1990–2012).

to the actors to carry unsustainable politics of construction—jeopardizing, in many occasions, the natural and unique heritage of the island. In fact, this phenomenon not only occurs in this particular insular territory; the same scenario has been described in the Canary Islands [24, 25]—or even generalizing further, we can pick some high touristic demand territory, and, unfortunately, similar results are easily found [26–30].

Through the analysis of **Table 4**, it is possible to verify the behavior of the agricultural areas; once again the highest values have been clearly found in one single land use, the 243—land principally occupied by agriculture, with significant areas of natural vegetation—which as decreased over the years (with an exception for the period of 2000 where it has decreased and recovered back in 2006). The second most representative land uses, considering artificial surfaces, is for the land use 242 (complex cultivation patterns), followed closely by the land uses 231, 241, and 222 (pastures, annual crops associated with permanent crops, and fruit trees and berry plantations). Those outcomes, as is the example of the decrease of the surface occupied by agriculture with significant natural vegetation, are not unexpected. Once again the territorial governance may play a key role in these results once the politics carried out toward the preservation of natural vegetation and traditional agriculture are not so profitable for the land owners, contrary to the reconversion of the land for a different use or agricultural technique—tendency that seems to be dangerous not only for the local heritage preservation as well as for a long-term sustainable development and growth.

In **Table 5** the behavior of the forest and seminatural areas is shown, with the highest values found in one land use, the 311—broad-leaved forest—which decreased over the years (with some oscillations in the 2000–2006 period). The second most representative land uses, considering artificial surfaces, is for the land use 313 (beaches, dunes, sands), followed closely by the land uses 322 and 321 (moors and heathland and natural grasslands). Here, the change of positions—in the period of 2006–2012—of the land use 324 (transitional woodland-shrub) replaced by the land use 334 (burned areas) should be highlighted; this results may be explained by the natural events as wildfires occur in the island—i.e., in fact, in 2016 wildfire events occurred once again in Madeira Island consuming several hectares of natural vegetation, agricultural areas, and urban areas leading to numbers that are even getting worst in future studies related to land use changes in Madeira Island. In this regard, the high values of the land use 313 (beaches, dunes, sands) are expected—considering the territorial features (island). In fact, where could be identified a barrier—once this use typology practically impossible to reconvert; nevertheless, and analyzed from another perspective, there is an opportunity to develop a touristic sustainable activity over this land use.

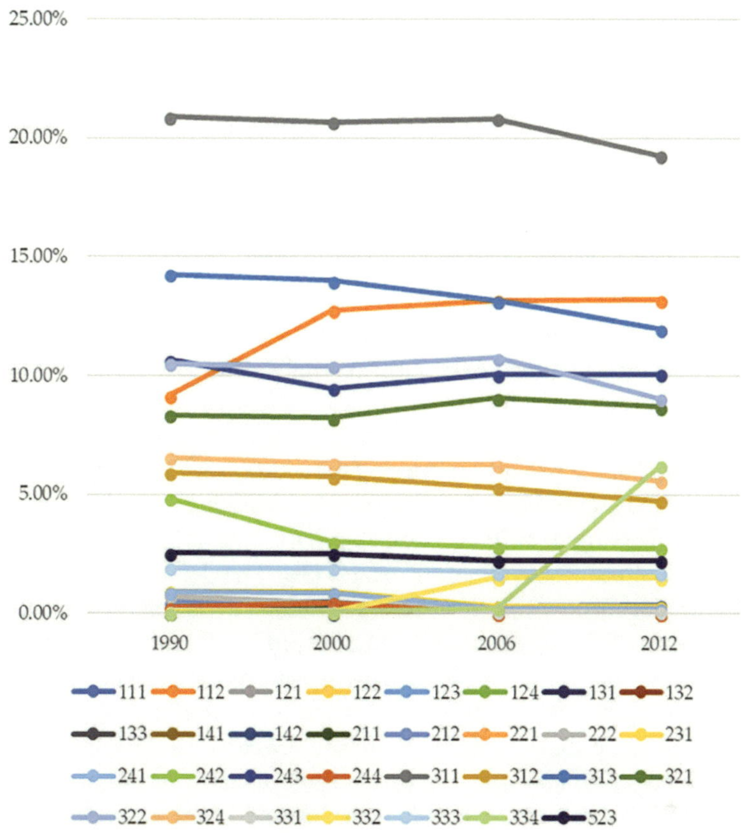

Figure 2. Evolution of the land cover through the years.

Finally, the land uses regarding water bodies have decreased over the last decades; once again expansionary and unsustainable politics may be at the core of the answer. In **Figure 2**, it is possible to verify that in most of the cases there is stability in the land uses. However, the land use 334 (burned areas), which has significantly increased, should be highlighted, within the last decade, and consequently the land use 311 (broad-leaved forest) has decreased. Also, the increases of the land uses 112 and 322 (discontinuous urban fabric and moors and heathland) should be highlighted.

In **Table 7**, the land use changes over the last decades according to levels 1, 2, 3, and 5 (artificial surfaces, agricultural areas, forests and seminatural areas, and water bodies) are presented. By far, the most representative level is the third (forests and seminatural areas), followed by the artificial surfaces (1), which became more representative in the last years in Madeira Island territory than the land use 2 (agricultural areas).

3.1. Barriers and opportunities

The changes in the land use could be understood as a direct manifestation of human activity over natural environments [31, 32]. Therefore, the natural factors and features—i.e., geomorphology, slope, relief, soil, and vegetation, among many others— are critical for the proper organization and distribution of the territory and their consequent land uses [31]. The lack of knowledge aligned with an existence of planning conducts to the destruction of the natural resources causing relevant (negative) impact on the local communities [33]. Thus, the proper identification and defining of risk areas—considering the planning and territorial management—are pivotal conditions for the prevention and minimization of the damages resulting from the phenomena and dangerous activities [34]. The uncontrolled growth (due to the lack of well-planning process) to built-up areas contributes to increasing the soil vulnerability and increasing the risk of natural disasters [35, 36], as is the case of erosion or landslides [37]—considering the local geomorphic features. In this regard, the urban expansion toward topographically "more" inclined and geologically unstable ground can cause problems and affect the population, the environment, and the local economy. The slope is assumed to be fundamental for the occurrence of slope movements, mainly due to the higher slope and the greater influence of gravity forces on the existing materials in the slopes that, if they are fragile, easily will disintegrate and move along the slope—which is their case in RAM [38]. Therefore, and considering the geomorphological risks, the slope assumes the main role, since it interacts with and for the erosion increasing in a geomorphological context and in the lithology allowing to define critical slopes for landslides—even in the vegetal cover ground, eliminating natural resources of the island [39]. The definition of land uses consistent with the risk degree that characterizes it and the prohibition or limitation to the urban expansion in the unstable areas are some of the options pointed out by Zêzere [37] to avoid such risks. In fact, the crossing of the constraints to urban growth with risk areas leads to the determination of land suitability for each category of use and respective infrastructure implementation. The limitation groups lead to the analysis of the urban land use capacity at the level of the existing one and at the level of areas of urban expansion (urbanizable land), consolidation, and reconversion [39–41]. Contextually, it is important to define classes and levels for specific land uses, based on urban and spatial planning criteria as well as in accordance with urban growth limitations, including the possibility of occurrence of natural hazards and disasters in specific areas [39].

On the other hand, the relief could also influence the urban growth and development [42]—once it forces the city to grow in a dispersed or apparently disorganized way, creating urban voids. In this case, it works as a topographical barrier and constitutes a natural element of obstruction to the urban expansion and as a barrier protecting (somehow) the fragmentation of the natural habitats. Such phenomenon is more relevant in cities located in hilly areas where the variation of altitude is large and there are steep slopes. Thus, the land uses of Madeira Island have been assessed from different perspectives and methods at RAM according to the abovementioned; in fact, the geomorphology of the territory strongly affects its development and growth.

Nevertheless, despite all the limitations inherent to the territorial relief, such topographic barriers at RAM will value the environmental dimension—considering as fundamental in a perspective of social and economic well-being, promoting the full exploitation of the values and endogenous natural resources. For this, the relationship between the economic activities and biodiversity and nature conservation is strongly influenced, namely, by the *unique fauna* and *flora* of the island as well as their ecosystems, natural landscapes, and humanized landscapes; in fact, these factors could also be seen as opportunities to promote sustainable development.

The valorization of the agricultural heritage of RAM (despite the stagnation of land use for agriculture) should be preserved and protected in a sustainable way; once in Madeira Island, there are several typical crops—i.e., banana, sugarcane, and vineyards, among many others.

http://dx.doi.org/10.5772/intechopen.80827

In this way, and if sustainable development policy will carry out all the involved actors and depending population, it could be valued by the so-called barriers existing in this "ultra-peripheral" territory.

4. Final remarks

Through the present study, it is possible to understand the impact of the land use changes and their dynamics on the specific insular territory. Also, throughout the analyses of the land use change patterns along with empirical knowledge of the territory, barriers and opportunities for a sustainable development and growth have been identified. Moreover, the limitations of such "ultra-peripheral" territories are evident, not only by the physical spatial dimensions presented but also by the difficulty to proceed to the reconversion of the uses. Considering such remarks and the particularity of these territories, the main actors/decision-makers and their policies and action over the territory are even more relevant and need to be conducted in a more reasonable way—considering the fragility of this region; such policies and actions present higher impact on the territory as well as on their inhabitants' life's quality standards and finally on the long-term sustainable development.

Therefore, the study of the land use change patterns is seen as pivotal to understand the dynamics and tendencies of these territories as well as to provide clues for the main actors to where the efforts toward a sustainable development and growth should be placed.

As the final remarks, the land uses could be understood as another tool for the knowledge of the territory—assessing the past and envisioning the future.

Author details

Rui Alexandre Castanho[1,2,3,4]*, Sérgio Lousada[5], José Manuel Naranjo Gómez[4,6], Patrícia Escórcio[5], José Cabezas[3,4], Luis Fernández-Pozo[3] and Luís Loures[4,7,8]

*Address all correspondence to: alexdiazbrown@gmail.com

1 Faculty of Applied Sciences, University of Dąbrowa Górnicza, Dąbrowa Górnicza, Poland

2 ICAAM—Institute for Agrarian and Environmental Sciences, Évora, Portugal

3 Environmental Resources Analysis Research Group (ARAM), University of Extremadura, Badajoz, Spain

4 VALORIZA—Research Centre for Endogenous Resource Valorization, Portalegre, Portugal

5 Faculty of Exact Sciences and Engineering (FCEE), Department of Civil Engineering and Geology (DECG), University of Madeira (UMa), Funchal, Portugal

6 Polytechnic School, University of Extremadura, Caceres, Spain

7 Polytechnic Institute of Portalegre (IPP), Portalegre, Portugal

8 Research Centre for Tourism, Sustainability and Well-being (CinTurs), University of Algarve, Faro, Portugal

References

[1] García-Álvarez D, Camacho Olmedo MT. Changes in the methodology used in the production of the Spanish CORINE: Uncertainty analysis of the new maps. International Journal of Applied Earth Observation and Geoinformation. 2017;**63**:55-67. DOI: 10.1016/j.jag.2017.07.001

[2] Bartholomé E, Belward AS. GLC2000: A new approach to global land cover mapping from earth observation data. International Journal of Remote Sensing. 2005;**26**(9):1959-1977. DOI: 10.1080/01431160412331291297

[3] Weber J-L. Implementation of land and ecosystem accounts at the European Environment Agency. Ecological Economics. 2007;**61**(4):695-707. DOI: 10.1016/j.ecolecon.2006.05.023

[4] Hazeu G, Büttner G, Arozarena A, Valcárcel N, Feranec J, Smith G. Detailed CLC Data: Member states with CLC level 4/level 5 and (Semi-)automated solutions. European Landscape Dynamics CORINE Land Cover Data. 2016:275-304. DOI: 10.1201/9781315 372860-27

[5] Pinto-Correia T, Kristensen L. Linking research to practice: The landscape as the basis for integrating social and ecological perspectives of the rural. Landscape and Urban Planning. 2013;**120**:248-256. DOI: 10.1016/j.landurbplan.2013.07.005

[6] Maul, George (2005) – Small islands. In: Maurice L. Schwartz (editor) Encyclopedia of Coastal Science. USA: Springer; pp. 883-887

[7] Amado, M. (2009): Planeamento Urbano Sustentável. Editora: Caleidoscópio Lisboa, Portugal . ISBN 9789728801748

[8] Baptista T et al. IDE-OTALEX C. The first crossborder SDI between Portugal and Spain: Background and development. Journal of Earth Sciences and Engineering. 2013;**3**(6):6-14

[9] Castanho R, Loures L, Fernández J, Fernández-Pozo L. Cross Border Cooperation (CBC) in Southern Europe—An Iberian case study. The Eurocity Elvas-Badajoz. Sustainability. 2017;**9**:360. DOI: 10.3390/SU9030360

[10] Castanho R, Loures L, Fernández J, Pozo L. Identifying critical factors for success in Cross Border Cooperation (CBC) development projects. Habitat International. 2016;**72**:92-99. DOI: 10.1016/j.habitatint.2016.10.004

[11] World Commission for Environment and Development WCED. Our Common Future. United Kingdom: Oxford; 1987

[12] Banos-González I, Martínez-Fernández J, Esteve MÁ. Tools for sustainability assessment in island socio-ecological systems: An application to the Canary Islands. Island Studies Journal. 2016;**11**(1):9-34

[13] Nielsen SN, Jørgensen SE. Sustainability analysis of a society based on exergy studies—A case study of the island of Samsø (Denmark). Journal of Cleaner Production. 2015;**96**:12-29. DOI: 10.1016/j.jclepro.2014.08.035

[14] Banos-González I, Martínez-Fernández J, Esteve-Selma MÁ. Dynamic integration of sustainability indicators in insular socio-ecological systems. Ecological Modelling. 2015;**306**:130-144. DOI: 10.1016/j.ecolmodel.2014.08.014

[15] EEA. EAGLE-Related Projects. 19 May 2018. [Online]. Available: https://land.copernicus.eu/eagle/files/eagle-related-projects/pt_clc-conversion-to-fao-lccs3_dec2010

[16] França JA, Almeida AB. Plano regional de água da Madeira. Síntese do diagnóstico e dos objectivos. In: 6° SILUSBA – Simpósio de Hidráulica e Recursos Hídricos dos Países de Língua Oficial Portuguesa. Praia, Cabo Verde; 2003. pp. 751-818

[17] Ramalho R, Brum da Silveira A, Fonseca P, Madeira J, Cosca M, Cachão M, et al. The emergence of volcanic oceanic islands on a slow-moving plate: The example of Madeira Island, NE Atlantic. Geochemistry, Geophysics, Geosystems. 2015;**16**:522-537. DOI: 10.1002/2014GC005657

[18] Pullen J, Caldeira R, Doyle JD, May P, Tomé R. Modeling the air-sea feedback system of Madeira Island. Journal of Advances in Modeling Earth Systems. 2017;**9**:1-24. DOI: 10.1002/2016MS000861

[19] Gonçalves R, Camacho R, Lousada S, Castanho R. Modeling of maritime agitation for the design of maritime infrastructures: The case study of madeira archipelago. Revista Brasileira de Planejamento e Desenvolvimento, Curitiba. 2018;**7**(1):29-50. DOI: 10.3895/rbpd.v7n1.7136

[20] Miranda D, Camacho R, Lousada S, Castanho R. Hydraulic studies and their influence for regional urban planning: A pratical approach to Fynchal's rivers. Revista Brasileira de Planejamento e Desenvolvimento, Curitiba. 2018;**7**(1):145-164. DOI: 10.3895/rbpd.v7n1.7179

[21] Camacho R, Lousada S, Castanho RA. Escoamento em Canais Artificiais. Àreas Urbanas em Situação de Cheia. Aplicação À Zona Baixa do Funchal [Flow in Artificial Channels. Urban Areas in Flood Situations. Applied to Funchal ́s Downtown]. Latvia: Novas Edições Académicas; 2018. ISBN: 978-613-9-61139-3

[22] DREM. Direção Regional de Estatística da Madeira. 2015. 01 June 2018. [Online]. Available: www.estatistica.madeira.gov.pt

[23] Lopes S. A utilização do SIG na estimativa da precipitação e escoamento fluvial na ilha da Madeira. Funchal, Portugal: LREC; 2011

[24] Joher S, Ballesteros E, Rodriguez- Prieto C. Contribution to the study of deep coastal detritic bottoms: the algal communities of the continental shelf off the Balearic Islands, Western Mediterranean. Mediterranean Marine Science. 2015;**16**:573-590

[25] Blazquez-Salom M, Bonet AAA, Cadena IY. Crisis and neoliberal tourist territorial planning in the Balearic Islands. Investigaciones Turisticas. 2015;**9**:24-49

[26] Morote AF, Sauri D, Hernandez M. Residential tourism, swimming pools, and water demand in the Western Mediterranean. The Professional Geographer. 2017;**69**(1):1-11

[27] Cantos JO, Vera-Rebollo JF. Climate change and tourism policy in Spain: Diagnosis in the Spanish Mediterranean coast. Cuadernos de Turismo. 2016;**38**:323-359

[28] Morote-Seguido AF, Hernandez- Hernandez M. Green areas and water management in residential developments in the European Western Mediterranean. A case study of Alicante, Spain. Geografisk Tidsskrift-Danish Journal of Geography. 2016;**116**(2):190-201

[29] Rodriguez-Rodriguez D, Malak DA, Soukissian T, Sanchez-Espinosa A. Achieving blue gowth through maritime spatial planning: Offshore wind energy optimization and biodiversity conservation in Spain. Marine Policy. 2016;**73**:8-14

[30] Rosado D, Usero J, Morillo J. Application of a new integrated sediment quality assessment method to Huelva estuary and its littoral of influence (Southwestern Spain). Marine Pollution Bulletin. 2015;**98**:106-114

[31] Gao P, Niu X, Wang B, Zheng Y. Land use changes and its driving forces in hilly ecological restoration area based on GIS and RS of northern China. Scientific Reports. 2015;**5**:11038

[32] Bertrand N, Vanpeene-Bruhier S. Periurban landscapes in mountain areas. At the crossroads of ecological and socio-economic studies. Journal of Alpine Research—Revue de géographie alpine. 2007;**95-4**:69-80

[33] IME. Natural Disasters Saving Lives Today, Building Resilience For Tomorrow. Westminster, London: 1 Birdcage Walk; 2013

[34] Zêzere J, Pereira A, Morgado P. Perigos naturais e tecnológicos no território de Portugal Continental. Lisboa: Centro de Estudos Geográficos, Universidade de Lisboa; 2006

[35] United Nations. Indicators of Sustainable Development Guidelines and Methodologies. New York: United Nations; 2001

[36] United Nations. Indicators of Sustainable Development: Guidelines and Methodologies. 3rd ed. New York: United Nations; 2007

[37] Zêzere JL. Dinâmica de vertentes e riscos geomorfológicos. In: Programa. Lisboa: Centro de Estudos Geográfico—Universidade de Lisboa; 2005

[38] Teixeira M. Movimentos de Vertente. Factores de Ocorrência e Metodologia de Inventariação. Portugal: Geonovas N.° 20; 2006. pp. 95-106

[39] Santos L d, Fortuna J. Modelo de exigências para uso urbano do solo. Critérios urbanísticos e riscos naturais—um exemplo em Coimbra. Territorium. 2005;**12**(12):69-95

[40] Codosero Rodas J, Naranjo Gómez J, Castanho RA, Cabezas J. Land valuation sustainable model on urban planning development: A case study in Badajoz, Spain. Sustainability; Special Issue Smart Cities and Villages. 2018;**10**:1450. DOI: 10.3390/su10051450

[41] Castanho RA, Vulevic A, Loures L, Cabezas J, Naranjo Gómez J. Redeveloping derelict landscapes on transboundary areas—fostering cross—cooperation as a possible solution. In: 24th APDR CONGRESS. UBI, Covilhã-Portugal. July 6-7. Proceedings Book. 2017. ISBN 978-989-8780-05-8

[42] Meneses FJ. O Urban Sprawl em Cidades Portuguesas de Média Dimensão—Análise da década de 1991 a 2001. Lisboa: Universidade Técnica de Lisboa; 2010

Chapter 3

Effects of Agricultural Land Use on the Ecohydrology of Small-Medium Mediterranean River Basins: Insights from a Case Study in the South of Portugal

Paula Matono, Teresa Batista, Elsa Sampaio and Maria Ilhéu

Additional information is available at the end of the chapter

http://dx.doi.org/10.5772/intechopen.79756

Abstract

Southern Europe has been experiencing an accelerated intensification of agricultural systems in the last decades with consequent environmental effects. This study aimed to evaluate the effects of agricultural land use in two small-medium river basins in the South of Portugal, regarding: (i) water quality and stream habitat; (ii) fish fauna; and (iii) soil. Sampling included fish captures, water, and soil sample collection. Hydromorphological habitat features were also assessed. Land use was quantified at the basin and local scales. Results showed that the most negative effects were associated with intensive, heavily irrigated, fertilized, and pastured local systems, mostly represented at the basin scale by olive groves, irrigated crops, and pastures. Conversely, local agricultural intensity did not prove to be a threat to the integrity and quality of the soil, seeming to ensure the sustainability of the local uses and their systems. Negative effects were observed on water quality and instream habitat and degradation of riparian vegetation, resulting in fish assemblages' impoverishment. This study contributes to a comprehensive approach to the effects of agricultural land use, highlighting the need to integrate the results of different natural resources to efficiently support policy and decision makers toward a sustainable agriculture, water management, and land use planning.

Keywords: agricultural intensity, agricultural systems, water quality, stream habitat, fish assemblages, soil quality, sustainable agriculture, water management, land use planning

1. Introduction

Land use includes different landscape elements that are very dynamic, having low temporal and structural stability [1, 2]. Land use/land cover changes (LULCC) are playing an important role on the global and local change phenomena with major impacts on the environmental and sociocultural sustainability. The European landscapes have been subject to rapid changes in land use throughout the second half of the twentieth century arising from developments in technology and management driven by socioeconomic and political forces [3]. Southern Europe has been shaped by human activity and maintained by traditional farming practices for centuries, but important changes in land use have occurred following the strong production incentives launched by the European Common Agriculture Policy Framework. Under this scenario, the South of Portugal has experienced a rather accelerated change in farming systems in the last three decades, due to the perspective of a profitable intensive and irrigated agriculture. Many of the agro-silvopastoral systems have been severely reduced due to intensification trends, sometimes followed in other areas by extensification or even abandonment [4–6]. Simultaneously, artificial water bodies have increased due to the construction of many dams (e.g., Alqueva dam) mainly for irrigation purposes. As a result, the irrigated area has increased considerably in the last decade and is currently about 30% of the cultivated land in Portugal [7].

The intensification of agricultural land use has raised the question of the long-term sustainability of agroecosystems [8]. The growing expansion of intensive agrosystems is expected to promote environmental degradation, including soil erosion, water resources depletion, risk of floods and landslides, water and soil contamination, and biodiversity loss [9, 10]. It is urgent to reverse this trend by encouraging farmers to adopt more sustainable practices that optimize the use of natural resources on which they depend, so that future generations will be able to meet their needs, while maintaining biodiversity.

Intensive farming, such as irrigated arable crops, pastures, and orchards, can result in several different types of stress, which alone or together affect the structure and functioning of aquatic ecosystems and biodiversity [11, 12]. For instance, decreased river discharge due to water overexploitation for irrigation purposes may change river hydrology (both groundwater and surface), increasing siltation and reducing habitat heterogeneity, with negative effects on the aquatic biota [13–16]. Soil and water can be contaminated by the random uses and the overdoses of synthetic fertilizers and other agrochemicals used to increase land productivity. The downstream effects of runoff from these systems may result in the increase of nutrient concentration in the water bodies leading to water eutrophication, dissolved oxygen depletion, and the loss of fish fauna integrity [12]. This phenomenon is particularly aggravated in Mediterranean climate regions, where floods alternate with long dry and hot periods, promoting the conditions to increase soil erosion and nutrient leaching, particularly in watersheds with high LULCC.

Soil erosion is also one of the most serious environmental problems associated with farming intensification, as well as loss of soil structure and stability [17]. Changes in soil aggregate stability may largely influence soil susceptibility to degradation [18, 19]. Soil structure stability has a key role in the functioning of soil, its capacity to support plant and animal life, water availability, and therefore is a good indicator of land integrity [18, 19]. The conversion of natural forest to other forms of land use can lead to a reduction in soil organic content, loss of soil quantity, and modification of soil structure [20]. Soil characteristics negatively affected by intensive agrosystems with tillage practices are soil organic matter, total porosity, aggregate stability, and bulk density [21]. Land use change also may affect water retention at field capacity in the soil; lower water content at the field capacity would be expected upon conversion of natural to cultivated lands [22]. Many examples of activities related with agrosystems acting as sources of soil change can be referred: biomass burning, fertilizer application, species transfer, plowing, irrigation, drainage, livestock grazing, pasture improvement [17], deforestation and site abandonment [23], breaking up of large tracts of grassland, expansion of cultures which promote erosion (e.g., maize and sugar beet), and farming of fields in the fall line [24]. The sustainability of cropping systems demands a focused attention to monitor soil quality because of the growing concern about the decline in soil productivity and the impoverishment of soil organic carbon caused by intensive agriculture practices [25].

Although considerable research has been conducted on the effects of land use on terrestrial environments in Mediterranean regions, studies on aquatic habitats are more limited. Furthermore, even though soil and aquatic degradation are widely recognized as major environmental problems resulting from land use intensification, integrated and comprehensive approaches considering the possible soil and water interconnections have received far less research.

Therefore, this study aimed to assess the effects of agricultural land use on water quality, stream habitat, structure and functionality of fish assemblages, and soil quality, based on a case study developed in two small-medium Mediterranean river basins in the South of Portugal.

2. Methods

2.1. Study area

The study was conducted in two small-medium river basins located in the South of Portugal (Alentejo region): Azambuja (261.92 km^2), a sub-basin of the Guadiana River, and Alcáçovas (429.64 km^2), a sub-basin of the Sado River (**Figure 1**).

This region is influenced by the Mediterranean climate, presenting high susceptibility to drought events [26]. The hydrological regime is very variable, with severe droughts and floods. Flow is strongly dependent on the seasonal distribution of rain, mainly concentrated in October–March. Small-medium river basins are particularly affected during the summer dry season (June–September), when streams became completely dry or reduced to isolated pools where fish fauna has to survive until the reestablishment of river continuity in the following rains [27]. Fish assemblages generally present low species richness and include many

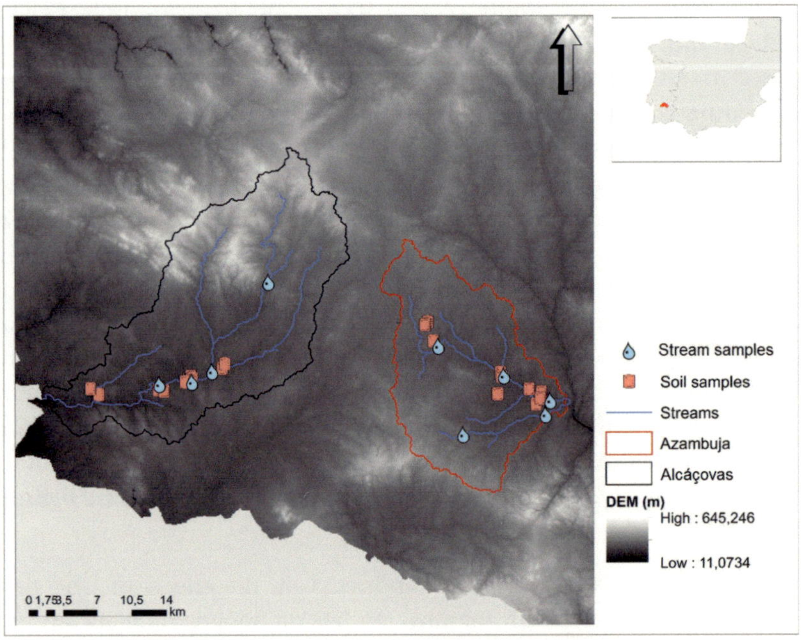

Figure 1. Location of the sampling sites in the Azambuja and Alcáçovas river basins in the South of Portugal.

endemic species with high conservation status, particularly in the Guadiana basin [28]. The most abundant and frequent species is roach (*Squalius alburnoides* Steindachner), followed by different species of barbels and nase.

The geomorphology is characterized by large lowland extensions (a mean altitude of 200 meters) with some more prominent areas, but without mountainous characteristics. In general, soils are incipient and slender with more aptitude for forest-pastoralism. The most representative types are lithosols, litholic soils, Mediterranean brown soils, and podzols, primarily derived from shale, clay, and limestone [29].

The Alentejo region is markedly rural, representing more than half of the agricultural area used in the country [7]. This landscape is dominated by olive groves and montado, an agroforestry system that has evolved from the Mediterranean forest by the planting of large areas of cork and holm oak, combined with low shrubs, and integrating livestock production. Extensive rainfed systems are also a mark of the Alentejo landscape and are generally present in large areas, combined with other productions [7].

The landscape has been changing with the intensification of the agricultural practices, cattle production, and the introduction of new crops. The irrigated land has registered a significant increase, mostly in permanent and temporary crops, olive groves, and vineyards, with the possibility of expansion in the future [7, 30]. Livestock production maintained high important in the agricultural activity, with an increase in cattle heads and a decrease in sheep and goats [30]. Consequently, pastures are mainly used as grazing area for livestock production, and this use has increased by about 42% in the last decade. Although livestock is mostly produced in extensive systems, there is often a high number of animals per farm (>300) [7]. The management of livestock is based on the rotational grazing of cattle on partitioned pasture areas, so that at each moment, a high stock rating may be observed. During the summer, when pastures become dry, livestock is often found in paddocks with free access to streams.

2.2. Site selection and sampling

2.2.1. *Stream habitat, water quality, and fish*

Sampling was held during 2012 in two small-medium river basins located in the South of Portugal: Azambuja (N = 5 sites) and Alcáçovas (N = 4 sites) (**Figure 1**). Sampling sites were selected based on a preliminary GIS analysis using ArcGIS software and considering a digital elevation model, slope, and exposure maps as well as flow direction and accumulation.

Fish captures took place in early spring, following the protocol developed and adopted for Portuguese rivers under the implementation of the Water Framework Directive [31]. Samplings were always carried out in flowing water conditions, immediately after the floods and previously to the strong reduction of flow during the summer period, in order to ensure high habitat diversity in the streams. Fishes were collected using backpack battery-powered electrofishing equipment (IG 200/2B, PDC Hans-Grassl GmbH, Schonau am Konigssee, Germany). All the necessary fishing permits were provided by the National Institute for the Conservation of Nature and Forests (ICNF). Captured fishes were identified to the species level and measured (total length, mm). Individuals of native species were returned alive to the water, whereas individuals of non-native species were removed and euthanized by thermal shock (freezing), in compliance with the Portuguese legislation and following the ethical guidelines of the Directive 2010/63/EU on animal welfare [32].

The environmental characterization of sites was carried out during the sampling procedure and included:

i. physicochemical parameters measured with a multiparameter probe—water temperature (°C), conductivity (μs cm^{-1}), pH, and dissolved oxygen (mg L^{-1});

ii. water nutrients evaluated through laboratory analyses according to the Standard Methods for the Examination of Water and Wastewater [33], after water sample collection—5-day biological oxygen demand—BOD_5 (mg L^{-1}), orthophosphate—PO_4^{3-} (mg L^{-1}), nitrite—NO_2^- (mg L^{-1}), nitrate—NO_3^- (mg L^{-1}), ammonia—NH_4^+ (mg L^{-1}), and total suspended solids—TSS (mg L^{-1}); and

iii. anthropogenic disturbance variables reflecting hydromorphological alterations induced by local land use practices—riparian vegetation, sediment load, hydrological regime, and morphological condition. Each variable was scored from 1 (minimum disturbance) to 5 (maximum disturbance) following [12] and based on [34].

Physicochemical parameters and nutrient concentrations were used to classify the water quality of the sampled sites into five classes, based on the water features for multiple uses, according to the Portuguese Environmental Agency guidelines [35]: excellent, good, reasonable, poor, and very poor.

2.2.2. Soil

Soil samples were collected in the dominant soil types from the slopes that drain into the stream section immediately upstream the stream sampling sites. The specific location of the soil samples depended both on soil maps [36] and land use/land cover maps, in order to achieve the highest number of possible combinations and replicates, in a total of 19 samples (Azambuja = 10 samples; Alcáçovas = 9 samples) (**Figure 1**).

Samples were collected at the root zones of 0–30 cm using a soil auger. This superficial soil layer is the most sensitive to material entraining through runoff, to washing and leaching of nutrients into deeper layers and groundwater, to the accumulation of nutrients from fertilization, to the accumulation of organic matter, and to structure degradation that influences infiltration and runoff, all these with potential consequences on the water quality of adjacent streams.

Each soil sample was air dried and passed through a 2-mm screen, and the coarse fraction (>2 mm) was separated. The fine soil fraction (<2 mm) was then subjected to laboratory analysis using standard procedures, and a wide range of parameters was analyzed, giving preference to the chemical characteristics, since it was intended to evaluate the relationship between soil results and water quality: textural class (relative proportion of sand, silt, and clay), pH (soil reaction), percentage of organic matter in the soil (OM), concentration of exchange bases ($cmol^{+}\cdot kg^{-1}$) (Ca^{2+} (calcium), Mg^{2+} (magnesium), Na^{+} (sodium), and K^{+} (potassium)), S (sum of bases) ($cmol^{+}\cdot kg^{-1}$), T (cation exchange capacity) ($cmol^{+}\cdot Kg^{-1}$), V (percentage of base saturation), P (available phosphorus) ($mg\cdot kg^{-1}$), and H^{+} (exchange acidity) ($cmol^{+}\cdot kg^{-1}$). Particle size distribution was determined by the hydrometer method [37]; soil pH was determined using a pH meter in a soil/liquid suspension of 1:2.5 [38]; organic carbon was determined using a chromic wet oxidation method [39]; available phosphorus was determined using the Bray II solution method [39]; exchangeable Mg^{2+} and Ca^{2+} were determined using ethylenediaminetetraacetic acid (EDTA) [40], while exchangeable K^{+} and Na^{+} were extracted using 1 N Neutral $C_2H_7NO_2$ and then determined using a flame photometer [40]; exchangeable acidity was measured titrimetrically using 1 M KCl per 0.05 M of NaOH [41], and effective cation exchange capacity was calculated from the sum of all exchangeable bases and total exchangeable acidity; percentage base saturation was calculated by the sum of the total exchangeable bases divided by the effective cation exchange capacity and then multiplied by 100.

2.2.3. Land use

The land use was assessed both at the catchment and local scales. At the catchment scale, the proportion of the most representative land use classes was calculated for the influence area of each sampling site, based on the land cover map applying the CORINE Land Cover Legend Level 5 at a scale of 1:10000 and aggregated to the level 3 [42]: irrigated arable land, nonirrigated arable land, vineyards, olive groves (mostly irrigated), pastures, annual and permanent crops, agroforestry (montado), hardwood forest, resin forest, mixed forests, and water plains. At the local scale, the dominant local land use observed at the surrounding area of each stream site was registered.

As in most sites, different land uses were present with possible effects on streams, and the overall level of agricultural intensity was also evaluated, based on the presence of irrigation, fertilization, and animal grazing. For each site, these three variables were scored between 0 (null impact) and 2 (high impact). The sum of these scores represented the total agricultural intensity, and three classes were established to classify the sampled sites: low (0–1), moderate (2–3), and high (≥4). For soil samplings, the exact local land use of the sample site was registered.

2.3. Data analysis

Fish captures were quantified as density (the number of individuals·100 m^{-2}) and fish assemblages were analyzed considering: (i) structural metrics (proportion of native and non-native species); (ii) functional guilds related to habitat (relative abundance of limnophilic, eurytopic, benthic, and water column species), breeding (relative abundance of lithophilic and phytophilic species), feeding (relative abundance of omnivorous and insectivorous species), and tolerance to degradation (relative abundance of tolerant species). Captured species were assigned to functional guilds according to the published literature [34, 43, 44] and expert judgment based on the available knowledge.

Soil quality was assessed regarding the performance of three ecological functions related to environmental regulation, biomass production, and reserves of water and biodiversity, through the soil quality index (SQI) using five indicators: chemical fertility, drainage, reaction (pH), organic matter (OM), and phosphorus (P). These five qualitative and quantitative indicators were scored from 1 to 3, following [45], and the SQI was calculated using the formula.

$$\mathrm{SQI} = \Sigma\, Ind/n \tag{1}$$

where *Ind* represents the score of each indicator and **n** the number of indicators. The SQI ranges between 0 (complete inability to perform the functions of environmental regulation, production of biomass, and ensure biodiversity) and 1 (full ability to perform the functions of environmental regulation, production of biomass, and ensure biodiversity).

Redundancy analysis (RDA) [46] was used to explore relationships between land use variables (both at local and basin scales) and: (i) water parameters and stream habitat features; (ii) structure and functionality of fish assemblages. A linear ordination method was selected after a preliminary detrended correspondence analysis that has shown a gradient length smaller than 3 SD [47]. A stepwise forward selection of the variables was used, and the final model was tested with Monte Carlo test under 999 permutations. Correlations higher than |0.4| were used in gradient interpretation.

A bivariate approach was used to analyze the proportion of water quality classes along the local agricultural intensity gradient, and the Friedman test was performed to search for significant differences. This approach was also used to evaluate the pattern of chemical soil parameters under different local land uses and along the local agricultural intensity gradient.

Prior to multivariate analyses, data were either log (x + 1) (linear measurements) or arcsin [sqrt (x)] (percentages) transformed to improve normality [48]. Statistical analyses were performed using the software Statistica 10 and Canoco 4.5.

3. Results and discussion

3.1. Land use vs. stream habitat and water quality

Seven significant variables ($p < 0.05$) were selected for the ordination model, including both local (N = 3) and basin scale (N = 4) land use variables (**Figure 2**). According to the canonical coefficients and inter-set correlations, axis 1 was mainly defined by irrigated arable land percentage (r = −0.63), pastures percentage (r = −0.70), and agroforestry (montado) percentage (r = 0.56). Axis 2 was mostly related with olive grove percentage (r = 0.63) and local levels of irrigation (r = 0.70), fertilization (r = 0.69), and animal grazing (r = −0.52) (**Figure 2**).

The ordination diagram (triplot) showed a good spatial segregation of sites along the first two axes, revealing also a good association of land use types with different water parameters and stream habitat features (76.3%). The first axis evidenced a clear association of extensive areas of irrigated olive groves and irrigated arable land in the drainage basin, as well as high local levels of irrigation and fertilization, with the degradation of most of the

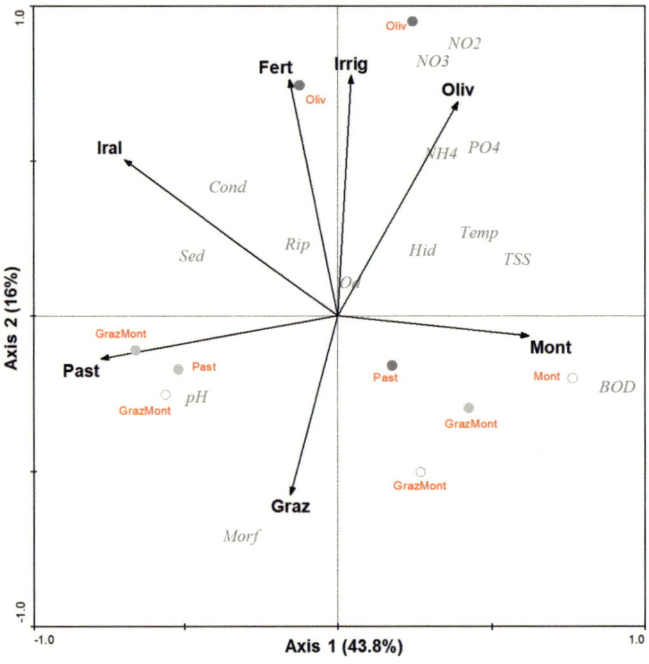

Figure 2. Ordination diagram (triplot) of the redundancy analysis between land use variables (at the local and basin scales) and aquatic habitat features in sampling sites located in the Azambuja and Alcáçovas river basins (N = 9). Sampling sites were identified by the dominant local land use (red) and were coded according to the classes of the agricultural intensification level: low (white dots), moderate (gray dots), and high (black dots). Abbreviations of land use variables: irrigated arable land (Iral), olive groves (Oliv), agroforestry-montado (Mont), pastures (Past), local level of animal grazing (Graz), local fertilization level (Fert), and local irrigation level (Irrig). Abbreviations of water parameters and stream habitat features: ammonia (NH_4), nitrate (NO_3), nitrite (NO_2), orthophosphate (PO_4), dissolved oxygen (Od), conductivity (Cond), water temperature (Temp), total suspended solids (TSS), biological oxygen demand (BOD), sediment load (Sed), riparian vegetation (Rip), hydrological regime (Hyd), and morphological condition (Morf). Abbreviations of local land uses: olive groves (Oliv), pastures (Past), grazed montado (GrazMont), and agroforestry-montado (Mont).

http://dx.doi.org/10.5772/intechopen.79756

water parameters and habitat features: increase in nutrient concentrations, high conductivity and water temperature, hydrological alterations, degradation of riparian vegetation, and high sediment loads. These results are strongly related to soil erosion, high surface runoff, and high levels of fertilization and irrigation commonly associated with intensive agriculture practices, as the high amount of organic material produced represents a considerable input of nitrates and phosphates leaching into superficial and groundwater very easily [49, 50].

Soil erosion is cited as one of the principal environmental problems associated with olive farming in Mediterranean regions [51]. In intensive olive plantations, farmers usually keep the soil bare of vegetation throughout the year, such that severe erosion occurs during heavy rains. Soil erosion and water runoff into nearby streams can be a major source of suspended sediments, nutrients, and pesticides in watersheds dominated by agricultural land [52, 53]. Natural stream flow can also suffer alterations [54, 55], by exacerbating the effects of seasonal or longer term droughts through water abstraction [56, 57], or by augmenting flows through irrigation returns, in some cases maintaining flowing rivers which would normally dry [58, 59].

The removal or degradation of riparian vegetation associated to the land use [60, 61] can further aggravate all the problems mentioned. Due to its position at the interface between terrestrial and aquatic ecosystems, riparian vegetation has the ability to prevent sediment runoff and to hold excess nutrients and modify their inputs to the stream [62], preventing negative consequences in overall water quality [63], and the increase of water temperature [64].

Large upstream areas of pastures and high local levels of animal grazing (mostly cattle) were mainly associated to higher pH values and instream morphological alterations. Livestock grazing with unrestricted access to streams has negative impacts on aquatic ecosystems [65, 66]. This practice increases instream trampling, habitat disturbance, and erosion from overgrazed stream banks, as well as reducing sediment trapping by riparian and instream vegetation and decreasing bank stability [67, 68].

Sites located in drainage basins with large areas of agroforestry (montado) did not show stream habitat alterations, although they were associated with the high values of BOD, TSS, and water temperature. This was possibly due to the fact that this site is located near an urban area and may be influenced by the drain treatment plant.

The water quality of the sampled sites ranged between moderate and very poor (**Figure 3**). Moreover, the relative proportion of the water quality classes of sites showed significant differences along the agricultural intensification gradient ($p < 0.05$), revealing a trend toward a progressive degradation. Nevertheless, even in sites with a low agricultural intensity, the water quality was predominantly moderate, with some percentage of poor quality. These results are in accordance with the fact that both local and basin scale variables were selected for the RDA (**Figure 2**), demonstrating that the effects of land use on water quality and stream habitat depend not only on the degree of intensity of the main local culture but also on cumulative multi-pressures acting at the basin scale. Scale is a particularly important issue in analyses of land use impacts on ecosystems because different perturbations, processes, and responses operate at different scales [69–72].

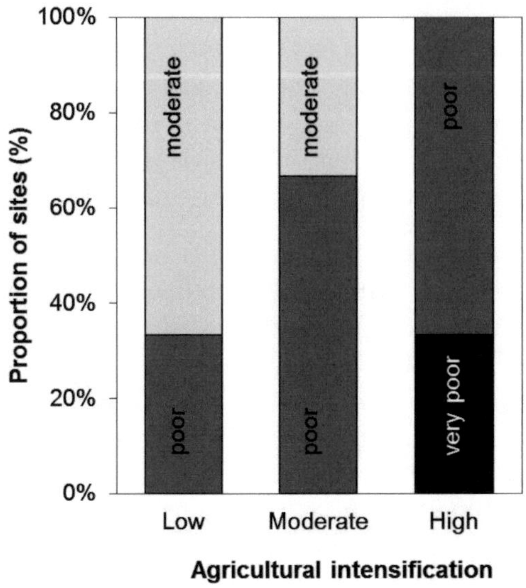

Figure 3. Relative proportion of the water quality classes along the agricultural intensification gradient in the sites sampled in the Azambuja and Alcáçovas river basins (N = 9).

3.2. Land use vs. fish assemblages

A total of 10 fish species were captured, including six native and four non-native species. On average, non-native species represented nearly 60% of the mean density per site. Native species only dominated fish assemblages in three sampling sites and are all endemic to the Iberian Peninsula, with high conservation status [28].

The RDA showed a reasonable segregation of the sites along the first two axes, which together explained 44% of data variability (**Figure 4**). Nevertheless, these axes revealed a good association between the relevant variables and fish metrics/guilds, explaining most of the species-variable relation (80.8%), thus supporting the interpretation of the results. Five variables were selected for the model, all reflecting land use at the catchment scale. Axis 1 was mainly defined by pasture percentage ($r = -0.48$), agroforestry (montado) percentage ($r = 0.4$), and olive grove percentage ($r = 0.55$). Axis 2 was mostly related with olive grove percentage ($r = -0.44$), and irrigated arable land ($r = 0.42$).

The ordination diagram (biplot) revealed different groups of fish metrics/guilds with which different land uses were particularly related. Sites under the upstream influence of large pasture areas were dominated by non-native, tolerant, omnivorous, and limnophilic species. Eurytopic and benthic species were associated with large areas of olive groves in upstream catchment areas. These results showed a clear impoverishment of the fish assemblages with the increase of the area occupied by more intensive agricultural uses in the drainage basin. Fish responds to the changes in the water parameters resulting from land use alterations [73] through interrelated impacts on water quality, hydrology, and habitat [54, 74], as seen in **Figures 2** and **3**. These impacts have been shown to substantially change fish assemblages [75], decrease species richness/diversity and sensitive species, while increasing tolerant and non-native species,

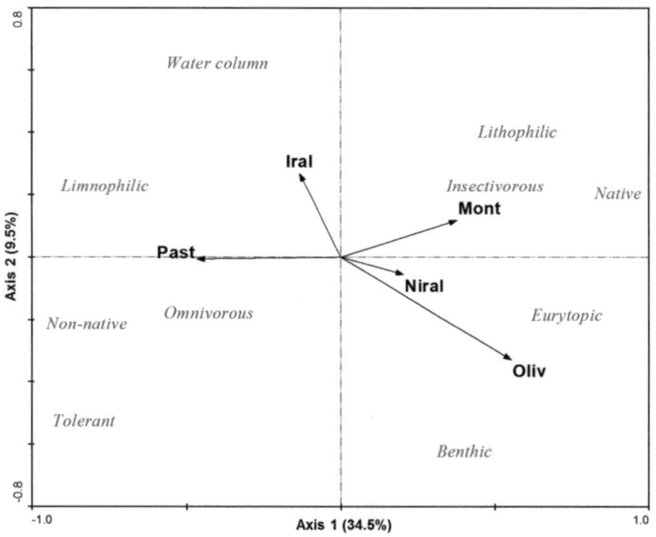

Figure 4. Ordination diagram (biplot) of the redundancy analysis between the land use variables (at the local and basin scales) and the fish metrics/guilds in sampling sites located in the Azambuja and Alcáçovas river basins (N = 9). Abbreviations of land use variables: irrigated arable land (Iral), nonirrigated arable land (Niral), olive groves (Oliv), agroforestry-montado (Mont), and pastures (Past).

and ultimately influencing the integrity of fish assemblages [12, 76, 77]. Conversely, extensive areas of agroforestry (montado) were related to high abundance of native, lithophilic, and insectivorous species, evidencing a lower disturbance level and enabling the support of more native specialist species [78].

Only basin scale variables were selected for the model, suggesting that the effects of land use on fish assemblages' structure and functionality operate at a larger spatial scale, as reported in other studies [79]. In fact, most agricultural and urban land uses occur at the larger watershed scale and their impacts cannot be fully understood by looking at adjacent riparian lands [80].

3.3. Land use vs. soil

Based on the most relevant soil variables, a consistent pattern was observed between the quality and chemical fertility of the soil and the local land uses (**Figure 5**), as well as regarding the intensity of the agricultural practices associated with them (**Figure 6**).

The organic matter content (OM) increased gradually along the agricultural intensity gradient, leading to an improvement in soil quality and health, since OM improves the soil structure and consequently increases the total porosity and friability, thus enhancing the overall resilience [81]. Similar results were obtained by Havaee et al. [82], even though different findings were reported by several authors, who concluded that there is a degradation of the soil by changing the uses for more intensive systems [83–88]. This was probably due to the inputs of organic matter and water irrigation used in the local land uses studied, which resort to these external factors in order to increase the land productivity. Simultaneously with the increase in the OM in most intensive local land uses, high data dispersion was also observed, probably

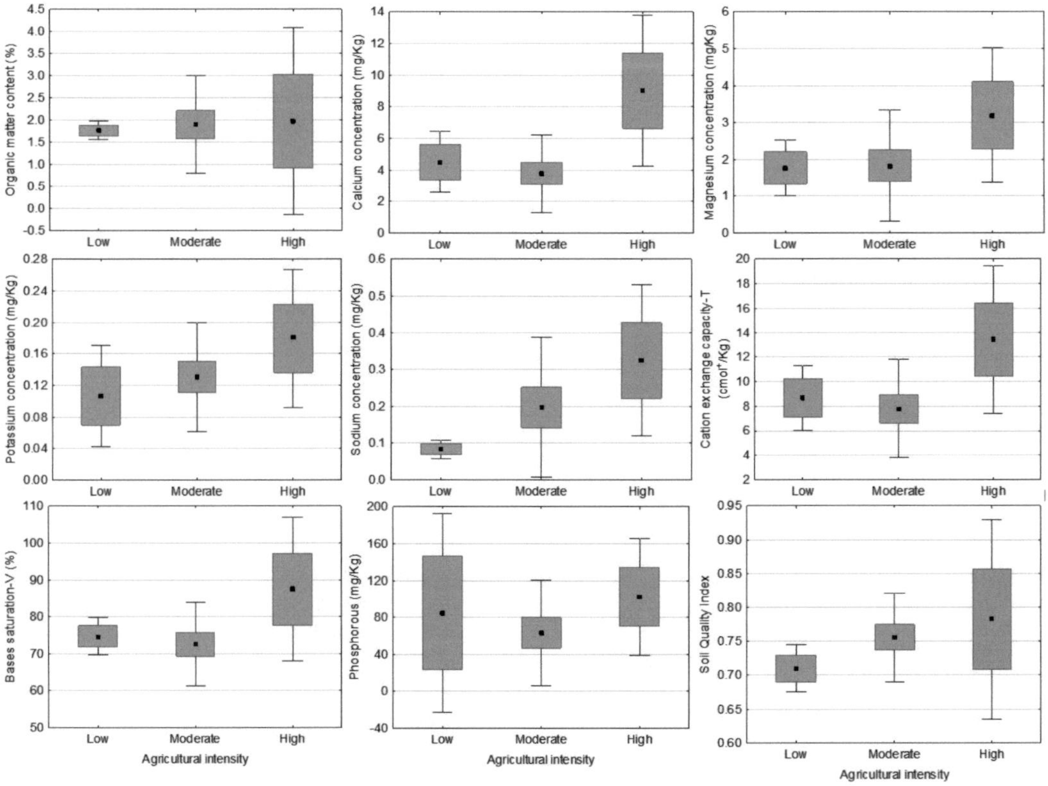

Figure 5. Box plots of the most significant results for soil characteristics along the agricultural intensity gradient. (▪): mean; box: ±SE; and whisker: ±SD.

due to the different sources of inputs in each of the local uses (**Figure 5**). The values of OM were higher in olive groves and in irrigated arable crops, possibly due to the input of external OM, whereas in pastures and in grazed olive groves, the origin of OM may be internal to the system, resulting from the presence of cattle. On the other hand, maize crop, which is also intensive, does not use OM supplementation.

The physical behavior of soil is controlled by the OM that is complexed with clay [89]. Although bulk density has not been evaluated in this study, this characteristic can be deduced from the OM/Clay ratio [82]. So, since the textural classes of the studied soils were always coarse, except in a single case in which it was median, it can be affirmed that the behavior of the bulk density, in face of the different land uses and corresponding intensity of the agricultural systems, was opposite to that of OM. Therefore, the higher the OM content, the higher the OM/Clay ratio, the lower the bulk density and, consequently, the lower the compaction, the higher the total porosity and the infiltration rate. In this way, the water that infiltrates the soil is retained by the high retention power of the OM [90, 91].

Several studies reported a reduction in the cation exchange capacity (T), which reflects a reduced chemical soil fertility, as the greatest impact of land use intensification on soil [83, 87, 88].

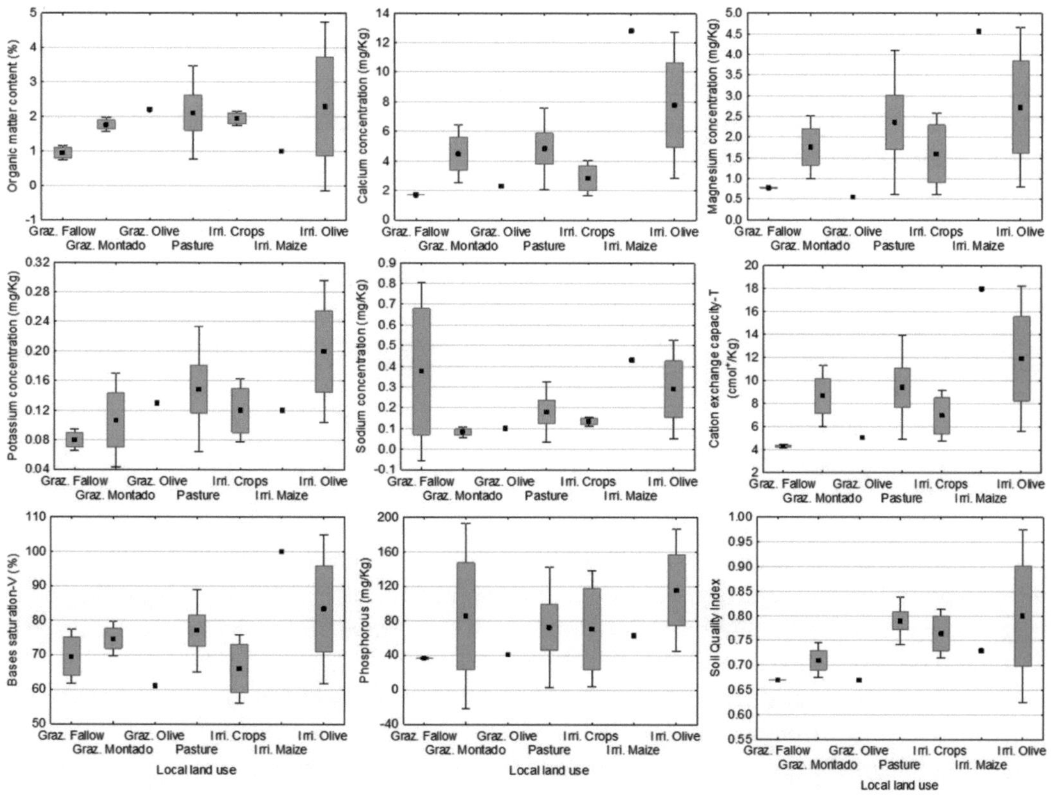

Figure 6. Box plots of the most significant results for soil characteristics in different local land use types. (▪): mean; box: ±SE; and whisker: ±SD.

However, in this study, the T, the bases of exchange (Ca^{2+}, Mg^{2+}, K^{+}, and Na^{+}), as well as the degree of saturation with bases (V), showed an increase along the gradient of agricultural intensity. This was possibly a consequence of the influence of the OM, since the texture of most soil samples was coarse, and therefore, in these cases, clay had little influence on soil properties. As shown in **Figure 6**, the bases of exchange showed different increases, certainly due to the practices of liming, fertilization, and irrigation proper of the uses to which they were associated.

Regarding the available phosphorous (P), as for all the other chemical characteristics, it increased in the soil in land uses with more intensive systems of exploitation (**Figure 6**). These results contradict other studies registering a phosphorous decrease when a change in the land use for more intensive systems occurs [84, 88, 92], due to the crop mining and the removal of waste and erosion. This unexpected pattern was better understood when observing the results obtained for each local land use (**Figure 5**), perceiving that they derived from the fertilization used in systems associated with more intensive uses, where productivity and yield are important [93].

The SQI revealed a pattern similar to the other analyzed characteristics, increasing with the agricultural intensity of the systems, despite showing high data dispersion (**Figure 6**), and explained when analyzing the results obtained for each local land use (**Figure 5**). Although the SQI was evaluated based on the parameters presented here, to which information about the drainage was added (considering the texture, the shape of the terrain, and the average annual precipitation), this did not alter the trend of the results, and some conclusions can be drawn about the consequences of different land uses on the quality of the soil and, in turn, on the adjacent aquatic ecosystems.

The obtained results showed that the higher the OM, the higher the rate of infiltration and the water retention in the soil, apparently with little influence of soil chemical constituents on aquatic ecosystems, since the water that solubilizes or suspends them is mostly retained in the soil. Only in situations where land use practices implies leaving the soil bare during the beginning of the rainy season, it will be possible to verify this type of consequences through the surface runoff [94].

4. Conclusions

Agricultural land use has shown to have strong negative effects on water quality, stream habitat, and degradation of riparian vegetation, ultimately resulting in fish assemblages' impoverishment and clearly benefiting non-native species, which thrive under altered conditions [95]. The most negative effects were associated with intensive, heavily irrigated, fertilized, and pastured agricultural systems, mostly represented at the basin scale by olive groves, irrigated crops, and pastures. Conversely, agroforestry (montado) results emphasize the potential contribution of this stable production system to biodiversity conservation.

Since Mediterranean rivers exhibit high levels of fish fauna endemicity, human impacts on these systems have the potential to extirpate native species and reduce local, regional, and global native biodiversity [96]. It should be further highlighted that considering the forecasted climate changes and their possible joint effects with land use changes, far reaching effects are likely to occur on ecological communities in Mediterranean regions in the future [97, 98].

Regarding soil, local agricultural intensity did not prove to be a threat to the integrity and quality of the soil, seeming to ensure the sustainability of the local uses and practices, contrary to what is usually found. The intensification of agricultural systems, by means of a high consumption of water or energy, can be carefully planned, thus preventing soil degradation through the known threats defined by Thematic Strategy for Soil Protection [99], namely, decline in organic matter, compaction, floods, soil erosion, salinization, contamination, landslides, and sealing. The careful planning and execution of agricultural practices that intensify the production systems (but not involving degradation) are the recommendations of several authors [88, 100, 101].

These are preliminary findings based on a case study, and more detailed research is further required to substantiate the results and assess the direct relationship between soil and the aquatic ecosystem, namely, by considering more soil characteristics and diversity, covering larger spatial and temporal scales, and considering climate data. This would allow a better understanding of the complex pathways underpinning the interaction among the processes and factors involved.

Nevertheless, this study emphasizes the tight interaction between streams and the terrestrial ecosystem and shows that both direct and indirect aspects of this linkage are relevant to stream ecosystem functioning. Moreover, the effects of the agricultural practices do not have the same spatial and temporal expression on the natural resources involved, namely, soil and water. As such, river basin management must integrate a vision of compromise between the intensification of agricultural systems and the conservation of different natural resources and ecosystems. Planners and policy makers should bring stakeholders together, based on the understanding of land-water relationships in a watershed, to plan for a sustainable agriculture, targeting and balancing locally specific environmental and socioeconomic needs.

Acknowledgements

This work is funded by National Funds through FCT - Foundation for Science and Technology, I.P., within the framework of the project UIDMulti044632016. We thank the Institute of Mediterranean Agrarian and Environmental Sciences (ICAAM) and the Center for Interdisciplinary Development and Research on Environment, Applied Management and Space (DREAMS). Special thanks to Maria José Barão for laboratory analyzes of water samples. We are also grateful to all the colleagues and volunteers for their valuable help in the fieldwork.

Conflict of interest

The authors declare no conflict of interest.

Author details

Paula Matono[1,2], Teresa Batista[1,2,3], Elsa Sampaio[1,4,5] and Maria Ilhéu[1,2*]

*Address all correspondence to: milheu@uevora.pt

1 Institute of Mediterranean Agricultural and Environmental Sciences (ICAAM), University of Évora, Évora, Portugal

2 Department of Landscape Environment and Planning, Science and Technology School, University of Évora, Évora, Portugal

3 Intermunicipal Community of Central Alentejo (CIMAC), Évora, Portugal

4 Department of Geosciences, Science and Technology School, University of Évora, Évora, Portugal

5 Centre for Interdisciplinary Development and Research on Environment, Applied Management and Space (DREAMS), Lusófona University of Humanities and Technologies, Lisbon, Portugal

References

[1] Guiomar N, Batista T, Fernandes JP, Cruz CS. Corine Land Cover nível 5—Contribuição para a carta de uso do solo em Portugal Continental. Évora: AMDE; 2011

[2] Brown DG, Duh JD. Spatial simulation for translating between land use and land cover. International Journal of Geographical Information Science. 2004;**18**(1):35-60

[3] Rounsevell MDA, Annetts JE, Audsley E, Mayr T, Reginster I. Modelling the spatial distribution of agricultural land use at the regional scale. Agriculture Ecosystems and Environment. 2003;**95**:465-479

[4] Pinto-Correia T, Mascarenhas J. Contribution for the extensification/intensification debate: What is happening to the Portuguese Montado? Landscape and Urban Planning. 1999; **46**:125-131

[5] Pinto-Correia T, Breman B. The new roles of farming in a differentiated European countryside: Contribution to a typology of rural areas according to their multifunctionality. Application to Portugal. Regional Environmental Change. 2009;**3**(9):143-152

[6] Pinto-Correia T, Vos W. Multifunctionality in Mediterranean landscapes—Past and future. In: Jongman R, editor. The New Dimension of the European Landscapes. Wageningen FRONTIS Series. The Netherlands: Springer; 2004. pp. 135-164

[7] INE. Recenciamento Agrícola 2009—Análise dos primeiros resultados. IP, Lisboa, Portugal: Instituto Nacional de Estatística; 2011

[8] Liebig MA, Tanaka DL, Wienhold BJ. Tillage and cropping effects on soil quality indicators in the northern Great Plains. Soil and Tillage Research. 2004;**78**:131-141

[9] Beaufoy G, Pienkowski M. The environmental impact of olive oil production in the European Union: practical options for improving the environmental impact. European Forum on Nature Conservation and Pastoralism, Brussels, European Commission; 2000

[10] Beaufoy G. EU Policies for Olive Farming: Unsustainable at all Counts. Report Produced Jointly by WWF Europe and BIRDLIFE International. 2001

[11] Reidsma P, Tekelenburg T, Van Den Berg M, Alkemade R. Impacts of land use change on biodiversity in the European Union. Agriculture, Ecosystems and Environment. 2006;**114**:86-102

[12] Matono P, Sousa D, Ilheu M. Effects of land use intensification on fish assemblages in Mediterranean climate streams. Environmental Management. 2013;**52**:1213-1229

[13] Resh VH, Brown AV, Covich AP, Gurtz ME, Li HW, Minshall W, et al. The role of disturbance in stream ecology. Journal of the North American Benthological Society. 1988;**7**(4):433-455

[14] Bernardo JM, Ilhéu M, Matono P, Costa AM. Interannual variation of fish assemblage structure in a Mediterranean river: Implications of streamflow on the dominance of native or exotic species. River Research and Applications. 2003;**19**:1-12

[15] Bunn SE, Arthington AH. Basic principles and ecological consequences of altered flow regimes for aquatic biodiversity. Environmental Management. 2002;**30**:492-507

[16] Meador MR, Carlisle DM. Relations between altered streamflow variability and fish assemblages in eastern USA streams. River Research and Applications. 2011;**28**(9):1358-1368

[17] Turner BL II, Meyer WB. Global land use and land cover change: An overview. In: Meyer WB, Turner BL II, editors. Changes in Land Use and Land Cover: A Global Perspective. Cambridge: Cambridge University Press; 1994. pp. 3-10

[18] Lipiec J, Kuoe J, Nosalewicz A, Turski M. Tillage system effects on stability and sorptivity of soil aggregates. International Agrophysics. 2006;**20**:189-193

[19] Shein EV, Umarova AB, Milanovski EY, Sokolova IV. Preferential water flow, local soil biota and structure degradation in chernozem 20 years after land-reclamation. International Agrophysics. 2010;**24**:75-80

[20] Jiang Y, Yuan D, Zhang C, Kuang M, Wang J, Xie S, et al. Impact of land use change on soil properties in a typical karst agricultural region of Southwest China: A case study of Xiaojiang watershed, Yunan. Environmental Geology. 2006;**50**:911-918

[21] Kizilkaya R, Dengiz O. Variation of land use and land cover effects on some soil physico-chemical characteristics and soil enzyme activity. Zemdirbyste-Agriculture. 2010;**97**(2): 15-24

[22] Haghighi H, Kheirkhah M, Saghafian B. 2011: Evaluation of Soil Hydraulic Parameters in Soils and Land Use Changes. Earth and Environmental Science. Chapter 20 Open Acess, In-Tech Book: Rijeka; 2011. 10.5772/19059

[23] Skole DL. 1994: Data on global land-cover change: Acquisition, assessment, and analysis. In: WB Meyer, Turner II BL, editors. Changes in Land Use and Land Cover: A Global Perspective. Cambridge: Cambridge University Press; 1994. p. 437- 471

[24] Briassoulis H. Analysis of land use change: Theoretical and modeling approaches. In: Loveridge S, editor. Web Book of Regional Science. Morgantown, WV: Regional Research Institute, West Virginia University; 2000. Available from: http://www.rri.wvu.edu/WebBook/Briassoulis/contents.htm

[25] Bhattacharyya R, Tuti MD, Kundu S, Bisht JK, Bhatt JC. Conservation tillage impacts on soil aggregation and carbon pools in a sandy clay loam soil of the Indian Himalayas. Soil Science Society of America Journal. 2012;**76**:617-627

[26] Miranda P, Coelho FS, Tomé AR, Valente MA. 20th century Portuguese climate and climate scenarios. In: Santos FD, Forbes K, Moniz R, editors. Climate Change in Portugal: Scenarios, Impacts and Adaptation Measures—SIAM Project. Lisboa: Gradiva; 2002. pp. 23-84. Available from: http://www.siam.fc.ul.pt/SIAM_Book

[27] Ilhéu M. Patterns of habitat use by freshwater fishes in Mediterranean rivers [Ph.D. thesis]. Évora, Portugal: University of Évora; 2004

[28] Cabral MJ, Almeida J, Almeida PR, Dellinger TR, Ferrand de Almeida N, Oliveira ME, et al. Livro Vermelho dos Vertebrados de Portugal. Lisboa, Portugal: Instituto da Conservação da Natureza; 2005

[29] IUSS Working Group WRB. World reference base for soil resources 2006. In: World Soil Resources Reports No. 103. 2nd ed. Rome: FAO; 2006

[30] DRAPAL. Caracterização Agrícola do Alentejo Central. Direcção Regional de Agricultura e Pescas do Alentejo; 2013

[31] INAG I.P. Manual Para a Avaliação Biológica da Qualidade da Água em Sistemas Fluviais Segundo a Directiva Quadro da Água Protocolo de Amostragem e Análise Para a Fauna Piscícola. Lisboa, Portugal: Ministério do Ambiente, do Ordenamento do Território e do Desenvolvimento Regional, Instituto da Água, I.P; 2008

[32] European Commission. Directive 2000/60/EC of the European Parliament and of the Council of. Establishing a framework for community action in the field of water policy. Official Journal of the European Communities. 2000;**L327**:2000. Available from: http://ec.europa.eu/environment/water/water-framework/index_en.html

[33] Clesceri LS, Greenberg AE, Eaton AD. Standard Methods for the Examination of Water and Wastewater. 20th ed. Washington, DC, USA: American Public Health Association, American Water Works Association, Water Environmental Federation; 1998

[34] Fame Group. Development, Evaluation & Implementation of a Standardized Fish-Based Assessment Method for the Ecological Status of European Rivers—A Contribution to the Water Framework Directive; Final Report, Scientific Achievements (Sections 5 & 6) (Co-Ordinator: Stefan Schmutz). Vienna, Austria: Institute for Hydrobiology and Aquatic Ecosystem Management, University of Natural Resources and Applied Life Sciences; 2004. Available from: http://fame.boku.ac.at

[35] APA. Agência Portuguesa do Ambiente. Available from: http://snirh.pt/snirh/_dadossintese/qualidadeanuario/boletim/tabela_classes.php

[36] Cardoso JC. Os Solos de Portugal. Sua Classificação, Caracterização e Génese. 1—A Sul do Rio Tejo. Lisboa: Direcção-Geral dos Serviços Agrícolas, Secretaria de Estado da Agricultura; 1965

[37] Gee GW, OR D. Particle size analysis. In: Dane JH, Topp GC, editors. Methods of Soil Analysis. Physical Methods. Part 4. Madison: Soil Science Society of America; 2002. pp. 255-293

[38] Herdershot WH, Lalande H, Duquette M. Soil reaction and exchangeable acidity. In: Carter MR, editor. Soil Sampling and Methods of Analysis. Boca Raton, FL: Lewis Publishers; 1993. pp. 141-145

[39] Nelson DW, Sommers LE. Total organic carbon and matter. In: Page AL, editor. Methods of Soil Analysis, Part 2: Chemical and Microbiological Properties. 2nd ed. Kentucky, USA: American Society of Agronomy; 1982. p. 570

[40] Thomas GW. Exchangeable cations. In: Page AL, editor. Methods of Soil Analysis, Part 2: Chemical and Microbiological Properties. 2nd ed. Kentucky, USA: American Society of Agronomy; 1982

[41] Mclean ED. 1982. Soil pH and lime requirements. In: Page AL, editors. Methods of Soil Analysis, Part 2: Chemical and Microbiological Properties. 2nd ed. American Society of Agronomy: Kentucky, USA; 1982. p. 199-234

[42] Batista T. Carta de ocupação e uso do solo do Distrito de Évora e Município de Sousel—Legenda Corine Land Cover Nível 5. Évora: CIMAC; 2011

[43] Holzer S. European fish species: Taxa and guilds classification regarding fish-based assessment methods [Ph.D. thesis]. Universität für Bodenkultur; 2008

[44] Magalhães MF, Ramalho CE, Collares-Pereira MJ. Assessing biotic integrity in a Mediterranean watershed: Development and evaluation of a fish-based index. Fisheries Management and Ecology. 2008;**15**:273-289

[45] Ilhéu M, Batista T, Matono P, Corte-Real J, Sampaio E. Efeitos da ocupação e uso de solo na ecohidrologia de pequenas bacias hidrográficas. XV Encontro da Rede de Estudos Ambientais dos Países de Língua Portuguesa, Luanda; 2013

[46] Jongman RHG, ter Braak CJF, Van Tongeren OFR. Data Analysis in Community and Landscape Ecology. Wageningen, The Netherlands: PUDOC; 1987

[47] Ter Braak CJF, Smilauer P. CANOCO Reference Manual and User's Guide to Canoco for Windows: Software for Canonical Community Ordination (Version 4.0). Ithaca, NY, USA: Microcomputer Power; 1998

[48] Legendre P, Legendre L. Numerical Ecology. 2nd ed. Amsterdam, The Netherlands: Elsevier; 1998

[49] Carpenter SR, Caraco NF, Correll DL, Howarth RW, Sharpley AN, Smith VH. Nonpoint pollution of surface waters with phosphorous and nitrogen. Ecological Applications. 1998;**8**(3):559-568

[50] Mallin MA, Johnson VL, Ensign SH, Macpherson TA. Factors contributing to hypoxia in rivers, lakes, and streams. Limnology and Oceanography. 2006;**51**:690-701

[51] Graaff J, Eppink LAAJ. Olive oil production and soil conservation in southern Spain, in relation to EU subsidy policies. Land Use Policy. 1999;**16**:259-267

[52] Kuhnle R, Bennett S, Alonso C, Bingner R, Langendoen E. Sediment transport processes in agricultural watersheds. International Journal of Sediment Research. 2000;**15**:182-197

[53] Vanni M, Renwick W, Headworth J, Auch J, Schaus M. Dissolved and particulate nutrient flux from three adjacent agricultural watersheds: A five-year study. Biogeochemistry. 2001;**54**:85-114

[54] Allan JD. Landscapes and riverscapes: The influence of land use on stream ecosystems. Annual Review of Ecology, Evolution, and Systematics. 2004;**35**:257-284

[55] Stohlgren T, Chase T, Pielke RA, Kittels TGF, Baron JS. Evidence that local use practices influence regional climate, vegetation, and stream flow pattern in adjacent natural areas. Global Change Biology. 1998;**4**:495-504

[56] Sophocleous M. Interactions between groundwater and surface water: The state of the science. Hydrogeology Journal. 2002;**10**:52-67

[57] Menció A, Mas-Pla J. Assessment by multivariate analysis of groundwater–surface interactions in urbanized Mediterranean streams. Journal of Hydrology. 2008;**352**:355-366

[58] Prat N, Munné A. Water use and quality and streamflow in a Mediterranean stream. Water Research. 2000;**34**:3876-3881

[59] Klose K, Cooper SD, Leydecker AD, Kreitler J. Relationships among catchment land use and concentrations of nutrients, algae, and dissolved oxygen in a southern California river. Freshwater Science. 2012;**31**:908-927

[60] Aguiar F, Ferreira MT. Human-disturbed landscapes: effects on composition and integrity of riparian woody vegetation in the Tagus River basin, Portugal. Environmental Conservation. 2005;**32**:1-12

[61] Salinas MJ, Casas JJ. Riparian vegetation of two semi-arid Mediterranean rivers: Basin-scale responses of woody and herbaceous plants to environmental gradients. Wetlands. 2007;**27**:831-845

[62] Muenz TK, Golladay SW, Vellidis G, Smith LL. Stream buffer effectiveness in an agriculturally influenced area, southwestern Georgia: Responses of water quality, macroinvertebrates, and amphibians. Journal of Environmental Quality. 2006;**35**:1924-1938

[63] Sekely AC, Mulla DJ, Bauer DW. Streambank slumping and its contribution to the phosphorus and suspended sediment loads of the Blue Earth River, Minnesota. Journal of Soil and Water Conservation. 2002;**57**(5):243-250

[64] Wohl NE, Carline RF. Relations among riparian grazing, sediment loads, macroinvertebrates, and fishes in three central Pennsylvania streams. Canadian Journal of Fisheries and Aquatic Sciences. 1996;**53**(suppl 1):260-266

[65] Lyons J, Weigel BM, Paine LK, Undersander DJ. Influence of intensive rotational grazing on bank erosion, fish habitat quality, and fish communities in southwestern Wisconsin trout streams. Journal of Soil and Water Conservation. 2000;**55**:271-276

[66] Magner JA, Vondracek B, Brooks KN. Channel stability, habitat and water quality in South-eastern Minnesota (USA) streams: assessing managed grazing practices. Environmental Management. 2008;**42**:377-390

[67] Kaufmann JB, Kreuger WC. Livestock impacts on riparian ecosystems and streamside management implications: A review. Journal of Range Management. 1984;**37**:430-438

[68] Vidon P, Campbell MA, Gray M. Unrestricted cattle access to streams and water quality in till landscape of the Midwest. Agricultural Water Management. 2008;**95**:322-330

[69] Allan JD, Johnson LB. Catchment-scale analysis of aquatic ecosystems. Freshwater Biology. 1997;**37**:107-111

[70] Allan JD, Erickson DL, Fay J. The influence of catchment land use on stream integrity across multiple spatial scales. Freshwater Biology. 1997;**37**:149-161

[71] Cooper SD, Diehl S, Kratz K, Sarnelle O. Implications of scale for patterns and processes in stream ecology. Australian Journal of Ecology. 1998;**23**:27-40

[72] Townsend CR, Downes BJ, Peacock K, Arbuckle CJ. Scale and the detection of land-use effects on morphology, vegetation and macroinvertebrate communities of grassland streams. Freshwater Biology. 2004;**49**:448-462

[73] Morgan DL, Thorburn DD, Gill H. Salinization of southwestern Western Australian rivers and the implications for the inland fish fauna—The Blackwood River, a case study. Pacific Conservation Biology. 2003;**9**:161-171

[74] Paul MJ, Meyer JL. Streams in the urban landscape. Annual Review of Ecology, Evolution, and Systematics. 2001;**32**:333-365

[75] Argent DG, Carline RF. Fish assemblages changes in relation to watershed landuse disturbance. Aquatic Ecosystem Health and Management. 2004;**7**(1):101-114

[76] Fischer JR, Quist MC, Wigen SL, Schaefer AJ, Stewart TW, Isenhart TM. Assemblage and population-level responses of stream fish to riparian buffers at multiple spatial scales. Transactions of the American Fisheries Society. 2009;**139**:185-200

[77] Roth NE, Allan JD, Erickson DL. Landscape influences on stream biotic integrity assessed at multiple spatial scales. Landscape Ecology. 1996;**11**:141-156

[78] Poff NL, Allan JD. Functional organization of stream fish assemblages in relation to hydrologic variability. Ecology. 1995;**76**:606-627

[79] Richards C, Johnson LB, Host GE. Landscape-scale influences on stream habitats and biota. Canadian Journal of Fisheries and Aquatic Sciences. 1996;**53**:295-311

[80] Wang L, Lyons J, Kanehl O, Gatti R. Influences of watershed land use on habitat quality and biotic integrity in Wisconsin streams. Fisheries. 1997;**22**:6-12

[81] Karchegani MP, Ayoubi S, Mosaddeghi MR, Honarjoo N. Soil organic carbon pools in particle size fractions as affected by slope gradient and land use changes in hilly regions, western Iran. Journal of Mountain Science. 2012;**9**:87-95

[82] Havaee S, Ayoubi S, Mosaddeghi MR, Keller T. Impacts of land use on soil organic matter and degree of compactness in calcareous soils of central Iran. Soil Use and Management. 2014;**30**:2-9

[83] Mulugeta L, Karltun E, Olsson M. Assessing soil chemical and physical property responses to deforestation and subsequent cultivation in smallholders farming system in Ethiopia. Agriculture, Ecosystems and Environment. 2005;**105**:373-386

[84] Eyayu M, Heluf G, Tekaliign M, Mohammed A. Effects of land-use change on selected soil properties in the Tera Gedam Catchment and adjacent agroecosystems, north-west Ethiopia. Ethiopian Journal of Natural Resources. 2009;**11**(1):35-62

[85] Nega E, Heluf G. Influence of land use changes and soil depth on cation exchange capacity and contents of exchangeable bases in the soils of Senbat Watershed, western Ethiopia. Ethiopian Journal of Natural Resources. 2009;**11**(2):195-206

[86] Kumar M, Babel AL. Available micronutrient status and their relationship with soil properties of Jhunjhunu Tehsil, District Jhunjhunu, Rajasthan, India. Journal of Agricultural Science. 2011;**3**(2):97-106

[87] Mojiri A, Aziz HA, Ramaji A. Potential decline in soil quality attributes as a result of land use change in a hillslope in Lordegan, Western Iran. African Journal of Agricultural Research. 2012;**7**(4):577-582

[88] Yitbarek T, Heluf G, Kibebew K, Sheleme B. Impacts of land use on selected physicochemical properties of soils of Abobo area, Western Ethiopia. Agriculture, Forestry and Fisheries. 2013;**2**(5):177-183

[89] Dexter AR, Richard G, Arrouays D, Czyz EA, Jolivet C, Duval O. Complexed organic matter controls soil physical properties. Geoderma. 2008;**144**:620-627

[90] Warburton ML, Schulze RE, Jewitt GPW. Hydrological impacts of land-use change in three diverse South African catchments. Journal of Hydrology. 2009;**414-415**:118-135

[91] Zégre N, Skaugset AE, Som NA, McDonnell JJ, Ganio LM. In lieu of the paired catchment approach: Hydrologic model change detection at the catchment scale. Water Resources Research. 2010;**46**(11):W11544

[92] Wakene N, Heluf G. Forms of phosphorus and status of available micronutrients under different land use systems of Alfisols in Bako area of Ethiopia. Ethiopian Journal of Natural Resource. 2003;**5**(1):17-37

[93] Fageria NK, Moreira A, dos Santos AB. Phosphorous uptake and use efficiency in field crops. Journal of Plant Nutrition. 2013;**36**(13):2013-2022

[94] Weatherheada EK, Howden NJK. The relationship between land-use and surface water resourcesin the UK. Land Use Policy. 2009;**265**:243-250

[95] Hermoso V, Clavero M, Kennard MJ. Determinants of fine-scale homogenization and differentiation of native freshwater fish faunas in a Mediterranean Basin: Implications for conservation. Diversity and Distributions. 2012;**18**:236-247

[96] Clavero M, Hermoso V, Levin N, Kark S. Geographical linkages between threats and imperilment in freshwater fish in the Mediterranean Basin. Diversity and Distributions. 2010;**16**:744-754

[97] Bêche LA, Connors PG, Resh VH, Merenlender AM. Resilience of fishes and invertebrates to prolonged drought in two California streams. Ecography. 2009;**32**:778-788

[98] Klausmeyer KR, Shaw MR. Climate change, habitat loss, protected areas and the climate adaptation potential of species in Mediterranean ecosystems worldwide. PLoS One. 2009;**4**(7):e6392

[99] EC. COM 2006/231 2006. Communication from the Commission to the Council, the European Parliament, the European Economic and Social Committee and the Committee of the Regions—Thematic Strategy for Soil Protection. Brussels: Commission of the European Communities; 2006

[100] Onwudike SU. Effect of land use types on vulnerability potential and degradation rate of soils of similar lithology in a tropical soil of Owerri, Southeastern Nigeria. International Journal of Soil Science. 2015;**10**(4):177-185

[101] Gutzler C, Helming K, Balla D, Dannowski R, Deumlich D, Glemnitz M, et al. Agricultural land use changes—A scenario-based sustainability impact assessment for Brandenburg, Germany. Ecological Indicators. 2015;**48**:505-517

Chapter 4

Agricultural Zoning and Policy Conflict: Thailand's Experience

Nararuk Boonyanam

Additional information is available at the end of the chapter

http://dx.doi.org/10.5772/intechopen.80262

Abstract

Agricultural zoning is a regulatory approach that redefines property rights. The technique has been used to preserve agricultural area and ensure food security of a country. This chapter described a historically grounded approach to establish the agricultural zoning of a country and its applications for Thailand, including the analysis of its performance using the historical research method. Reviews of historical literatures, research papers, agricultural acts, government office reports, and news have identified the most crucial factor that causes agricultural zoning in Thailand to fail both at the farm level and at the country level, as the agricultural commodity price. The findings suggest that agricultural zoning in Thailand should undergo major revision by taking the agricultural commodity price into account. The agricultural commodity price set based on its origin was proposed to be appended in the existing agricultural zoning program. The method will address the two critical issues that the agricultural zoning programs in Thailand try to solve: the mismatch land-use problem and the crop price instability.

Keywords: property right, agricultural zoning, Thailand, agricultural transformation, agricultural policy

1. Introduction

The tragedy of the commons is the ruin of a common-property resource by uncontrollable use when each rational person pursuing his own best interest in a society that believes in the common-property regimes. As individuals increase their use of the resource, they gain the full benefits of the increased use but share only a proportion of cost. Since there is no guarantee on the property right, there is no incentive to preserve or increase the production of the resource for future use. One obvious example of the tragedy of the commons happens in Ethiopia where there is a huge increase in the demand of fuel wood. When the harvest of fuel wood exceeds the rate of the wood production, farmers use straw and dung which are used to maintain soil fertility for fuel resulting in dramatic soil erosion. At the end, all trees are wiped out, and farmers abandon their land as there is a total collapse in soil fertility. The country does not have sufficed food for its population and has to rely on the food aid from international institution. As Hardin [1] stated "Freedom in a commons brings ruin to all."

The establishment of property rights was seen as a tool to avoid the tragedy of the commons especially in land resource. However, the full ownership land right establishment led us to the same route in which all land resources are converted to nonagricultural use. Property rights form the basis for all market exchange and lead to the efficiency in land-use allocation. Unfortunately in the case of land use, most of the efficiency uses are in the nonagricultural activities. As landholder weighs the value of land in different uses, the benefit from agricultural use is typically a lot lower than other types of use even in extensive farming. Therefore, the full ownership land right turns out to be another nightmare. We then fall into the same situation that food security is in crisis due to dramatic reduction in agricultural land. Every country tries to solve this problem by introducing different types of land right in between the two extremes. Considering **Figure 1**, two extremes of property right are presented at the end of the line. Common right is property that is owned by a group of individuals. Access, use, and exclusion are controlled by the joint owners, while the full ownership right is the property that is owned by private owner or a group of legal owners. Access, use, exclusion, and management are controlled by the private owner or a group of legal owners.

In the middle of the line, it is the partial formal right such as preemptive right and usufruct right. These types of right restrict transfer of holding to only by inheritance and prohibit the transfer of ownership or rental. The restrictions aim to preserve agricultural area and ensure that the landless farmers cannot sell their land for a conversion purpose. Therefore, its occupancy level is in between the two extremes of land right type. However, the partial formal land right gains limited successes as it reduces incentives to invest in agricultural land and pose constraints to agricultural growth and natural resource management [2, 3]. Furthermore, it is found that rich people grabbed these lands with partial formal right especially prime plots with commercial prospect although the restriction is in place. Therefore, policy maker tries to find other methods to preserve agricultural land. The list of regulatory method varies which includes such things as purchase development right, purchase in fee, gifting to a trust, use-value taxation, growth-management acts, agricultural protection zoning, and many others. Among these long list methods, the agricultural protection zoning is the most persistent method despite many criticisms of its shortcomings [4].

The agricultural protection zoning or agricultural zoning is a regulatory approach that redefines property rights. It has been widely used since the publication of The Economics of Zoning

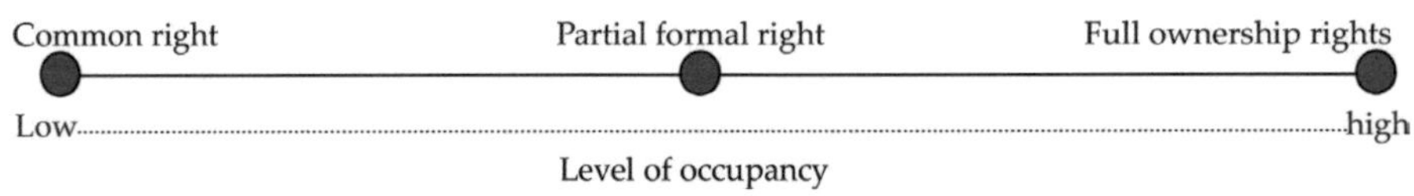

Figure 1. Land right map.

http://dx.doi.org/10.5772/intechopen.80262

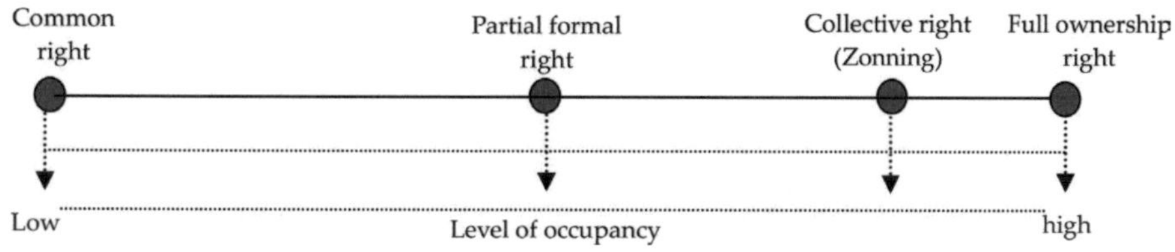

Figure 2. Agricultural zoning in land right map.

Laws [5]. The method was subtitled a property right approach to American land-use controls [6]. Nelson [7] stated that zoning creates collective property rights that are held by the local government. The entitlement itself is shared between the local government and the individual owner, and by this the individual right declines. The method is intended for protecting agricultural land and farming activities from incompatible nonfarm uses by specifying many factors, such as the uses allowed, minimum farm size, the number of nonfarm dwellings allowed, or the size of a buffer separating farm and nonfarm properties. If we take a look at the land right map again, the collective right or zoning should be located at the right-hand side of the line next to the full ownership right as shown in **Figure 2**. The level of occupancy is prone to the side of full ownership right as its restriction is weaker than that of the partial formal right.

Considering the location of zoning method on the land right map, zoning would appear to effectively address the issue of agricultural land reduction and food scarcity as it removes the decision to convert from the agricultural landowner, which is the shortcoming of the full ownership right, as well as address the occupancy level issue of partial formal right which creates uncertainty that leads to lower investment on agricultural land of farmer. There are a number of attempts to evaluate the effectiveness of agricultural zoning method; therefore, after discussing the methodological chapter in the next section, the review of agricultural zoning performance literature will be explored followed by the application of agricultural zoning in Thailand, its policy conflict, findings of the study, and a proposal to revise the agricultural zoning used, respectively.

2. Methodological chapter

This chapter was motivated by three major objectives: to examine the general application of the agricultural zoning to a country; to examine the adjusted application of agricultural zoning to Thailand; and to analyze the historical tendency of agricultural zoning policy and propose an additional tool that could bring success to the objectives of the agricultural zoning program used in Thailand.

The method employed for this chapter is historical research method. The method examines an account of what has happened in the past and presents events in the context of the present condition. It helps to reflect and provide possible answers to current issues and problems. The past events reveal the general and the particular in historical phenomena and give an understanding of various development stages of the event. The method helps to capture the nuance and idea of the event and helps to answer the questions: where have we come from, where are we now, and where are we going? [8].

The secondary sources which include historical literatures in agricultural zoning, research papers, Thailand agricultural acts, government office reports, and online news were used as a tool to gain an experiential sense of the past and provide policy design resource. In the previous section, the review of general application of agricultural zoning program reveals that the method is actually one of the kinds of property rights, and the general objectives to apply the agricultural zoning of a country are to preserve the agricultural area and ensure food security. In the next section, the review of agricultural zoning performance paper in various countries will be discussed. This step helped to establish a background of the historical analysis into the policy design process.

3. Effectiveness of agricultural zoning

Researchers have developed several techniques and models to capture the impact of agricultural zoning on land-use change. These techniques can be grouped into two main approaches which are the pricing approach and the outcome approach. The pricing approach is the way which researchers use market prices as a proxy for effects on land use as land price reflects the conversion potential of the zoning policy that embodies in the parcel. While the outcome approach focuses on observing the physical land-use change comparing to the use before the implementation of the policy.

Pricing approach has several features. One popular feature of pricing approach is based on the land rent theories of Von Thunen and Ricardo. Given respective attributes and location, any parcel of land can be modeled for use in a way that would earn the highest rent. A number of studies have attempted to estimate the extent to which support policies have increased farmland prices, but I found only one that examines the effect of zoning policy on farmland price which is the work of [9]. The work found that large lot zoning, total acreage, distance from a highway, distance from a town, and the presence of a house on the parcel are the determinants of the price per acre of farmland. In addition, weighted productivity index and greater soil value were found to be negatively correlated with land prices. Another popular feature of pricing approach is based on the concept of local government that tries to maximize profit through zoning policy. Zoning is a function of land attributes and land price. Land prices are also a function of parcel characteristics and the designed zoning. If zoning policy is effective, the method should produce an effect on land price. Many works examine the effect of the zoning on land price based on this concept. Vaillancourt and Monty [10] found that the land zoned for agriculture is 15–30% less in price than unrestricted land. Wallace [11] also found that zoning policy decreases the price of parcel zoned for large lots and the land price is found to be higher for other type of zoning. Logan and Zhou [12] indicated that zoning has modest effect on the changes in land rent, local population, family income, and the number of black population. The agricultural landowners perceive that the limitations imposed on their permitted uses will decrease their land value, but it is not always be the case. McMillen and McDonald [13] analyzed the correlation between zoning policy and land use, finding policy to have a strong effect on land utilization. Henneberry and Barrows [14] indicated that the effect of agricultural zoning on land price depends on the location and characteristics of the parcel. The agricultural zoned parcel that is located far from the city center is found to have a higher price than the agricultural zoned parcels close to the city. The result is in line with [15] who found that the land zoned for low-density development has a significant decrease in value, while the land zoned for high density or for community has a significant increase in value. More recent work of [16] also showed that agricultural zoning can be an effective policy for protecting productive farmland, while in the evidence from Japan, Nishihara [17] showed that agricultural land zoning policy has a great impact on land value increase and discourages the landowner from cultivating the land. A small gap in the anticipation of land policy to convert agricultural zone to residential development zone prevents landowner from selling or leasing the land to a more efficient farmer. Some studies based on this concept indicated that the zoning policy is not an effective tool for managing land use. Mark and Goldberg [18] examined the relationship between rezoning and changes in property values and the ability of zoning to mitigate externalities and found that rezoning does not necessarily lead to changes in land use and value. In the evidence from the United States, Pogodzinski and Sass [19] showed that zoning has no effect on house price but instead the minimum lot sizes, minimum side yard restriction, and maximum height restrictions affect house price, suggesting that the zoning method does not affect land use.

The other approach that uses to evaluate the effectiveness of zoning measure is the outcome approach. The approach is a direct proxy on land-use change as it compares the real physical change before and after the implementation of the policy. Not much research has been done on this approach. In England, Coughlin [20] found that agricultural zoning significantly reduces the sales of farmland to the conversion purpose, while in the evidence from the United States, Kline and Alig [21] found that the zoning reduces the likelihood of development on the land located within forest use, but the result for exclusive farm use zones is inconclusive. Levine [22] examines the effect of the growth-control enactment between 1979 and 1988 on net housing construction between 1980 and 1990. The study showed that local growth-management measures significantly displaced new construction such as rental housing and an expansion of the metropolitan areas into the interiors of the state. The measures have strong impact on low-income households and minorities, but not all growth-control measures were associated with this change, and the strong effect was found in the zoning measure. Cho and Wu [23] also adopts this concept to evaluate the interactions between residential development, land-use regulations, and public financial impacts in California, Idaho, Nevada, Oregon, and Washington. The study found that the land-use regulation decreases land development, long-run expenditure and property tax at the higher cost of housing prices, and property tax, while land-use regulations, land development, public expenditure, and property tax are affected by population, geographic location, land quality, housing prices, and the risks and costs of development. Mayer and Somerville [24] found that land-use regulation lowers the steady-state level of new construction. Metropolitan areas with more extensive regulations have up to 45% fewer starts and price elasticities that are more than 20% lower than those in less regulated markets. Kuminoff and Sumner [25] found that population growth and edge length of urban interface are statistically significant and positively correlated with conversion. Zoning and development restrictions were not significant explanatory variables for conversion. More strong result was found in the work of [26]. They evaluated the efforts of Utah County to discourage the farmland conversion to other uses and found that zoning has not been successful in agricultural land protection. Coxhead and Demeke [27] evaluated the determinants of land-use decisions made by farmers in an upland area of the Philippines. According to the findings, the most important predictors explaining substantial farmland allocations were crop prices and policy reform.

The past evidence and studies do not provide any conclusion, that is, if the agricultural zoning policy is actually helpful to protect agricultural land and ensure the food sustainability. The only best conclusion that can be drawn from these researches is that the performance of zoning policy depends on the circumstance and the context of each government. One must be tempted to think that the government that has similar context and circumstance will be able to refer to the research result of each other as a guideline for managing their land use, but in reality there is no absolute likeness in a context and circumstance of government. The agricultural zoning formulation of a country is, in fact, a complex issue. Therefore, there is a need for evaluating each government policy performance separately in order to improve and solve an individual measure while the policy is on process.

4. Agricultural zoning in Thailand

In this section, the history and the implementation of agricultural zone in Thailand will be discussed as a case study. Thailand has a unique circumstance and context which other countries cannot refer to its success or failure in agricultural zoning policy as their guideline for formulating their own agricultural zoning. What happened in Thailand will be a supporting evidence to the conclusion in the previous section that the effectiveness of agricultural zoning cannot be judged by referring to the result of other area and the evaluation must be done separately for each area. Again, the historically grounded approach was used by reviewing relevant Thailand agricultural acts and policies, various government office reports, agricultural economics crop zoning management manual, and online news.

Thailand is one of the Asian developing economies. As an agricultural country that is facing dynamic change in all aspects especially land use and land market, Thailand has introduced the Agricultural Economics Act which creates the agro-economic zone or in short "agrozoning" since 1979." The act defined an agro-zoning as "an area of agricultural production, to be established according to the soil type, rainfall, temperature, economic crop, farm type and main income of farmer by using the boundary line of the province as border zone." The objective of the agro-zoning is to plan the long-term development in agriculture, promote land-use type that match with its parcel suitability, control the agriculture data and statistic, and follow up the agricultural program with low budget and timeframe. In 1987, the government has divided the agro-zoning into 24 zones as shown in **Table 1** based on the factors prescribed in the act. Therefore, the provinces that have similar attribute and characteristics were included in the same zone, and the specific crop type was set and encouraged for production in each zone.

In 2013, the government adjusted the agro-zoning to become agricultural economic zone or in short "agricultural zoning" in order that it is possible to carry out in practice. The definition of agricultural zoning has been adjusted to "an area of agricultural production, including animal husbandry and reforestation to be established according to the market conditions and

http://dx.doi.org/10.5772/intechopen.80262

Zone number	Province	Important commodities
1	Udon Thani, Nong Bua Lamphu, Nong Khai	Cassava, rice, cow, buffalo
2	Sakon Nakhon, Nakhon Phanom, Mukdahan	Cassava, rice, cow, buffalo
3	Yasothon, Ubon Ratchathani, Amnat Charoen	Cassava, rice, cow, buffalo, jute
4	Khon Kaen, Kalasin, Maha Sarakham, Roi Et	Cassava, rice, cow, buffalo
5	Surin, Buriram, Sisaket	Rice, buffalo, corn
6	Nakhon Ratchasima, Chaiyaphum	Cassava, rice, cow, buffalo, jute
7	Phetchabun, Lopburi, Saraburi	Rice, corn, green beans, millet
8	Nakhon Sawan, Uthai Thani	Rice, corn, millet, buffalo, cassava
9	Tak, Kamphaeng Phet, Sukhothai	Rice, cow, soybean, peanut, fruit
10	Phichit, Phitsanulok	Rice, green bean, buffalo, cow, tobacco, corn
11	Nan, Phrae, Uttaradit	Rice, soybean, peanut, tobacco, buffalo
12	Chiang Rai, Phayao, Lampang	Buffalo, rice, tobacco, vegetable
13	Chiang Mai, Lamphun, Mae Hong Son	Vegetable, buffalo, cow, soybean
14	Chai Nat, Suphan Buri, Sing Buri, Ang Thong	Rice, cow, sugarcane
15	Ayutthaya, Pathum Thani, Nonthaburi, Bangkok	Rice, fruit, vegetable, flower
16	Kanchanaburi, Rajburi, Phetchaburi, Prachuap Khiri Khan	Sugarcane, cow, rice, corn, fruit
17	Samut Songkhram, Samut Sakhon, Nakhon Pathom	Coconut, fruit, fisheries, flower
18	Prachinburi, Sa Kaeo, Chachoengsao, Nakhon Nayok	Rice, buffalo, cassava
19	Samut Prakan, Chonburi, Rayong	Cassava, fisheries, rice, sugarcane, coconut
20	Chanthaburi, Trat	Fruit, cassava, rubber, corn, fisheries
21	Chumphon, Ranong, Surat Thani	Rubber, coffee, cow, coconut, fisheries, oil palm
22	Nakhon Sri Thammarat, Phatthalung, Songkhla, Satun	Rubber, cow, rice, fisheries, coconut
23	Phang Nga, Krabi, Trang, Phuket	Rubber, coffee, cow, cashew nut, fisheries, oil palm
24	Pattani, Yala, Narathiwat	Rubber, cow, coconut, fruit

Source: Office of Agricultural Economics, 2015 [28]

Table 1. Twenty-four agricultural economic zones in Thailand.

agricultural economy of the country by taking into consideration conditions similar to the main factors such as climate, water resources, crop area, animal feed, types of farming and income of farmers." The main objective of the agricultural zone in this era is to promote the land use that matches with its suitability; to balance the crop supply with the market demand, in the hope that the balance quantity will solve the crop price instability issue; and to develop a systematic control of the agricultural program at a provincial level. The government uses zoning to set specific area for specific crop production based on water, soil, rainfall level, temperature, location, and farmers' income. It is also anticipated that the zone set will help to increase the production productivity in each area as the crop set offers better suitability for the farmers' respective areas. Since farmers are not familiar with such crops, the government provides training for farmers on the production of new crop. In addition, a guaranteed selling price was implemented to help farmers who need to replace the crop type to comply with the zone set. By these actions the crop production will be matched with the physical characteristics of the area, and the farmers will have sufficed knowledge and experience for production. The government also moves and expands the existing factories to the most suitable agricultural areas, so that there will be markets to support such crops. Since the establishment of the program in 1979, the program appears to have limited success. Therefore, the government tries to catch the public attention by announcing the agricultural economics crop zoning of 20 agricultural commodities in 2013. These commodities comprise thirteen agricultural crops (rice, cassava, hevea, oil palm, sugarcane, maize, pineapple, longan, rambutan, durian, mangosteen, coconut, and coffee), five livestock (beef cattle, dairy cattle, swine, chicken, and hen), and two fisheries (sea prawn and freshwater animal). The Ministry of Agriculture and Cooperatives (MOAC) then has scheduled a road map to speed the development of the zoning in the period of 2013–2032 which is totally 20 years.

The crucial point here is that Thailand has established the agricultural zoning with different sets of objectives from other countries. The measure does not intend to prevent the loss of agricultural land or to control an expansion of urban area into agricultural area. The data from the Office of Agricultural Economics (OAE) in **Table 2** showed that Thailand has an increasing trend of agricultural land area. There was a drastic increase in agricultural land area between 1986 and 2015, and a quick surge of agricultural land area was found between 2001 and 2002. It was increased from 127,431,751 rai in 2001 to 146,768,298 rai in 2002 due to a great number of agricultural price support schemes and huge demand of economic crop in the global market. However, after 2002 its trend still appears to be increasing along the horizon even in the top ten most urbanized city in Thailand, except for Bangkok, the capital city, as shown in **Tables 3** and **4**.

The figures and facts indicated that Thailand uses zoning to address the issue of mismatching between land use and land suitability and the issue of crop price instability. Therefore, the agricultural zone has been adopted by adjusting it to be commodity-based format and not use-based format as in other countries. As mentioned previously, we can see that agricultural zoning in Thailand has been divided into two periods. In the beginning of the program which is during 1979–2012, the area was first set followed by deciding the suitable crop to be produced in each area, while the second period which is between 2013 and 2032, the crop was first to be set (how much the country wants the specific crop in the market) and then followed by matching the area with the announced crop. The MOAC has created the country's Agri-map which provides suitability information of land in each area, and the provincial office needs to confirm if the map complied with the real physical land of the province. In this period, the zoning program will be operated at a provincial level.

Year	Total land area (rai)	Agricultural area (rai)	(%)
1986	320,696,888	127,789,900	39.85
1987	320,696,888	128,062,343	39.93
1988	320,696,888	128,545,799	40.08
1989	320,696,888	128,546,022	40.08
1990	320,696,888	128,762,844	40.15
1991	320,696,888	129,621,724	40.42
1992	320,696,888	128,589,662	40.10
1993	320,696,888	127,794,556	39.85
1994	320,696,888	128,338,834	40.02
1995	320,696,888	128,959,887	40.21
1996	320,696,888	128,303,197	40.01
1997	320,696,888	127,602,084	39.79
1998	320,696,888	126,901,617	39.57
1999	320,696,888	127,762,512	39.84
2000	320,696,888	127,597,090	39.79
2001	320,696,888	127,431,751	39.74
2002	320,696,888	146,768,298	45.77
2003	320,696,888	146,541,409	45.69
2004	320,696,888	146,602,041	45.71
2005	320,696,888	146,673,114	45.74
2006	320,696,888	146,895,404	45.81
2007	320,696,888	147,119,966	45.88
2008	320,696,888	147,492,689	45.99
2009	320,696,888	147,588,846	46.02
2010	320,696,888	148,055,549	46.17
2011	320,696,888	148,067,113	46.17
2012	320,696,888	149,240,058	46.54
2013	320,696,888	149,236,233	46.53
2014	320,696,888	149,225,195	46.53
2015	320,696,888	149,242,393	46.54

Source: Office of Agricultural Economics, 2018 [29]

Table 2. Agricultural land area in Thailand from 1986 to 2015.

Year	Agricultural area (rai)				
	Bangkok	Chonburi	Songkhla	Samut Prakan	Nakhon Ratchasima
1986	395,075	1,861,900	1,879,848	311,154	7,857,931
1987	355,211	1,724,904	1,878,161	266,853	7,780,447
1988	323,837	1,716,842	1,895,693	245,732	7,746,955
1989	306,574	1,621,656	1,863,396	217,131	7,716,845
1990	313,599	1,446,529	1,913,696	179,865	7,720,889
1991	288,292	1,492,668	1,933,158	239,123	7,833,264
1992	253,871	1,474,013	1,919,411	198,333	7,832,394
1993	237,315	1,466,989	1,924,027	187,608	7,807,216
1994	221,131	1,416,303	1,975,175	197,666	7,747,841
1995	206,822	1,367,404	2,040,258	201,977	7,680,030
1996	181,099	1,336,578	2,035,236	211,205	7,626,904
1997	159,288	1,303,320	2,034,943	213,217	7,615,387
1998	139,394	1,267,573	2,034,351	221,155	7,573,219
1999	125,359	1,271,462	2,062,808	206,558	7,659,630
2000	120,828	1,287,038	2,066,468	200,875	7,676,804
2001	119,623	1,305,153	2,071,358	194,711	7,661,896
2002	322,363	1,631,771	2,428,149	217,151	8,392,312
2003	323,070	1,627,028	2,439,855	216,859	8,299,745
2004	323,361	1,620,896	2,444,975	216,424	8,293,541
2005	323,779	1,620,333	2,450,061	216,666	8,291,510
2006	324,046	1,621,946	2,458,414	216,668	8,310,461
2007	323,269	1,834,552	2,459,470	214,764	8,336,850
2008	323,661	1,624,954	2,463,045	213,389	8,349,161
2009	325,012	1,620,023	2,470,831	212,981	8,336,476
2010	324,645	1,622,629	2,452,728	212,346	8,379,865
2011	326,573	1,621,671	2,451,857	212,248	8,378,455
2012	233,995	1,720,845	2,266,426	211,838	8,382,871
2013	233,911	1,720,137	2,266,511	211,449	8,383,120
2014	233,266	1,718,785	2,267,234	211,421	8,382,551
2015	233,186	1,719,457	2,269,009	211,704	8,385,473

Source: Office of Agricultural Economics, 2018 [29]

Table 3. Agricultural land area of the top ten most urbanized cities: Bangkok, Chonburi, Songkhla, Samut Prakan, and Nakhon Ratchasima.

Year	Agricultural area (rai)				
	Nonthaburi	Chiang Mai	Udon Thani	Pathum Thani	Khon Kaen
1986	198,698	1,401,397	4,829,053	763,674	4,068,582
1987	209,468	1,370,415	4,986,930	780,590	4,158,015
1988	201,581	1,381,934	5,062,530	707,577	3,979,004
1989	196,101	1,377,851	5,080,723	686,146	4,112,119
1990	194,967	1,269,312	5,097,243	695,483	4,157,993
1991	179,941	1,279,141	5,163,631	716,225	4,200,254
1992	166,071	1,247,978	5,184,504	670,599	4,100,790
1993	157,400	1,234,921	3,743,486	613,007	4,073,572
1994	153,632	1,250,985	3,737,699	607,271	4,099,446
1995	148,965	1,265,266	3,726,665	601,844	4,130,219
1996	160,802	1,259,494	3,706,577	577,502	4,089,495
1997	171,899	1,257,296	3,678,779	545,589	4,049,981
1998	184,525	1,248,918	3,638,467	521,029	4,000,177
1999	180,115	1,300,222	3,650,181	502,336	4,044,045
2000	170,652	1,300,530	3,645,349	491,555	4,054,090
2001	165,677	1,308,829	3,632,734	484,043	4,054,701
2002	227,610	1,724,473	4,385,100	572,478	4,502,696
2003	227,992	1,735,925	4,347,167	573,399	4,471,391
2004	228,184	1,742,904	4,346,415	573,884	4,472,107
2005	228,438	1,744,032	4,348,154	574,508	4,474,476
2006	228,596	1,749,532	4,356,468	574,889	4,480,981
2007	228,034	1,758,322	4,371,859	573,489	4,495,596
2008	228,172	1,769,898	4,382,422	573,796	4,507,209
2009	229,089	1,775,998	4,376,284	574,765	4,500,321
2010	228,568	1,790,597	4,397,072	573,062	4,520,704
2011	228,437	1,788,971	4,398,996	572,401	4,519,542
2012	219,259	1,830,686	3,868,581	510,265	4,219,317
2013	219,160	1,830,291	3,867,979	510,192	4,219,427
2014	219,012	1,830,078	3,868,003	509,752	4,219,573
2015	219,185	1,829,976	3,868,703	510,443	4,219,853

Source: Office of Agricultural Economics, 2018 [29]

Table 4. Agricultural land area of the top ten most urbanized cities in Thailand: Nonthaburi, Chiang Mai, Udon Thani, Pathum Thani, Khon Kaen.

After the implementation of the agricultural zoning, the agricultural sector in Thailand has transformed in the dimensions of crop diversification. Using the formula of Gibbs and Martin with the data from OAE, crop diversification index was calculated and presented in **Figure 3**. The index showed that Thailand has moved from mono-cropping pattern to a more diversified cropping pattern especially during the 1980s and 1990s which after the agricultural zoning was implemented. In 1977 and 1978, before the implementation of the zone, the index was 0.51, but in 2015, it was increased to 0.68. More farmers switch from rice to perennial crops such as coffee, sugarcane, cassava, corn, oil palm, fruit, vegetables, and flowers, livestock, fisheries, etc.

Despite that agricultural zoning has caused a significant agricultural transformation in Thailand, the level of success is considered to be limited considering that it has already been implemented for over 35 years. Only my work [30] analyzes the root cause of failure of the agricultural zoning in Thailand. The work found that the agricultural zone in Thailand was set based on physical and environmental characteristics and failed to consider crop prices, input price, and property right, which farmer prioritizes as their top priorities when allocating production to their agricultural land. These missing factors except property rights are economic factors. The zoning control results in land-use efficiency fail because variable factors that the government considers are not in line with private demand. The government fails to adequately perceive the private demands existing in land use of landowner and responds instead to the political pressure of organizing land use at the country level. The analysis of the work is however limited to the farm-level variable. The external factors at the country level such as the government policies are not included in the analysis. In reality, such factors have tremendous effect on the zoning program. Therefore, in the next section, we will go over some of the governments' agricultural policies that conflict with the zoning program and cause it to fall short of expectation.

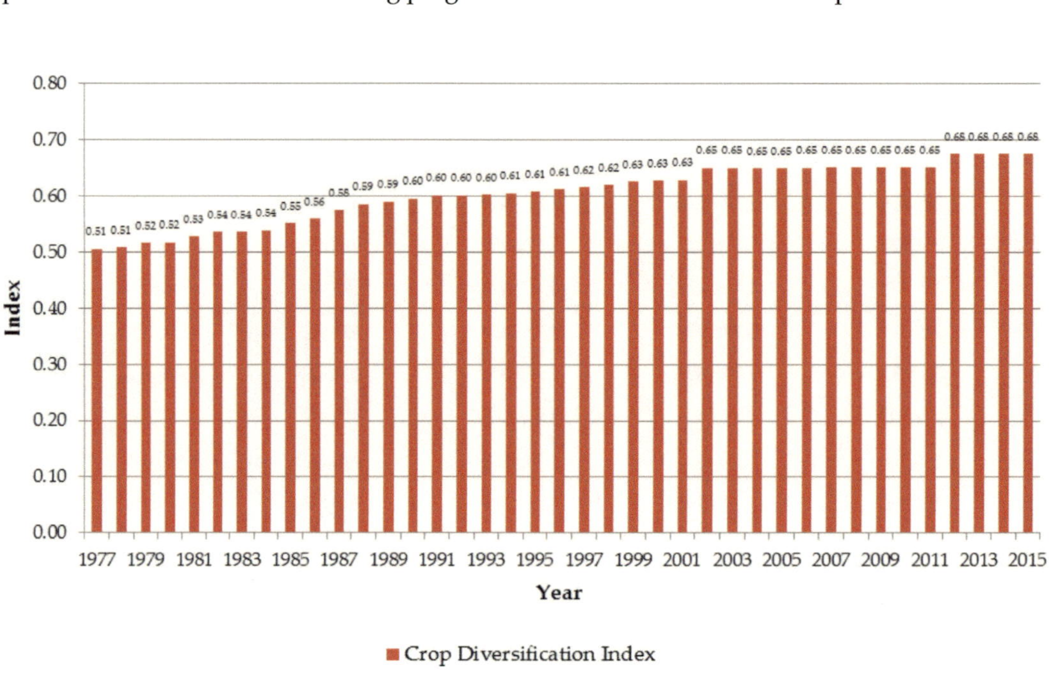

Figure 3. Crop diversification index of Thailand between 1977 and 2015.

5. Policy conflict

Most of the agricultural policies in Thailand were found to be conflicted with each other especially the policy of agricultural zoning and crop subsidy schemes. Agricultural commodity price in Thailand has been intervened by its government from time to time (more appropriate phase is "all the time"). There were mortgage schemes for rice, cassava, longan, rubber, oil palm, shallots, sunflower seed, and many more. On top of this, there are aid fund, act, strategic plan, and programs for many agricultural commodities such as rubber aid fund, the Sugarcane and Sugar Act of 1984, rubber price stability program, oil palm industry strategic plan, etc.

The most notorious program that ruins the agricultural zoning program is the rice mortgage scheme. Rice production is the largest portion of the Thai economy and labor force; therefore, any policy on rice will produce a widespread effect in the country. In 2011, while the agricultural zoning is functioning, the rice mortgage scheme was implemented with the objective to increase the rice price, create stabilize Thai economy through increased domestic consumption, increase Thai farmers' well-being, and control the world market price of rice. The attempt to replace rice paddy field with sugarcane field by agricultural zoning failed because most farmers switch from other crops to rice in order to obtain the guaranteed price. Likewise, this incident happens in the case of the mortgage scheme for cassava, longan, hevea, oil palm, etc. While one of the main objectives of agricultural zoning is to control the supply quantity of products, so that the product price is not fluctuated, the price is in fact the cause of the change in the quantity supply. This happens because the agricultural commodities are in fact in the perfectly competitive market. In microeconomics theory, equilibrium price in the perfectly competitive market is determined through a process of interaction between the market demand and market supply not the market supply alone. Therefore, by introducing the series of mortgage scheme for each crop, the zoning policy has failed to function properly.

Another program to be discussed is the government's strategic plans for palm oil year 2008–2012 which aim to expand the plantation area of oil palm from 2.5 million rai in 2007 to 5 million rai in 2012, and the strategic plans for palm oil year 2015–2026 which aim to expand the plantation area of oil palm from 5 million rai in 2017 to 7.5 million rai in 2026. The strategic plan is part of the government's alternative energy development plan for 2015–2036. Oil palm is a primary raw material for biodiesel production which is an alternative energy of the country with an increasing demand in recent year. In 2012, the Ministry of Energy has a target of biodiesel production of 3100 million liters per year to increase the country's' energy security. According to the Agri-map of the MOAC, the suitable area for oil palm production is in the south of Thailand where there is the highest rainfall in the country. The rainfall in the south of Thailand is regular and scattered throughout the region. Oil palm needs plenty of water and 6 hours of sunlight each day with the plantation area lower than 300 meters above sea level. Each tree needs 5–350 liters of water daily, monthly rainfall of 150 millimeters, and dry season not more than 60 days [31]. Therefore, the zone set for oil palm is in the southern area of Thailand. However, the high demand of domestic palm oil and the oil palm strategic plan creates an attractive economic incentive for farmer to switch the production from other crop to oil palm. In 2015, the Ministry of Agriculture and Cooperatives (MOAC) introduced the oil palm to northeast of Thailand through the pilot projects in Nong Khai province. Northeast of Thailand is not suitable for the production of oil palm as it has poor soil, low rainfall, and a long dry season. However, the expansion of oil palm plantation area has been promoted through various interventions, such as seedling, low-interest rate funding, input, mills, and refineries. As a result, oil palm plantation area has rapidly increased in the northeast of Thailand. The attempt to encourage land use according to its suitability of the zoning program has again failed due to an attractive economic incentive created by the government. Besides, the other objective of zoning which is to stabilize the crop price also failed. Consider **Figure 4**, it presented the price of oil palm during 1997–2018. The oil palm price has been fluctuated even during the period of both oil palm strategic plans (2008–2012 plan and 2015–2026 plan). In 2017, only 2 years after the implementation of the second plan, oil palm price starts a dramatic fall due to the excess supply of the crop. The government tries to solve the low-price issue by exporting the residual quantity to global market, but the attempt has not succeeded in solving the excess supply problem and low oil palm price in the country. As of March 2018, the oil palm price is as low as 3.02 baht per kilogram. The government is now considering an income compensation scheme for farmer as a short-term assistance and a cut down oil palm tree program with compensation as a long-term measure to combat the low-price issue.

The last instrument to discuss is the Sugarcane and Sugar Act of 1984. The act was enacted after the implementation of agricultural zoning. Since the enactment of the Sugarcane and Sugar Act of 1984, sugarcane production area has considerably expanded by almost 160% from 1982 to 1996 [32]. According to the act, sugarcane farmers and sugar factories share the profits from production at a rate of 70:30 under the condition that the sugarcane field must not be located more than 80 km from the factory. The objective of the act is to stabilize the sugarcane price for farmer. After the enactment of the act, the sugar factory has been expanded to 53 factories in 2015 from 17 factories in 1984. The sugar price remains quite high as sugarcane

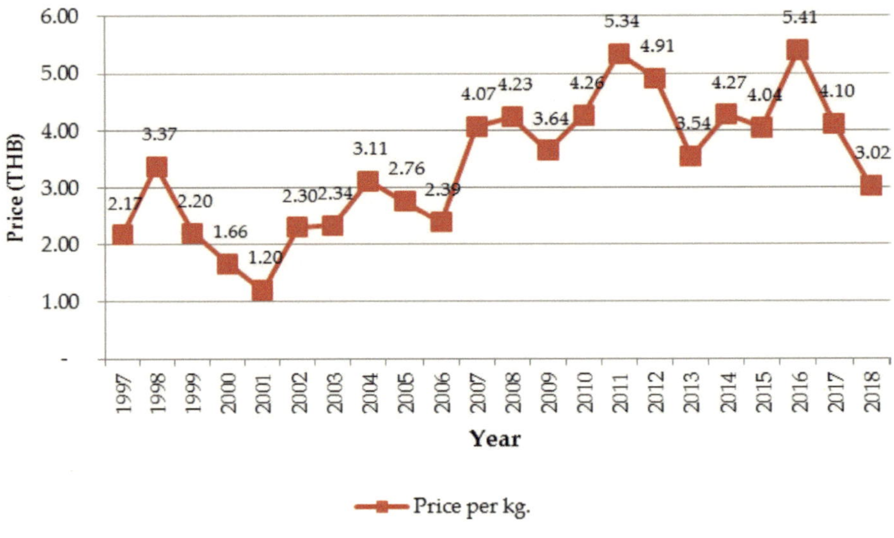

Figure 4. Oil palm price between 1977 and 2018.

http://dx.doi.org/10.5772/intechopen.80262

farmers and factories try to push the domestic price up to share the set benefit. The country exports approximately 70% of sugar product, and only 30% was distributed for domestic use. The agricultural zoning again could not resist the expansion of land use in sugarcane production because of the high economic incentive initiated by the act.

The government tends to implement the policy to tackle the issue at hand without considering the existing active policies. These actions make each agricultural issue more complicated to be handled and result in a high government budget spending in each incident. After the implementation of the agricultural zoning, the government should have a minimum intervention in the market, so that the instrument will be able to function properly.

6. Findings of the study

The most crucial factor that causes failure on agricultural zoning in Thailand at the farm level is the agricultural commodity price, while the governments' latter crop subsidy scheme is the most crucial factor that causes the failure of agricultural zoning at the country level.

Obviously, the finding reveals that the most crucial factor that causes agricultural zoning in Thailand to fail at the farm level and at the country level is actually agricultural commodity price. This suggests that agricultural zoning in Thailand should undergo major revision by taking the agricultural commodity price into account. Therefore, a proposal to revise the agricultural zoning used in Thailand based on this finding will be discussed in the next section.

7. A proposal for a revision of agricultural zoning in Thailand

The revision of agricultural zoning proposed here is based on the principles for dealing with two critical issues that agricultural zoning program in Thailand tries to solve: the mismatching between land use and land suitability and the crop price instability.

The proposal is to set agricultural commodity price based on its production area or in other words its origin. This zone pricing is to be included in the existing zoning program. The Land Development Department (LDD) has completed the country soil survey and matched the land quality with each crop growth requirement in order to help the government to determine the crop zone. They have divided land suitability into four categories: highly suitable (S1), moderately suitable (S2), marginally suitable (S3), and not suitable (N) [33]. The government will be able to set the crop price according to these criteria. The crop that is grown in the area with highly suitable category will have the highest price followed by the crop that is grown in the moderately suitable area, marginally suitable area, and not suitable area, respectively. The proof of the commodity origin is the certificate of origin issued by the local government office. By setting the agricultural commodity price based on its production area, an incentive for farmer to allocate their land to the most suitable crop set by the government is initiated. The method will eliminate the price risk for farmer and the mismatch land-use problem at the same time. The factory and the mill are assured of the quality of the product as it was produced on the most suitable area for such crop. To "set" the price is different from to "control" and to "fix" the price. The proposal here is not to control the market mechanism of demand and supply. The price based on the commodity origin can be set in several ways such as the percentage difference between the crop grown in highly suitable area and the crop grown in moderately suitable area.

One concern of this method is that the agricultural commodities are in the perfectly competitive market, so the country cannot prevent imported agricultural commodities especially their price set. However, when the use of agricultural land is complied with its suitability, it can be anticipated that the commodity produced can be competed with imported product in terms of quality. The origin declaration will also be applied for imported agricultural commodities. Typically, all types of the certificate of origin (Form D, CO, Form FTA, etc.) are included in the import document set. The importer only needs to put the tag on their commodity showing where the origin of their product is, and this practice has already been done recently in Thailand for agricultural product sale in modern trade store. The consumer will benefit from the information on the commodity origin as some imported product prices are much lower than that of the domestic product but with great poor quality especially in terms of safety. Consumer behavior is changing as there is a raised concern on safety food; therefore, the product origin information is a crucial factor for making a purchase decision instead of price. The agricultural commodities are price inelastic of demand as it is a basic necessity product. In the short run, the quantity purchase is not very responsive to price change. Therefore, the concern of the lower price of imported product is quite inconsiderable. In the worst case, the impact from imported product can be dealt through price protection scheme, but it is not suggested as Thailand has an obligation under the Doha Round of WTO negotiations in November 2001 to reduce the domestic agricultural support and protection. Therefore, the price protection will not be a sustainable way to deal with the problem.

8. Conclusion

Agricultural zoning is a regulatory approach that redefines property rights. The technique has been used to preserve agricultural area and ensure food security. Many empirical evaluations of its effectiveness in several countries have been made, but they cannot be used as a reference to other countries as each country has unique context and circumstance.

Thailand is one of the Asian developing countries, which has adopted the agricultural zoning, but the method is adjusted to be the commodity-based format and not use-based format. The main objectives of agricultural zoning in Thailand are to address the issue of mismatching between land use and land suitability and the issue of crop price instability. Agricultural zoning in Thailand has been established since 1979, but it showed limited success mainly due to the variable factors that the government considers which are not in line with private demand. In addition, the implementations of latter policies are contradicted to the zoning policy. The program has caused an agricultural transformation in Thailand in the dimension of crop diversification, but it needs a major revision in order to achieve its goals.

The revision proposed is to set agricultural commodity price based on its origin. There are two strategies for meeting Thailand's agricultural zoning objectives. First, after setting the price based on its origin, the government should also play a small role as possible in intervening the agricultural commodity market. Second, the government should integrate a crop production information system with the Agri-map of the MOAC for farmer to check the production area quantity of each crop in the country. This current production quantity information should be updated at least monthly, so that farmer will have sufficient information to make a decision on the quantity supply of the agricultural commodity apart from the price.

Acknowledgements

I would like to express my appreciation to Professor Tongroj Onchan and Assistant Professor Supachat Sukharom for giving useful comments on my chapter draft and the Faculty of Economics at Si Racha, Kasetsart University, Chonburi, Thailand, for financial support.

Author details

Nararuk Boonyanam

Address all correspondence to: noaratc@gmail.com

Faculty of Economics at Sri-Racha, Kasetsart University, Sriracha Campus, Chonburi, Thailand

References

[1] Hardin G. The Tragedy of the Commons. Science. 1968;**162**(3859):1243-1248. DOI: 10.1126/science.162.3859.1243. PMID 5699198

[2] Gine X. Land security in rural Thailand: Evidence from a property rights reform. World Bank Policy Research Working Paper No. 3734; 2005

[3] The United States Agency for International Development. Land Tenure and Property Right Assessment for Angola. Washington, DC: ARD Inc.; 2005. Contract No. LAG-I-00-98-00031-00

[4] Fischel WA. The Evolution of Zoning Since the 1980s: The Persistence of Localism; September 1, 2010. Available at SSRN: https://ssrn.com/abstract=1686009 or http://dx.doi.org/10.2139/ssrn.1686009

[5] Fischel WA. The Economics of Zoning Laws: A Property Rights Approach to American Land Use Controls. Baltimore, MD: Johns Hopkins University Press; 1985. 400 p

[6] Cole HD, Ostrom E. Property in Land and Other Resources. Lincoln Institute of Land Policy: United States; 2012. 504 p

[7] Nelson RH. Zoning and Property Rights: An Analysis of the American System of Land-Use Regulation. Massachusetts: The MIT Press; 1977. 259 p

[8] Marshall C, Rossman GB. Designing Qualitative Research. Newbury Park, DA: Sage Publications, Inc.; 1998. 261 p

[9] Stewart PA, Libby LW. Determinants of farmland value: The case of DeKalb County, Illinois. Review of Agricultural Economics. 1998;**20**(1):80-95

[10] Vaillancourt F, Monty L. The effect of agricultural zoning on land prices, Quebec, 1975-1981. Land Economics. 1985;**61**(1):35-42

[11] Wallace NW. The market effects of zoning undeveloped land: Does zoning follow the market? Journal of Urban Economics. 1988;**23**:307-326

[12] Logan JR, Zhou M. Do suburban growth controls control growth? American Sociological Review. 1989;**54**(3):461-471

[13] McMillen DP, McDonald JF. Urban land value functions with endogenous zoning. Journal of Urban Economics. 1991;**29**:14-27

[14] Henneberry DM, Barrows RL. Capitalization of exclusive agricultural zoning into farmland price. Land Economics. 1990;**66**:249-258

[15] Bockstael NE. Modeling economics and ecology: The importance of a spatial perspective. American Journal of Agricultural Economics. 1996;**78**:1168-1180

[16] Esseks JD, Long LM. The political viability of agricultural protection zoning to prevent premature conversion of farmland. In: Libby LW, Abdalla C, editors. Protecting Farmland at the Fringe: Do Regulations Work? Strengthening. The Research Agenda, Conference Proceedings, Baltimore, MD; 5-7 September, 2001 [Internet]; 2001. Available from: http://aede.ag.ohio-state.edu/Programs/Swank/pdfs/protecting_farmland_at_the_fringe.pdf [Accessed: April 5, 2018]

[17] Nishihara M. Real option valuation of abandoned farmland. Review of Financial Economics. 2012;**21**(4):188-192

[18] Mark JH, Goldberg MA. Land use controls: The case of zoning in the Vancouver area. American Real Estate and Urban Economics Association Journal. 1981;**9**(4):418-435

[19] Pogodzinski JM, Sass TR. The theory and estimation of endogenous zoning. Regional Science and Urban Economics. 1994;**24**:601-630

[20] Coughlin RE. Formulating and evaluating agricultural zoning programs. Journal of the American Planning Association. 1991;**57**(2):183-192

[21] Kline JD, Alig RJ. Does land use planning slow the conversion of forest and farm lands? Growth and Change. 1999;**30**:3-22

[22] Levine N. The effects of local growth controls on regional housing production and population redistribution in California. Urban Studies. 1999;**36**:2047-2068

http://dx.doi.org/10.5772/intechopen.80262

[23] Cho SH, Wu J. Measuring interactions among urban development, land use regulations, and public finance. In: Selected Paper, American Agricultural Economics Association Annual Meeting, Chicago, IL; August 5-8, 2001 at Agricultural Economics Search [Internet]; 2001. Available from: http://agecon.lib.umn.edu/cgi-bin/pdf_view.pl?paperid=2552 [Accessed: March 15, 2018]

[24] Mayer CJ, Somerville CT. Land use regulation and new construction. Regional Science and Urban Economics. 2000;**30**:639-662

[25] Kuminoff NV, Sumner DA. Modelling farmland conversion with new GIS data. In: Proceeding of the American Agricultural Economics Association Meeting; 5-8 August 2001; Chicago; 2006. pp. 31-45

[26] Nellis L, Maca MN. The effectiveness of zoning for agricultural lands protection: A case study from Cache County, Utah. Landscape and Urban Planning. 1986;**13**(1986):45-54

[27] Coxhead I, Demeke B. Panel data evidence on upland agricultural land use in the Philippines: Can economic policy reforms reduce environmental damage? American Journal of Agricultural Economics. 2004:1354-1360

[28] Office of Agricultural Economics. Agricultural Information Center: Agro-Economic Zone; 2018. Available from: http://www.oae.go.th/main.php?filename=index [Accessed: April 4, 2018]

[29] Office of Agricultural Economics. Report of Agricultural Land Use in Thailand: Provincial Level Year 1986-2015; 2018. Available from: http://oldweb.oae.go.th/download/use_soilNew/soiNew/landused2558.html [Accessed: May 8, 2018]

[30] Boonyanam N. Agricultural economic zones in Thailand. Land Use Policy. 2017. DOI: 10.1016/j.landusepol.2017.05.023. Available from: https://www.sciencedirect.com/science/article/pii/S0264837716305889

[31] Yangdee B. Ten million rai of oil palm plantation: A catastrophe for the Thai people. In: World Rainforest Movement; 2011 [Internet]. n.d. Available from: www.wrm.org.uy/countries/Thailand/Catastrophe.pdf [Accessed: April 18, 2018]

[32] Chanhom K. WTO Ruling to Strengthen World's Export on Sugar: Thailand as a Case Study [Internet]. n.d. Available from: http://www.thailawforum.com/articles/wto-ruling-strengthen-world-export-sugar-thailand6.html [Accessed: May 20, 2018]

[33] Ministry of Agriculture and Cooperatives. Agricultural Zoning Management Manual [Internet]; 2014. Available from: www.oic.go.th/FILEWEB/CABINFO CENTER3/DRAWER073/.../00000109.PDF [Accessed: March 1, 2018]

Chapter 5

Consequences from Land Use and Indirect/Direct Land Use Change for CO_2 Emissions Related to Agricultural Commodities

Stefan J. Hörtenhuber, Michaela C. Theurl, Gerhard Piringer and Werner J. Zollitsch

Additional information is available at the end of the chapter

http://dx.doi.org/10.5772/intechopen.80346

Abstract

Increasing demand for food, feed, and fuels adds pressure on ecosystems through land use and land use change (LULUC), with greenhouse gas emissions among the most significant environmental impacts. Large regional variation in LULUC and indirect driving forces may not be adequately addressed by a one-size-fits-all approach that assigns equal LULUC emissions per unit of area, and by a focus on direct d(LU) LUC impacts only. Hence, our method integrates effects from international agricultural commodity trade as indirect emissions (iLULUC) of the demand of food and feed. In most countries, the majority of foods and feedstuffs (70% of global calories) are produced for the domestic market and the rest is exported and contributes to a hypothetical global pool of iLULUC emissions. Total LULUC emissions are calculated for individual countries, accounting for LULUC from increased domestic agricultural production for domestic consumption and for emissions imported from the global market's iLULUC pool. Furthermore, we estimate consumption-based emission factors for specific product groups per country. Results show that vegetable oils, oil crops, and cereals account for the majority of global LULUC emissions and iLULUC results derived with the presented method cannot be compared directly to dLULUC results; however, their orders of magnitude are similar.

Keywords: land use, land use change, LULUC, CO_2, greenhouse gas emissions, global warming potential, carbon footprint, food, consumption-based accounting

IntechOpen

1. Introduction

Increased global demand for food energy and protein is a major driver for the growing environmental impacts of food and feed production. Impacts include both land use (LU) emissions on already cultivated agricultural areas through intensification and land use change (LUC) emissions from newly converted areas such as primary and secondary forests, fallow land, and savannahs [1, 2]. The increased demands for livestock products and bioenergy are major causes of increases in agricultural LU [3]. This increased land use leads to LUC. Over the past 50 years, livestock and bioenergy accounted for 65 and 36% of LUC, respectively [3]. Other socioeconomic drivers of emissions from LULUC are population growth, economic development, and changing consumption patterns [4–7]. An accurate accounting for LULUC impacts is critical for life cycle assessment (LCA) frameworks and other assessment methods that quantify agricultural greenhouse gas (GHG) emissions.

LU and LUC (LULUC) are major contributors to global CO_2 emissions, especially in the tropical regions of South-America, Asia, and Africa. Emissions from LULUC contributed approximately 20% of total global CO_2 emissions during the last two decades of the twentieth century [8]. From 2000 to 2010, the proportion of CO_2 emissions originating from LUC substantially decreased, but still contributed about 10–12% of global CO_2 emissions [9, 10]. Simulations of the development of atmospheric CO_2 concentrations, which were used to determine the impact of LULUC since preindustrial times (i.e., the last 250 years), showed that almost a quarter (23%) of the increase in the CO_2 concentration originates from LULUC [11].

Emissions from the conversion of known and defined regions of origin are coined as "direct" (dLUC; see [1]). dLUC emissions consider carbon released when a specific area is transformed, e.g., from forest to cropland or builtup land (i.e., land for infrastructure, buildings). Although region-specific dLUC emission accounts are useful, they fail to account for the effects of international agricultural commodity trading. The concept of indirect LUC (iLUC) increasingly became an issue in the life cycle analysis of biofuels that substitute fossil fuel and often were discussed as climate-neutral alternatives [12]. Additionally, iLUC emissions have wide-ranging policy implications [13, 14]. Indirect effects not only apply to LUC, but also to LU emissions. Consequently, market-induced or policy-driven incentives to transfer and expand land (i.e., forest clearance) to meet increased market demands for bioenergy plants and biofuels, food and feed distant in countries are related to and responsible of iLULUC. However, iLULUC emissions from shifts in international agricultural commodity trading have, so far, been rarely estimated. The studies found in the literature strongly focus on the iLUC debate in the context of bioenergy plant cultivation [15].

In the globalized world, many countries are exporters of food, feedstuffs, and bioenergy fuels actually, and cause domestic (i) LULUC emissions on behalf of the countries buying their commodities on the global markets. We hypothesize that countries with increasing net agricultural exports will tend to emit more CO_2 from LULUC as well, because they are forced to increase production through conversion of previously unused land (i.e., LUC) and intensification of cultivation on existing land (causing LU emissions due to soil carbon losses). These developments are of course subject to other factors; for example, a growing domestic population will exacerbate LULUC emissions.

Consequently, the objective of this work is to provide a deterministic, top-down method which accounts for the effects of iLULUC linked to international agricultural commodity trade on country-specific LULUC emissions. The aim is to provide a consistent and scientifically robust method that allows for the inclusion of consumption-based iLULUC emission factors into LCA and carbon footprint of different agricultural commodities consumed in the different countries.

2. Methods

In this section, we describe the conceptual background used for the development of our method, some of its key assumptions, and the computational steps involved as well as the empirical analysis of country- and product-specific LULUC emissions.

2.1. The conceptual background of country-specific shares of agriculture-related LULUC emissions

In general, agricultural commodities with increasing production volume exert stronger pressure on (currently unused) land than products with decreasing production volumes when accounting for environmental impacts of LULUC. Therefore, increasing production should be assigned a larger share of impacts. In our approach, we assume that agricultural exports can be linked to international iLULUC effects: if domestic production becomes more export-oriented, domestic supply will decrease and the unmet domestic demand will lead to increased commodity imports if economically feasible. Within our approach, we assume the existence of a (hypothetical) global pool for iLULUC emissions based on the commodities that are traded.

Aside from the global iLULUC emissions pool, the method presented takes a country-specific approach, since trends in agricultural production, imports, and exports differ by region (as well as by product type). A country-specific method allows a better consideration of large regional LULUC variations than a one-size-fits-all approach. The latter would assign equal LULUC emissions on an area basis (for every hectare used globally to produce food, feed, fuels, or fibers; see, e.g., [16, 17]), regardless of regional differences. Moreover, if regional LULUC data and regional agricultural statistics are available within a country, the approach could easily be adapted to a higher spatial resolution as well.

Countries with increasing agricultural exports will feed a proportional share of their total LULUC (i.e., LU-related as well as dLUC- and iLUC-) emissions into this pool, thereby reducing their burden of LULUC emissions, and countries with increasing net imports will import a proportional share of these global iLULUC pool emissions. It is important to note that this takes a dynamic rather than a static view: the yearly changes in exports and imports determine the flows of iLULUC emissions, and not the absolute export and import data (see Eqs. (6) and (8)).

In order to allow an aggregation of the wide variety of agricultural commodities produced by a given country and traded internationally, we convert commodity masses obtained from the FAO statistics [18] to their energy equivalent, based on lower heating value (LHV) data from [19, 20]. Furthermore, all calculations in this study include CO_2 emissions only and other

GHGs such as methane and nitrous oxide are excluded, since they typically contribute little to total LULUC emissions change [21].

2.2. Empirical analysis of country-specific shares of agriculture-related LULUC emissions

We calculated net LULUC emissions for 175 nations based on data reported in Refs. [18–21]. As a starting point, we use the CO_2 emissions L_k from Ref. [21] that are caused by LULUC for each country.

$$L_{glo} = \sum_k L_k \tag{1}$$

where L_{glo} is the annual worldwide LULUC emissions (excluding those countries for which no suitable data are available), Tg a^{-1} and L_k is the annual LULUC emissions from country k, Tg a^{-1}.

Each country's LULUC emissions have to be reduced by those LULUC emissions that are caused by the expansion of infrastructure areas (including builtup areas based on [22]) in order to allocate the remaining LULUC emissions to agricultural commodities that enter the economy of each country. Thus, we split the infrastructure LULUC emissions and the agricultural LULUC emissions (based on 2013 areas in [18]) in proportion to their countrywide area.

$$AG_k = L_k - INF_k \tag{2}$$

where AG_k is the annual agriculture-related LULUC CO_2 emissions from country k, Tg a^{-1} and INF_k is the annual infrastructure-related LULUC CO_2 emissions from country k, Tg a^{-1}.

In the model presented here, agriculture-related LULUC emissions are in principle allocated to the emitting country, but we correct this number by accounting for iLULUC-causing increases of net agricultural imports (imports minus exports) into each country, thus obtaining the agriculture-related LULUC emissions due to the domestic consumption of agricultural commodities in a given country k:

$$NL_k = AG_k + NI_k \tag{3}$$

where NL_k is the net annual agriculture-related LULUC emissions due to domestic consumption of agricultural commodities in country k, Tg a^{-1} and NI_k is the LULUC-related emissions due to net agricultural import increases into country k, Tg a^{-1}.

The following equations illustrate how the net import emissions are calculated. We first calculate the global iLULUC pool (Eq. (4)) and then distribute the iLULUC pool's emissions to countries proportional to their net import increases during the selected accounting period (Eq. (6)).

The global iLULUC pool is established by adding all export increase-related LULUC emissions EX_k:

$$iLULUC_{glo} = \sum_k EX_k \tag{4}$$

EX_k in turn are defined as the share of a country's agriculture-LULUC emissions that is proportional to a country's export increases:

$$EX_k = AG_k \times e_k \tag{5}$$

where EX_k is the LULUC emissions due to agricultural commodity export increases of country k, Tg a^{-1} and e_k is the export-increase allocation factor (nondimensional, Eq. (6)).

The export-increase allocation factor e_k relates a country's agricultural export increases over the selected time period to its domestic agricultural production increase, both converted to annual energy equivalents based on the exports' mass-weighted LHV:

$$e_k = \frac{\Delta E_k}{\Delta D_k} \tag{6}$$

where ΔE_k is the average annual export increases of agricultural commodities expressed as annual energy equivalents, TJ a^{-1} and ΔD_k is the average annual domestic production of agricultural commodities expressed as annual energy equivalents, TJ a^{-1}.

In our analysis, the increases ΔE_k and ΔD_k are both calculated as average annual differences between a final (i.e., 2007–2009) and an initial 3-year period (i.e., 1998–2000).

Now that the global iLULUC emissions pool has been established, its emissions are distributed among all countries in proportion to their individual net import increases ni_k:

$$NI_k = iLULUC_{glo} \times ni_k \tag{7}$$

where ni_k is the net-import-increase allocation factor (nondimensional), based on energy equivalents (Eq. (8)).

The net-import-increase allocation factor, ni_k, is defined as the difference between a country's share of global import increases and a country's share of global export increases:

$$ni_k = \frac{\Delta I_k}{\sum_k \Delta I_k} - \frac{\Delta E_k}{\sum_k \Delta E_k} \tag{8}$$

where ΔI_k is the average annual import increase of agricultural commodities expressed as annual energy equivalents, TJ a^{-1}, $\Sigma_k \Delta I_k$ is the global sum of average annual import increases, TJ a^{-1}, ΔE_k is the average annual export increase of agricultural commodities expressed as annual energy equivalents, TJ a^{-1}, and $\Sigma_k \Delta E_k$ is the global sum of average annual export increases, TJ a^{-1}.

2.3. Empirical analysis of product group-specific shares of agriculture-related LULUC emissions for a given country

In a next step, net LULUC emissions can also be calculated specifically for a product group p that is consumed in a country k. The approach follows largely that for countries as described in the previous section.

For aggregating the various flows of agricultural commodities (i.e., imports, exports, domestic production, and domestic demand), we again use the average energy content of each product group, aggregated based on the mass-weighted single-commodity LHVs. The following product groups in Ref. [18] are considered here: alcoholic beverages, cereals (excluding beer), fruits (excluding wine), oil crops, pulses, spices, starchy roots, sugar and sweeteners, sugar crops, tree nuts, vegetable oils, vegetables, animal fats, eggs, meat, milk (excluding butter), offal, stimulants; no data are available for the groups "tobacco and rubber" and "miscellaneous."

Each product group in a country is assigned a share of the countrywide agricultural LULUC AG_k in proportion to its energy-equivalent share of the total agricultural production:

$$AG_{k,p} = AG_k \times a_{k,p} \tag{9}$$

where $AG_{k,p}$ is the LULUC emissions of agricultural product group p in country k, Tg a^{-1} and $a_{k,p}$ is the production allocation factor (nondimensional, Eq. (10)).

The production allocation factor, $a_{k,p}$, relates a product group's production increases in country k to that country's total domestic agricultural production increase, both converted to annual energy equivalents based on LHV:

$$a_{k,p} = \frac{\Delta P_{k,p}}{\sum_p \Delta P_{k,p}} \tag{10}$$

where $\Delta P_{k,p}$ is the average annual production increase of product group p in country k, expressed as annual energy equivalents, TJ a^{-1}.

As was done with countrywide emissions, product-specific LULUC emissions, $AG_{k,p}$, are adjusted with additional iLULUC emissions from the global iLULUC pool, $NI_{k,p}$, in proportion to their net import increases, $ni_{k,p}$. The expression for net LULUC emissions due to domestic consumption of product p is similar to that for the respective country as a whole:

$$NL_{k,p} = AG_{k,p} + NI_{k,p} \tag{11}$$

where $NL_{k,p}$ is the net annual agriculture-related LULUC emissions due to domestic consumption of product p in country k, Tg a^{-1} and $NI_{k,p}$ is the net import emissions due to net import increases of product group p into country k, Tg a^{-1}.

The net import emissions for product group p in country k are calculated as:

$$NI_{k,p} = iLULUC_{glo}{}^{*}\, ni_{k,p} \tag{12}$$

where $ni_{k,p}$ is the net-import-increase allocation factor for product group p in country k (LHV-based and nondimensional).

The net-import-increase allocation factor, $ni_{k,p}$, is defined as the difference between a country- and product-specific share of global import increases and the share of global export increases:

http://dx.doi.org/10.5772/intechopen.80346

$$ni_{k,p} = \frac{\Delta I_{k,p}}{\sum_k \left(\sum_p \Delta I_{k,p}\right)} - \frac{\Delta E_{k,p}}{\sum_k \left(\sum_p \Delta E_{k,p}\right)} \tag{13}$$

where $\Delta I_{k,p}$ is the average annual import increase of product group p in country k, expressed as annual energy equivalents, TJ a^{-1}, $\Sigma_k(\Sigma_p \Delta I_{k,p})$ is the global sum of average annual import increases of product group p, expressed as annual energy equivalents, TJ a^{-1}, $\Delta E_{k,p}$ is the average annual export increase of product group p in country k, expressed as annual energy equivalents, TJ a^{-1}, and $\Sigma_k(\Sigma_p \Delta E_{k,p})$ is the global sum of average annual export increases of product group p, expressed as annual energy equivalents, TJ a^{-1}, which is equivalent to the global sum of average annual import increases.

As a last optional step of the method, the net LULUC emissions due to domestic consumption of product group p in country k can be converted from countrywide amounts to emissions per unit mass consumed:

$$nl_{k,p} = NL_{k,p}/C_{k,p} \tag{14}$$

where $nl_{k,p}$ is the average annual net agricultural LULUC emissions per unit mass of product group p consumed in country k, Tg Tg^{-1} (kg kg^{-1}) and $C_{k,p}$ is the average consumption (average over the last 3 years of the period 2007–2009) of product group p in country k, Tg a^{-1}.

3. Results

3.1. LULUC-related emissions on a spatial basis

The average global iLULUC emissions pool was calculated at 1.2 Pg CO_2 per year. This is equivalent to approximately 30of all LULUC-related CO_2 emissions from the 175 countries analyzed in this study. **Figure 1** shows the average annual net agriculture-related LULUC emissions (NL_k) per ha of agricultural land, which is a combination of a country's agricultural LULUC emissions (AG_k) and the balance NI_k of (a) imported (positive) iLUC emissions and (b) exported (negative) dLUC emissions (see Eq. (3)).

In specific countries such as Australia and Japan, no net LULUC emissions were assigned (value 0; see also **Table 2**) due to two reasons: (i) neither imports nor exports increased, i.e., no national LULUC emissions are exported to the global iLULUC pool, nor is iLULUC imported from the pool, and (ii) national LULUC emissions are fully attributed to settlement (infrastructure) area expansion while agricultural land areas declined (compare Eq. (2)).

Net exporting countries such as Argentina or the USA even show (theoretically) negative net LULUC results per ha (**Figure 1**). This is a consequence of rapidly increasing (LHV-energy) net export volumes and little or no LULUC import increases (resulting in a negative net import increase balance NI_k), combined with low national LULUC emissions (AG_k).

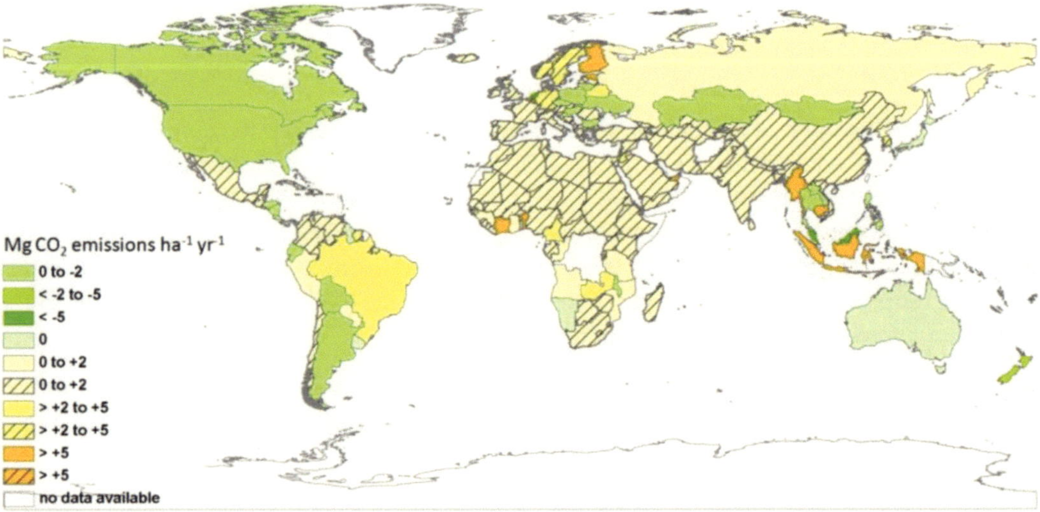

Figure 1. Average annual net agriculture-related LULUC emissions per ha of agricultural land (Mg ha^{-1} $year^{-1}$) corresponding to "NL_k" in Eq. (3) divided by agricultural land area. Hatched areas designate countries where iLULUC due to net import increases is more than half of total net agricultural LULUC emissions.

The highest average annual net agriculture-related LULUC emissions in **Table 1** were computed for Indonesia. Of the Indonesian LULUC-related CO_2 emissions, 53% are attributed to peat fires, 20% to peat drainage/oxidation, 22% to deforestation, and only 5% to palm oil and timber plantation establishment [23]. This illustrates that emissions may stem not only from deforestation and agricultural activities but also from other LULUC effects.

Agricultural LULUC emissions AG_k of 0 and 993 Tg a^{-1} were calculated for the USA and Brazil, respectively. These national LULUC emissions were corrected by −185 and − 110 Tg a^{-1} LULUC emission for the USA and Brazil, respectively, due to increased exports to the global

Product group	Allocation factor $ni_{k,p}$ (%)	Product group	Allocation factor $ni_{k,p}$ (%)
Alcoholic beverages	−1.1	Sugar crops	0.0
Cereals—excluding beer	−19.6	Tobacco and rubber	0.0
Fruits—excluding wine	+0.9	Tree nuts	−0.1
Miscellaneous	0.0	Vegetable oils	−3.6
Oil crops	−42.7	Vegetables	−0.1
Pulses	0.0	Animal fats	−0.2
Spices	0.0	Eggs	0.0
Starchy roots	+0.1	Meat	−4.3
Stimulants	−1.0	Milk—excluding butter	−0.5
Sugar and sweeteners	−27.3	Offals	−0.4

Table 1. Allocation factors for specific product groups' net import-increases for the example of Brazil.

iLULUC pool. Dividing by the domestic agricultural area (414*10^6 ha for USA and 276*10^6 ha for Brazil), we arrived at net LULUC emissions NL_k per average ha of agricultural land of about −0.3 and + 3.0 Mg a^{-1} ha^{-1} for USA and Brazil, respectively.

All country-specific emission factors for average hectares as well as product groups (see Section 3.2) are presented in the supplementary material (https://www.fibl.org/de/oesterreich/schwerpunkte-at/klimaschutz/klimaschutz-projekte/land-use-change.html).

3.2. Product group-specific LULUC emissions

In addition to countrywide net agricultural LULUC emissions, we calculated net LULUC emissions specifically for 3150 commodity groups that are consumed within the 175 specific countries of our analysis.

Figure 2 shows the global LULUC emissions of selected plant-based products, plotted over their global consumption. All product groups above the diagonal line (vegetable oils, oil crops, pulses, and tree nuts) are burdened with higher total LULUC emissions (a consequence of high production increases) relative to the proportions of their global consumption. Together, vegetable oils and oil crops account for 43% of all LULUC emissions, of which the larger part is attributable to bioenergy and food oil production. The other product groups in **Figure 2** (starchy roots, fruits, spices, and vegetables) have comparably low LULUC emissions per kg consumed. The highest absolute global average LULUC emissions per kg of product were found for vegetable oils (7.78 kg CO_2 kg^{-1}), followed by tree nuts (3.94 kg CO_2 kg^{-1}), pulses (1.96 kg CO_2 kg^{-1}), vegetables (1.42 kg CO_2 kg^{-1}), and oil crops (1.15 kg CO_2 kg^{-1}).

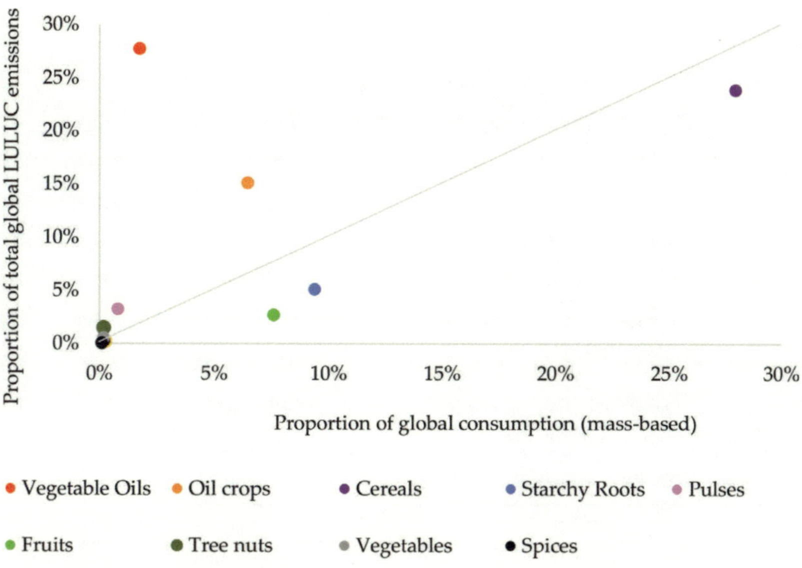

Figure 2. Proportions of global consumption and global LULUC emissions for selected plant-based products.

Table 1 illustrates the allocation factors ($ni_{k,p}$, Eq. (13)) for net-import increase-related iLULUC emissions for the example of Brazil. Some of the allocation factors are negative, indicating net export increases that shift emissions into the global iLULUC pool. The product groups with the largest export increases and therefore with the largest negative allocation factors are oil crops (mostly soy), sugars (from sugar cane), and cereals (mostly wheat and maize).

To complete the picture, the product-specific net LULUC emissions, $nl_{k,p}$, are shown in **Table 2** for selected countries. Interestingly, for Brazil, the strong export growth of oil crops, sugar/ sweeteners, and cereals (negative contribution to net LULUC emissions) is masked by a larger increase in domestic production $AG_{k,p}$ (Eqs. (9) and (10)) that causes high LULUC emissions of 3.66, 3.17, and 1.70 kg CO_2 per kg product consumed domestically. However, only for the product group offals are the export increases large enough to result in negative overall LULUC emissions. In contrast, for tree nuts, vegetable oils, spices, and oil crops, large net LULUC emissions are assigned per kg of product, pointing to domestic production increases outweighing the effects of export increases, or even net import increases exacerbating the domestic production increases.

Australia and Japan are not listed in **Table 2**, since they have no net agricultural LULUC emissions for any product group—in these countries, agricultural land use is decreasing or constant, and thus, all land expansion is assigned to infrastructure growth. In addition, both agricultural exports and imports from Australia and Japan decreased during the accounting period. In contrast, export-dominated countries such as Argentina, Canada, and the USA show

	AR	BR	CA	CN	FR	GER	ID	UK	USA
Cereals	−0.71	1.70	−0.05	0.03	0.22	0.41	4.55	0.55	−0.05
Oil crops	−0.30	3.66	−2.49	1.77	0.50	0.47	8.70	−0.39	−0.95
Sugar and sweeteners	−0.67	3.17	0.36	0.11	1.03	1.43	1.16	0.93	0
Sugar crops	0.02	0.67	0.01	0.00	0	0.01	0.15	0.21	0
Pulses	−0.42	0.97	−3.41	−0.17	1.76	−0.75	1.44	0.09	−0.7
Tree nuts	5.97	4.19	1.89	0.18	0.03	1.41	28.43	2.29	−1.26
Vegetable oils	−32.36	4.88	−3.21	2.64	0.98	5.04	131.77	1.09	0.95
Animal fats	−4.00	1.92	−0.40	−0.07	−0.08	−0.12	3.64	0.04	0.01
Eggs	0	0.44	0.04	−0.00	0.07	0.39	2.13	0.42	−0.12
Meat	−0.19	0.72	−0.18	0.01	0.04	−0.18	3.99	0.13	−0.06
Milk—excluding butter	−0.10	0.35	0.02	0.08	−0.03	0.04	12.49	0.38	−0.08
Offals	−0.29	−0.32	−0.88	1.03	−0.86	−5.44	2.79	0.18	−0.04

*AR = Argentina, BR = Brazil, CA = Canada, CN = China, FR = France, GER = Germany, ID = Indonesia, UK = United Kingdom, and USA = United States of America.

Table 2. Average net LULUC emissions for domestically consumed products in kg CO_2 per kg product for selected countries*.

mostly negative net LULUC emissions; in the case of the USA, this applies to fewer product groups than for Argentina. Countries like France and the United Kingdom show positive net agricultural LULUC emissions for most product groups, mainly due to import increases. Emissions for Indonesia are much higher than for the other countries because of large domestic LULUC emissions AG_k, regardless of the product group, which are partially a consequence of a rapidly growing population and an improved food supply [23].

Figures 3 and **4** show product groups associated with large positive or negative net LULUC emissions for selected countries. Hatched bars indicate a majority from net import-related LULUC emissions, while fully colored bars indicate the majority of emissions originating from domestic agricultural LULUC.

Plant-based commodities with high net emissions include spices, stimulants, oil crops, vegetable oils, tree nuts, and cereals (**Figure 3**). With regard to vegetable oils, Argentina and China are clearly increasing net exporters, and Brazil generally has large positive net LULUC emissions due not to imports, but to large domestic production increases. This applies also to production of Argentinean and Brazilian tree nuts.

Concerning livestock products, **Figure 4** shows a general dominating export role for Argentina and a specific role of animal fats, while most Brazilian livestock products are dominated by domestic LULUC emissions. For instance, Chinese imports of offal increased and thus lead to positive net LULUC (**Table 2**).

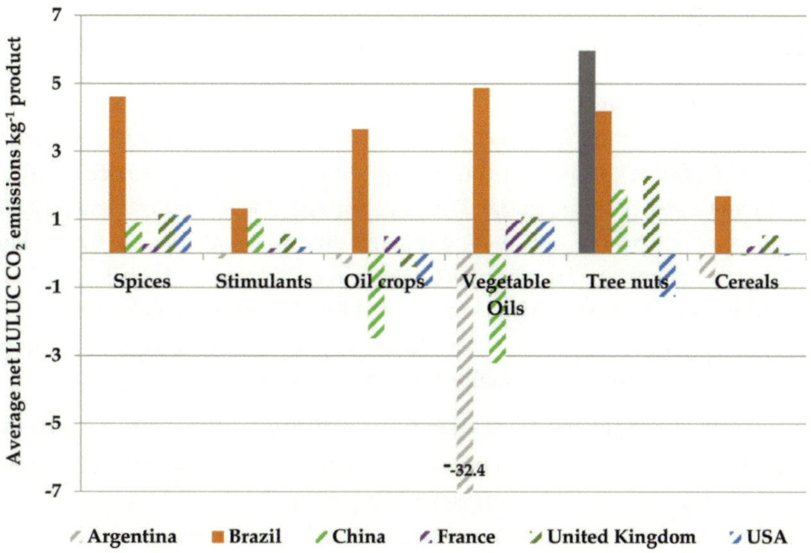

Figure 3. Average net LULUC emissions of specific vegetable product groups with comparably high emissions per kg of product (kg LULUC-CO_2 kg^{-1} product). Hatched columns represent a dominating contribution of iLULUC emissions to the net LULUC emissions per unit of product from different groups; solid columns indicate that net LULUC is dominated by emissions assigned to domestic production increases.

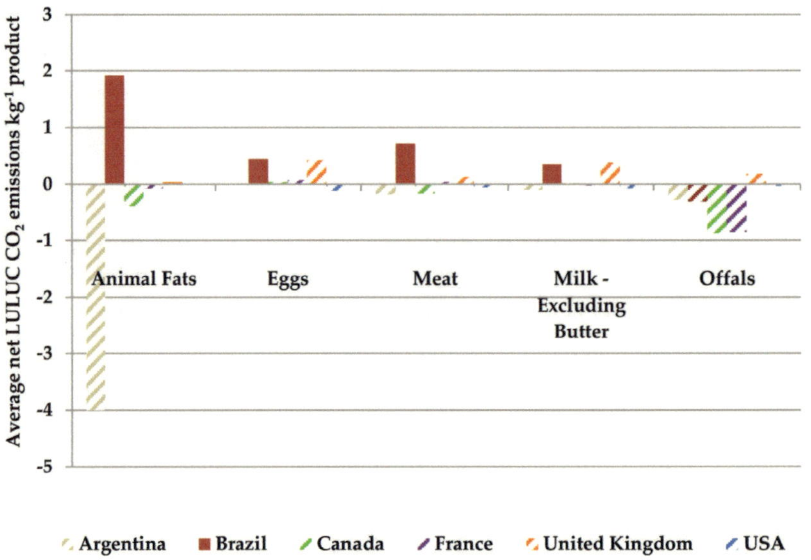

Figure 4. Average net LULUC emissions of specific livestock product groups with relatively high emissions per kg of product (kg LULUC-CO_2 kg^{-1} product).

4. Discussion

4.1. Novelty and limitations of the proposed method

Our method assumes that agricultural LULUC is a consequence of increasing demand for agricultural products and thus for land. We derive robust and globally consistent emission shares and emission factors based on the dynamic development of agricultural production, expressed in increases of produced (and net-imported) energy equivalents rather than on static, absolute shares of production (e.g., exported energy quantities as such). This focus on dynamic developments has the advantage of capturing the trends triggering LULUC impacts, but it also requires up-to-date information on rapidly changing global agricultural developments, making it difficult to extend the method to geographical entities smaller than countries (i.e., the level at which statistics are usually available; see [18]).

On the one hand, the method illustrated here is predicated on the principle of assigning an environmental burden (LULUC emissions) to an increase in commodity consumption, i.e., to the importing country, whose increased demand for the commodity is seen as causing the burden. On the other hand, one could also argue that it is the producer, not the consumer, who decides to satisfy a perceived demand, and therefore, the LULUC emissions should be assigned to the country of origin. Applied to LULUC, this shifted perspective would mean that export-related LULUC emissions are still assigned to the producing country. Hence, no "iLULUC emissions pool" would be necessary. A compromise approach would be to evenly divide the LULUC emissions from imports and exports between producer and consumer. Mathematically, this would correspond simply to cutting the size of the iLULUC emissions pool in half.

In most countries, the larger part of increased food and feedstuff production is for domestic purposes. Thus, most of a given country's LULUC emissions (globally approximately 70%) are assigned to the domestic territory. The remaining roughly 30% are exported or imported and are thus assigned to a global iLULUC emissions pool. In many countries though, LULUC from import increases accounts for more than half of the net LULUC (hatched areas in **Figure 1**).

For some countries, CO_2 emissions from LULUC could be overestimated because not all LULUC is linked to infrastructure, settlements, and agriculture, but also to, e.g., mining. The relatively undetailed allocation on the basis of the increase or decrease in areas for infrastructure, settlements, and agriculture introduces uncertainty. So far, the model also ignores the role of intensification as a cause of net export increases without causing LUC. Further studies could add such elements to the model, which is crucial for a correct assessment and allocation of agricultural LULUC emissions.

As stated above, emission shares are allocated in proportion to the energy content of agricultural product groups (based on their LHVs). As has long been debated (e.g., in LCA [24, 25]), allocation could also be based on commodity prices, but for the purposes of this study, the required data were not available. Such an economic aggregation would emphasize the role of monetary drivers for cultivation and agricultural management decisions, but on the other hand, it would be subject to confounding factors such as currency exchange rate fluctuations and fluctuations of auxiliary material prices (fuels, fertilizers, and pesticides).

Uncertainties may be introduced by input data from [18] concerning areas, yields, national consumption, or traded amounts. These data are reported by the national statistical authorities. In addition, the aggregation of single commodities into product groups such as "cereals" causes uncertainties, as different commodities within a group (e.g., types of cereal grains) will have different LHVs, which even further vary under practical conditions. For example, for the average LHV of the product group "cereals," we used the LHV of the globally dominant cereal commodity wheat as a default value. A comparison of the wheat LHV with the actual weighed average of the US cereal grain production mix shows a difference of 1.9% between the default value and the actual mix (US Department of Agriculture's statistical data sets for the years 1998–2000 and 2007–2009; http://quickstats.nass.usda.gov/). Additional uncertainty originates from the conversion of volume-based production information (bushels) to mass-based production data, as well as from the variability of published LHV values for grains.

From a global perspective, livestock products seem not to lead to particularly high LULUC emissions. However, the resulting numbers for $nl_{k,p}$ (see Eq. (14) and **Table 2**) are to some extent misleading, as they are based on production and net import increases. Those increases were rather low for livestock products over the observed period (e.g., in Brazil in **Table 1**), but arable land is increasingly used for livestock feed production, i.e., cereals or by-products from oil crops (oil cakes or solvent-extracted meal). The real LULUC emissions from livestock products are therefore likely to be higher than the numbers obtained with this method. Consequently, a part of the emissions linked to, e.g., oil crops have actually to be allocated to livestock products.

A limitation of our approach is that it does not consider historically grown and established bilateral trade connections between countries. For example, when the US corn is explicitly produced for the Chinese market, then US LULUC emissions end up in the global pool and obliterate the fact that China alone would be responsible for the LULUC change emissions. However, the focus of the study was the construction of a global iLULUC emissions pool in order to account for the changing global interrelationships of the agricultural commodity marketplace.

4.2. Direct (LU)LUC emissions versus results of the proposed method

Some studies (e.g., [1, 26]) computed direct LU emissions and dLUC emissions for specific oil crops from specific countries, e.g., Brazil and Argentina, and for the import mix of such crops used, e.g., in Austria [26]. For the latter, our results are comparable to those for Germany, as most oil crops imported into Austria are transported through Germany and they are influenced in both countries by the European markets.

For the example of oil crops, i.e., the basis for vegetable oils and by-products (mainly feed), which are consumed in Austria, the method proposed here assigns 1.99 kg CO_2 to 1 kg of product. Most of the oil crops or their products are imported into Austria and, in addition, no dLUC emissions are relevant for domestic oil crops. Thus, LULUC emissions are sourced exclusively from contributions to the iLULUC pool. Based on market information (e.g., Refs. [27, 28]), 50% each of the oil crops are estimated to come as soybeans from North America (no dLUC emissions) and South America. The resulting level of 1.61 kg of dLULUC emissions is in line with the 1.99 kg CO_2 stated above. The emissions are linked to imports from Brazil, which show 3.097 kg dLUC-CO_2 per kg of soybeans and LU-related emissions of 0.019 kg LU-CO_2 per kg of soybeans [1]. Together, dLULUC accounts for 3.22 kg CO_2 per kg of Brazilian soybeans, which is comparable to the 3.66 kg CO_2 derived with the method presented herein. It has to be noted that d(LU)LUC emission factors cannot be directly compared to the iLULUC emission factors presented here. While dLULUC estimates are close to the numbers from the presented method in specific cases such as of Austria, dLUC emission factors alone are insufficient and should be replaced or accompanied by emission factors which consider iLULUC effects in LCAs and carbon footprints.

5. Conclusion

We propose an integrated dynamic treatment of emissions from LULUC, caused by domestic agricultural production, and from iLULUC that is linked to international agricultural commodity trade, which may be used in LCA frameworks and other assessment methods that include GHG emissions accountings. iLULUC effects are accounted for which are induced by countries with increasing demand for certain agricultural commodities. LULUC emissions are not only caused by growing national agricultural land use, but also by the growth of builtup areas. Indirect LULUC emissions related to an increase in net agricultural imports represent the balance of (a) (positive) iLULUC emissions from import increases and (b) (negative) dLUC emissions from exported commodities. Our model thus reflects a dynamic rather than a static perspective of agricultural commodity production and trade—it uses the increases of production, exports, and imports in place of their absolute values.

Indirect LULUC factors are derived by converting data on agricultural commodity production and trade to the commodity's corresponding energy content on an LHV basis. A (hypothetical) global iLULUC pool reflects the global interconnectedness of agricultural commodity trade;

http://dx.doi.org/10.5772/intechopen.80346

national iLULUC emissions may be derived from it and represent the LULUC emissions inherent in the traded products.

Our results account for the allocation of emissions to specific product groups consumed in a country in proportion to their corresponding energy content on an LHV basis. This allows for the aggregation of agricultural product group data on different spatial levels, and it provides a more detailed focus compared to generic agricultural land-related emission estimates. With this approach, 3150 new results from 175 countries are provided with the respective indirect (LU)LUC effects. The results vary substantially between nations, with clear differences between producing and exporting countries versus importing countries. A similar differentiation applies to specific product groups within a country.

LUC-related GHG-accounting should rest on a well-documented computational basis as a prerequisite for a fair differentiation of "LULUC-emitting/exporting nations" versus "LULUC-importing nations" on the one hand and between (LU)LUC-driving product groups versus product groups with little or no effects on LULUC emissions on the other. Further work should address the validation and improvement of the model and its input data.

Acknowledgements

The authors are grateful to Dr. Christian Schader, Dr. Rainer Weisshaidinger, Theresia Markut, and Dr. Matthias Meier from the Research Institute of Organic Agriculture (FiBL) for their valuable input and for many stimulating discussions. We gratefully acknowledge support from the Austrian Science Fund (FWF) project GELUC (P29130-G27).

Conflict of interest

The authors declare no conflict of interest.

Author details

Stefan J. Hörtenhuber[1,2]*, Michaela C. Theurl[1,3], Gerhard Piringer[2] and Werner J. Zollitsch[2]

*Address all correspondence to: stefan.hoertenhuber@fibl.org

1 Research Institute of Organic Agriculture (Forschungsinstitut für biologischen Landbau FiBL), Vienna, Austria

2 Department of Sustainable Agricultural Systems, BOKU—University of Natural Resources and Life Sciences, Vienna, Vienna, Austria

3 Institute of Social Ecology (SEC), BOKU—University of Natural Resources and Life Sciences, Vienna, Vienna, Austria

References

[1] Hörtenhuber S, Piringer G, Zollitsch W, Lindenthal T, Winiwarter W. Land use and land use change in agricultural LCAs and carbon footprints—The case for regionally specific LUC versus other methods. Journal of Cleaner Production. 2014;**73**:31-39. DOI: 10.1016/j.jclepro.2013.12.027

[2] Fargione J, Hill J, Tilman D, Polasky S, Hawthorne P. Land clearing and the biofuel carbon debt. Science. 2008;**319**:1235-1238. DOI: 10.1126/science.1152747

[3] Haberl H. Net land-atmosphere flows of biogenic carbon related to bioenergy: Towards an understanding of systemic feedbacks. GCB Bioenergy. 2013;**5**:351-357. DOI: 10.1111/gcbb.12071

[4] Alexander P, Rounsevell MDA, Dislich C, Dodson JR, Engström K, Moran D. Drivers for global agricultural land use change: The nexus of diet, population, yield and bioenergy. Global Environmental Change. 2015;**35**:138-147. DOI: 10.1016/j.gloenvcha.2015.08.011

[5] Defries RS, Rudel T, Uriarte M, Hansen M. Deforestation driven by urban population growth and agricultural trade in the twenty-first century. Nature Geoscience. 2010;**3**:178-181. DOI: 10.1038/ngeo756

[6] Gerbens-Leenes W, Nonhebel S. Food and land use. The influence of consumption patterns on the use of agricultural resources. Appetite. 2005;**45**:24-31. DOI: 10.1016/j.appet.2005.01.011

[7] Stehfest E, Bouwman L, Van Vuuren DP, Elzen MGJD, Eickhout B, Kabat P. Climate benefits of changing diet. Climatic Change. 2009;**95**:83-102. DOI: 10.1007/s10584-008-9534-6

[8] Solomon S, Qin D, Manning M, Chen Z, Marquis M, Averyt KB, Tignor M, Miller HL, editors. Climate change. In: The Physical Science Basis. Contribution of the Working Group I to the Fourth Assessment Report of the Intergovernmental Panel on Climate Change. Cambridge, UK and New York, USA: Cambridge University Press; 2007. pp. 499-587

[9] Harris NL, Brown S, Hagen SC, Saatchi SS, Petrova S, Salas W, Hansen MC, Popatov PV, Lotsch A. Baseline map of carbon emissions from deforestation in tropical regions. Science. 2012;**336**:1573-1576. DOI: 10.1126/science.1217962

[10] Poruschi L, Dhakal S, Canadell J. Ten Years of Advancing Knowledge on the Global Carbon Cycle and Its Management. Global Carbon Project Report No. 7 [report]. Global Carbon Project, Tsukuba, Japan; 2010

[11] Hörtenhuber S. Greenhouse gas emissions from dairy production—Assessment and effects of important drivers [dissertation thesis]. Vienna: BOKU—University of Natural Resources and Life Sciences; 2011

[12] Farrell AE, Plevin RJ, Turner BT, Jones AD, O'Hare M, Kammen DM. Ethanol can contribute to energy and environmental goals. Science. 2006;**311**:506. DOI: 10.1126/science.1121416

[13] Finkbeiner M. Indirect Land Use Change (iLUC) within Life Cycle Assessment (LCA)—Scientific Robustness and Consistency with International Standards [Internet]. 2013. Available from: http://www.ufop.de/files/5213/6853/0753/RZ_VDB_0030_Studie_ENG_Komplett.pdf [Accessed: May 18, 2018]

[14] Fritsche U, Henneberg K, Hünecke K. The "iLUC Factor" as a Means to Hedge Risks of GHG Emissions from Indirect Land Use Change, Working Paper [Internet]. 2010. Available from: https://www.oeko.de/oekodoc/1030/2010-082-en.pdf [Accessed: May 18, 2018]

[15] Saez de Bikuña K, Hamelin L, Hauschild MZ, Pilegaard K, Ibrom A. A comparison of land use change accounting methods: Seeking common grounds for key modeling choices in biofuel assessments. Journal of Cleaner Production. 2017;**177**:52-61. DOI: 10.1016/j.jclepro.2017.12.180

[16] Williams AG, Audsley E, Chatterton JC, Murphy-Bokern D, Brander M. Greenhouse gas emissions from UK food and drink consumption by systems LCA: Current and possible futures. In: Proceedings of the LCA Food 2010 Conference on Life Cycle Assessment in the Agri-Food Sector, 22–24 September 2010, Bari, Italy. 2010. pp. 363-368

[17] Vellinga TV, Blonk H, Marinussen M, van Zeist WJ, de Boer IJM, Starmans D. Methodology Used in FeedPrint: A Tool Quantifying Greenhouse Gas Emissions of Feed Production and Utilization. Report 674 [report]. Wageningen UR Livestock Research; 108p

[18] FAO (Food and Agricultural Organisation of the United Nations). FAO Statistical Databases [Internet]. 2014. Available from: http://faostat.fao.org [Accessed: February 13, 2014]

[19] Fehrenbach H, Giegrich J, Reinhardt G, Schmitz J, Sayer, U, Gretz M, Seizinger E, Lanje K. Criteria for a Sustainable Use of Bioenergy on a Global Scale. Report on behalf of the German Federal Environment Agency [report]; 245p

[20] Beilicke G. Bautechnischer Brandschutz—Brandlastberechnung. Germany: Buchfabrik Halle; 2010. 265p. ISBN: 978-3-942578-00-4

[21] Olivier J, Janssens-Maenhout G. Total Greenhouse-Gas Emissions. Part III of: CO_2 Emissions from Fuel Combustion. Paris, France: International Energy Agency; 2011. pp. III.1-III.57

[22] Krausmann F, Erb K-H, Gingrich S, Haberl H, Bondeau A, Gaube V, Lauk C, Plutzar C, Searchinger TD. Global human appropriation of net primary production doubled in the 20th century. Proceedings of the National Academy of Sciences of the United States of America. 2013;**110**(25):10324-10329. DOI: 10.1073/pnas.1211349110

[23] Boer R, Sulistyowati IL, Zed F, Masripatin N, Kartakusuma DA, Hilman D, Mulyanto HS. Indonesia's Second National Communication under the United Nations Framework Convention on Climate Change (UNFCCC). Jarkarta: Ministry of the Environment; 2010

[24] Curran M-A. Co-product and input allocation approaches for creating life cycle inventory data: A literature review. The International Journal of Life Cycle Assessment. 2007;**12** (Special Issue 1):65. DOI: 10.1065/lca2006.08.268

[25] Ekvall T, Tillman A-M. Open-loop recycling: Criteria for allocation procedures. The International Journal of Life Cycle Assessment. 1997;**2**(3):155-162. DOI: 10.1007/BF02978810

[26] Hörtenhuber SJ, Lindenthal T, Zollitsch W. Reduction of greenhouse gas emissions from feed supply chains by utilizing regionally produced protein sources: The case of Austrian dairy production. Journal of Science and Food Agriculture. 2011;**91**(6):1118-1127. DOI: 10.1002/jsfa.4293

[27] FBuG (Forum Bio- und Gentechnologie e.V.) [Internet]. 2014. Available from: http://www.transgen.de/anbau/460.gentechnisch-veraenderte-sojabohnen-anbauflaechen-weltweit.html [Accessed: February 13, 2014]

[28] OVID (Verband der ölsaatenverarbeitenden Industrie in Deutschland) [Internet]. 2014. Available from: http://www.ovid-verband.de/unsere-branche/daten-und-grafiken/ [Accessed: February 13, 2014]

Chapter 6

Land Acquisition and Use in Nigeria: Implications for Sustainable Food and Livelihood Security

Isaac B. Oluwatayo, Omowunmi Timothy and
Ayodeji O. Ojo

Additional information is available at the end of the chapter

http://dx.doi.org/10.5772/intechopen.79997

Abstract

Land acquisition and use remain a critical issue of great policy relevance in developing countries such as Nigeria. This study therefore examined land acquisition and use in Nigeria within the context of food and livelihood security. The chapter used secondary data obtained from the World Bank website, National Bureau of Statistics (NBS) and other sources. It was found that there are gender, location and income-group considerations in the allocation of land in Nigeria. While the urban land market is relatively more formal, the rural land market is informal and the transactions were not documented in most cases. The study found that bureaucratic bottlenecks, high cost of registering land and long registration procedures, and inconsistent policy regimes impede the development of land market in Nigeria. Based on the findings of the study, it was recommended that the government should reduce and make the processes to be completed in registering lands in Nigeria easier. The Land Use Act 1978 should be amended to capture the prevailing realities around customary laws and informal markets. The government should reduce the cost of land registration in Nigeria. Multilateral organisations and government should co-create and co-finance innovative interventions to improve activities in the land market.

Keywords: conflict, food security, land acquisition, livelihood security, Nigeria, welfare

1. Introduction

Land acquisitions are broadly defined to include not only the purchase of ownership rights but also the acquisition of user rights, for instance through leases or concessions, whether short or long term [1]. Land tenure system can be defined as the rights and institution that governs access to and use of land [2]. Tenure system of land involves a system of rights, duties and responsibilities concerning the use, transfer, alienation and ownership security of land and its resources. From the ongoing, it is clear that land acquisition and use cannot be discussed extensively without incorporating land tenure system. This chapter therefore presents a comprehensive review of literature on the land acquisition and use in Nigeria and the implications for sustainable food and livelihood security in the country.

Land is a veritable ingredient of development especially in the agricultural and tourism sector of any economy. Nigeria has a total land mass of 924,768 sq.km with a population of 198 million and annual population growth rate of 2.8% [3]. Nigeria comprises over 250 ethnic groups located within the 36 states and the Federal Capital Territory [4]. Land is an asset and factor of production for households in Nigeria [5]. However, the level of access and title ownership is determined by the state [6]. Therefore, the land system is characterised by several actors including government, community leaders, families, lawyers, middle men and estate agents among others. All activities of the different actors are regulated by the government through policies and programmes. Generally, land systems thrive on clearly stated property rights. Two types of proprietary rights have been defined in literature-absolute or nonderivative interests and derivative interests [6]. The absolute or nonderivative interest is a nonrestrictive access and use of land conferred on the holder. The absolute interest on land has also been explained as inclusive of highest scope of proprietary decisions on the use and management of land. Derivative interest derives from a larger estates or superior estates. The derivative rights cover leaseholds, life interests, mortgage, rents and pledges among others. The two types of property rights (absolute or nonderivative interest and derivative interest) exist in Nigeria.

The Nigerian land system has evolved over the years as classified into precolonial, colonial and postcolonial periods in literature [7]. The three periods are explained below:

a. Precolonial land ownership structure: Prior to the colonial era, lands were solely owned by families and communities. The land was owned by the community and family heads who then allocate based on the needs of their subordinates [8]. The legal estate or authority existed at the community or family level. Thus, the leadership of communities and families had absolute interests, while constituents had derivative interests.

b. Ownership structure during colonial rule: The ownership of land was regulated by the colonial authorities before independence. The legislations included Treaty of Cession (1861), Land Proclamation Ordinance (1900), Land and Native Rights Act (1916), Public Lands Acquisition (1917), State Land Acts (1918) and Town and Country Planning Act (1947). The colonial legislations were meant to take property rights out of the reach community leaders. For instance, in 1900, the Land Proclamation Ordinance created by Lord Lugard regarded the principles of native law and custom and stipulated that the title of land can only be acquired through the high commissioner [6].

http://dx.doi.org/10.5772/intechopen.79997

c. Postcolonial ownership structure: As depicted earlier, the land ownership structure in Nigeria has evolved over the years. Two key legislations have been enacted since independence: Land Tenure Law of Northern Nigeria of 1962 and Land Use Act of 1978. The land tenure law of Northern Nigeria of 1962 stipulated that the minister responsible for land matter controls, holds and allocates land (unoccupied or occupied native lands) to natives of Northern Nigeria. This implies that non-natives except for the approval of the minister could not land titles. The law granted the natives of Northern Nigeria the right to own land for a limited number of years. The individual/native may sell, mortgage or transfer the land subject to the minister's approval. The Land Tenure Law of 1962 was repealed, and land use decree of 1978 was implemented.

Land tenure issues are important components of developmental discourse [9]. This is because unplanned or weak regulatory undermines development as informal settlements grow thereby stressing already inadequate urban infrastructure [5]. Therefore, poor land management affects security and growth as it induces, slums and suboptimal living conditions [10, 11]. The Nigerian land use decree of 1978 stipulates that all land belong to the government holding same in trust for the public [12]. This implies that the government allocates land to individuals and corporate entities based on the objectives of interested parties [13].

Objectives of Land Use Act 1978

- Make land accessible to all Nigerians
- Prevent speculative purchases of communal land
- Streamline and simplify the management and ownership of land
- Land available to governments at all levels for development
- Provide a system of government administration of rights towards improving tenure security

The aftermath was political considerations in the allocation of land, corruption and rise in lobbyist tendencies. The land use act avails the opportunities to own lands without recourse to families and communal land holdings. The process of obtaining certificates of occupancy is characterised by bureaucratic bottlenecks, high registration fees and perpetual payment of levies and taxes [10]. Till date, land tenure is governed by customary laws, especially in rural Nigeria. Therefore, tenure security is low as the transactions in the land market are largely informal.

Nigeria trails other African countries in the ease of registering land indices (see **Table 1**) [14]. Nigeria ranks 179th in the ease of registering land compared to Botswana (81st position), Morocco (86th position), South Africa (107th position) and Ghana (119th position). In terms of the number of procedures required to complete land title registration, there are 11 procedures in Nigeria compared to Botswana, Morocco, South Africa and Ghana with four procedures, six

S/N	Country	Registering of property (DTF)	Registering property index	Procedures (number)	Time (days)	Cost (% of property value)	Quality of land administration index (0–30)
1	New Zealand	94.97	1	2	1	0.1	26
2	United States	76.80	37	4.4	15.2	2.5	17.6
3	China	76.15	41	4	19.5	3.4	18.3
4	United Kingdom	74.51	47	6	21.5	4.8	24.5
5	Botswana	65.45	81	4	27	5.1	10
6	Morocco	64.35	86	6	22	6.4	15.5
7	South Africa	58.43	107	7	23	7.6	13.5
8	Cote d'Ivoire	57.56	113	6	30	7.4	10.5
9	Ghana	55.5	119	6	52	6.7	8.0
10	Burkina Faso	50.44	140	4	67	12	11.5
11	Algeria	43.83	163	10	5.5	7.1	7
12	Nigeria	34.08	179	11.3	68.9	10.5	7.4

Source: World Bank [14].

Table 1. Ease of registering land in selected countries (2017).

procedures, seven procedures and six procedures, respectively. This depicts low level of innovation and inefficiency in the land registration process in Nigeria. As expected from a country with one of the highest number of procedures for land title registration, it takes more days to register land title in Nigeria than elsewhere (see **Table 1**).

2. State of land acquisition and land market in Nigeria

Generally, lands are small and fragmented which imparts significantly on the mechanisation potentials of farming households. Similarly, fragmented nature of land systems inhibits the creation of an effective land market, which consequently hinders the emergence and development of an overall equitable distribution of wealth [15]. Land markets should be accessible to all categories of people including women and minority groups.

According to Dale et al. [15], land markets should possess the following features:

- Guaranteed security of land rights
- Low transaction cost for all users

- Access to credit
- Transparency, openness and ease of access to all
- Protection of minorities
- Support for social, economic and environmental sustainability
- Public/private partnerships to provide value-for-money services

Small-scale farmers dominate rural landholdings with average farm size ranging from 0.5 ha in the South to 4 ha in the North. About 50% of the Nigerian farms are less than 1 ha, while 15% are less than 5 ha [16]. There are three types of land markets in Nigeria. The types of markets are:

- Formal markets where certificates of occupancy from the government are allocated.
- Combination of formal and informal markets where transfer of land rights are certificate of occupancy.
- Informal market where the bulk of the transactions are not documented as title owners do not obtain certificate of occupancy [16].

The land can be used for different purposes including agricultural production, industrial production, buildings for households and establishment of parks among others [5]. There are location, gender, income-class considerations in the allocation and ownership of land in Nigeria [17]. In terms of ownership structure, men typically own more land compared to women in Nigeria (see **Table 2**). Majority of the land owners inherited it from their family, while only 7 and 2.2% of male and female, respectively, reported purchase of land (National Bureau of Statistics; World Bank and Federal Ministry of Agriculture and Rural Development, [18]).

In the study conducted by Eze et al. [19], they examined land tenure systems, farm sizes, agricultural productivity and innovation by looking at the socio-economic characteristics of farmers and identified the factors that affected agricultural productivity in the Imo State. Five communities were chosen randomly, and from each of these communities, twenty farmers were randomly chosen. Data were collected, collated and analysed using relevant techniques such as means, percentages, frequency distribution and multiple regression analysis. The results showed that 85% of the respondents practiced individual land tenure system alone. It was also revealed that the laws of inheritance and increase in population led to the subdivision and fragmentation of existing farmland in such a manner that the sizes of farm holdings

Outright purchase		**Rented**		**Used free of charge**		**Allocated by community**		**Family inheritance**	
Male (%)	Female (%)	Male (%)	Female (%)	Male (%)	Female (%)	Male (%)	Female (%)	Male (%)	Female (%)
7.0	2.2	6.8	11.8	7.9	11.8	7.1	5.9	71.2	68.4

Source: NBS et al. [18].

Table 2. Land ownership structure in Nigeria.

discouraged agricultural commercialization. Also, the study revealed that fragmentation widened the distance between plots that led to increased waste of man-hour and energy. It was also shown that mechanisation of agriculture was impracticable under land fragmentation and adoption of modern innovation was reduced since just 35.0% of the respondents claimed to have adopted other forms of innovation. From the analysis, it was found that lands were severally fragmented during the acquisition and sharing of either family or community lands, and this went a long way in discouraging adoption of laudable innovation like mechanised farming as a result of reduced farm size of land. Also farmers were not at liberty to sell a portion of land acquired through communal ownership and thus majority of them could not acquire extensive land for considerable agricultural productivity. The econometric result revealed those socioeconomic factors that significantly affected the farmers' productivity in the area to include planting materials, household size, farming experience, tenure system and labour cost.

Alarima et al. [12] examined the land rights and rental systems followed by sawah rice farmers in Nigeria. The study was conducted in six states in Nigeria that use sawah rice technology: Kwara, Ondo, Niger, Ebonyi, Kaduna and Abuja. A total of 124 sawah farmers were selected for participation based on their involvement in sawah-based rice production. Data used in this study were collected from October 2009 to January 2011 in all the sawah sites in Nigeria. T-test was used to determine significant differences in the yields and farm sizes of landlord and tenant farmers. The land tenure system practiced in the lowlands is governed primarily by inheritance (71.8%), with temporary arrangements made through rentals (37.1%). Tenants pay ₦12,000 ha^{-1} $year^{-1}$ in land-for-cash agreements and 5% of the total yield of rice in land-for-paddy agreements. Generally, agreements between landlords and tenants were verbal, binding and honoured by both parties. Land conflicts occurred when either party breaches the agreement and always result in the landlords taking over the land. Significant differences in the farm sizes and yields of landlord and tenant farmers were found. The findings of the study indicate the importance of secured land tenure, which affects the practice of sawah and the productivity of rice farming. Sustaining and improving sawah rice production in the study area would require addressing the land tenure issues of both landlords and tenants. Access to land for tenants will enhance their participation in sawah and increase their chances of increasing their income and emerging from poverty. Investment in more durable inputs such as power tillers, dykes and irrigation canals will decrease if the land is not secured. Therefore, tenants and landless people need more secure access to land to provide them with opportunities to manage their sawah plots so that they will have higher yields. Farmers whose land security was not guaranteed would be more inclined towards short-term investments in land, and the sustainability of sawah would not be expected to be their priority.

Also, the study by Ojo [20] examined the effects of land acquisition for large-scale farming on the performance, productivity and technical efficiency of small-scale farming in Nigeria. The farmers were grouped into two groups: farmers whose families donated land for large-scale oil palm project (A) and those who did not (B). Data collected were analysed using descriptive statistics, gross margin and stochastic frontier production function analyses. The study revealed that acquisition of land for large-scale farming in the study area had adverse effects on small-scale farming. This could further worsen the food security crisis in Nigeria because about 80% of farmers in Nigeria practice small-scale farming [21] that constitutes over 90% of food and agricultural production in the country [22]. It was revealed that group B farms were more productive in the allocation of resources and overall production as measured by the decreasing positive elasticity of production of most of the variables involved in the production function analysis. The group B farms were also more technically efficient than the group A farms.

The study therefore recommended that government in its drive to encourage large-scale agricultural production should not acquire land near the towns and people's settlements; rather, land on the highways/expressways should be opened up and allocated to prospective large-scale farmers.

Again, Twene [23] investigated the effects of the large-scale agricultural land grabbed for the Bui Dam project on the livelihoods of the affected people. Both quantitative and qualitative data were collected through the use of interviewer questionnaire administration, interview guide, focus group discussions and observations. A total of 142 household heads were interviewed, although some key informants such as chiefs and community development officers of the district assemblies were also interviewed. The descriptive statistical tools and the t-test were employed to analyse the quantitative data whilst content analysis was applied to qualitative data with the result presented in the form of direct quotations. The study revealed through the result of the t-test for difference in output within 2005/2006 and 2013/2014 production seasons that local food crop production and the quantity of fish catch have declined after the land grabs.

Equally, it was found that annual income levels of the local people had fallen after the Bui Dam project due to reduction in their productivity of both crop production and the quantity of fish caught. Thus, farming and fishing were the most affected occupations in the study area. These were attributed to loss of farmland and low level of experience for fishing in the newly created lake as well as the problem of proximity between the resettled communities and the river.

In addition, it was found that the land grabbing situation resulted in conflict between some communities and the Bui Power Authority. The study showed that the local people initially reacted by engaging in conflict with officials of the Bui Power Authority, particularly in Dokokyina. Equally, it was found that the land-grabbing incident in the study area was not accompanied by adequate compensation packages as promised which were the bases for the conflict because assets lost were not in commensurate measure with compensation packages received.

However, it was also revealed that the acquisition of the land for the Bui Dam project and its associated relocation of the affected people have adversely affected the natural capital base of the people (land, forest and water bodies). Thus, generally, the study discovered that the local people's access to the natural capital had worsened, while access to the physical capital had improved.

Finally, with regard to interventions and coping strategies, the study revealed that the main coping strategies adopted by both men and women in the study communities after the land was taken for the Bui Dam project are casual work (by-day) and petty trading, respectively.

The growing interest in petty trading was attributed to the view that the construction of the dam has caused influx of people into the study area, thereby providing market for consumable goods. Other coping strategies included farming, premixed fuel business, wood gathering and pito brewing. The local people in the study communities were unaware of any alternative livelihood interventions provided by the Bui Power Authority and the district assemblies to ensure the sustainability of their livelihoods.

The study therefore recommended the introduction of a comprehensive livelihood enhancement programme such as skill training for the youth and the landless group of people in the study communities by the Bui Power Authority and the district assemblies. This would enable them to promote their livelihood sustainability.

3. Land tenure, food security and livelihood security

Rural Nigeria is agricultural as 85% of the residents depend on agriculture for their livelihood. However, access to land is limited as families and community heads still manage to control land thereby determining access to land. Given the position of Land Use Act 1978, it implies that the beneficiaries of the communal land allocation system are not formally recognised as the legal holders of right to the land. Again, family and community heads rely on memory and reference to natural and artificial features to define plots of land that is prone to uncertainty regarding the location of boundaries [23]. This is because most communal land allocations are not documented [23].

Availability of land determines food and livelihood security given the level of agricultural development in Nigeria [17]. This is because farming operations will remain at subsistence level due to inadequate access to land. In fact, an estimated 95% of agricultural lands in Nigeria are not titled [24]. This undermines the capacity of farmers to present lands as collateral to access formal loans from financial institutions [24]. Again, the lack of absolute or nonderivative property interest constrains the ability of farming households to plant cash crops consequently limiting their income generation potentials [17]. Therefore, food security is difficult as the population continues to grow and agricultural land becomes scarce [25]. The challenges of agricultural production and food security including inadequate access to land, finance and technology, inconsistent policy regime, infrastructure deficit and adverse climate change impacts have been documented in literature [26, 27].

4. Land acquisition and livelihood security in Africa

Moreover, Onoja and Achike [28] reviewed cases of land grabbing by foreign investors in West Africa, identified the possible drivers of large-scale land acquisition by foreign investors in the region and discussed the implications of the findings for agricultural and land policy reforms in West Africa. Prior to the study, reports indicating that large portions of land (estimated 50–80 m ha) have been bought by international investors in middle- and low-income countries, with roughly two-thirds of those purchases occurring in sub-Saharan Africa, called for a cursory appraisal of the implications of the trend of land grabbing for West African food security (Sahel West Africa Countries, SWAC/OECD, [29]). Land transactions involving foreign investors had increased in the area over the years. Over 100,000 ha have been documented in Nigeria. Ghana and Mali have many significant transactions on land by foreign investors. Several investors have more than 100,000 ha. Burkina Faso has one significant land transaction (200,000 ha), while Niger and Senegal have relatively small land transactions. Most lands grabbed in West Africa were profit driven (by biofuel investors) and were made under the guise of using the lands acquired for agricultural investments. Land tenure and investment in land have far-reaching economic and social implications and are therefore key issues for small family–operated farms and their relations with agribusiness (Sahel West Africa Countries, SWAC/OECD, [29]). According to the Food and Agriculture Organisation [30], many problems that are now being recognised in natural and agricultural land systems have arisen out of the use of inadequate technologies for assessing and monitoring land resources, preventing land pollution and rehabilitating contaminated lands. According to Cotula, Vermeulen, Leonard, and Keeley, transactions labelled as "large-scale" involved between 1000 and 500,000 ha [31, 32]. Increasing evidence is emerging to affirm that the problem of large-scale land acquisition by foreign investors in Africa is following a dangerous trend, which needs to be monitored.

Global Development [33] reported that research findings have indicated that a million Chinese farmers have joined the rush to Africa and that some of the world's richest countries are buying or leasing land in some of the world's poorest to satisfy their insatiable appetites for food and fuel. In the new scramble for Africa, the report added, 2.5 m ha (6.2 m acres) of farmland in five sub-Saharan countries have been bought or rented in the past 5 years at a total cost of $920 m (£563 m). Recent high-profile land purchases encompassing thousands of hectares of prime agricultural land have raised concerns over equitable land access [30]. Sub-Saharan Africa, especially Nigeria and other West African countries, is not exempt from this development (see Cotula, Vermeulen, Leonard, and Keeley, [31, 32]). Such a trend is more disturbing when considered alongside the future of food production from SSA land, where FAO [34] put the estimated share of arable land in total agricultural land at only 15.6% as of 2000. Response indicators showed that the value of agricultural production per hectare of agricultural land is highest in South Asia, at I$ 720.6, while Sub-Saharan Africa trailed behind, globally ranking lowest with a value of I$ 71.8. Under this scenario, worrying over the growing trend of large-scale land acquisition by foreign investors-who are, at best, interested in growing crops that can only contribute to food security and economic growth of countries outside SSA-while the limited land available for African farmers is diminishing in the face of lingering hunger and poverty is justified.

Food security is a current issue in Nigeria, as it is across Africa. According to the review, it was noted that there are 307 million hungry people in Africa, most of whom live in Sub-Saharan Africa (265 million). A FAO statistic indicated that at least 9.4 million Nigerians were undernourished and that out of Nigeria's 147.7 million citizens, 6% were highly undernourished (2011). Instead of dealing with food supply or food security problems at such a critical time, the country is selling off arable lands to foreign investors prospecting in biofuels production to the extent of losing greater than 136,000 ha of land from only eight deals that could have been used in producing food crops. The implication of this is that the drive for food security will still be a far-fetched dream as long as attraction of foreign investment in agriculture only aims to produce biofuels for profit. The auctioning of fertile farmlands for this purpose also portends danger of losing job opportunities, increasing poverty in the country, and helping the growth of foreign companies to the disadvantage of poor land owners in Nigeria. The overall implication of these trends is to increase poverty, unfavourable terms of trade against Nigeria, desertification, increased global warming and the adverse consequences of climate change and the disempowerment of indigenous citizens, who will now be left with few pieces of land that may not reach even 1 ha. The study therefore recommended that a regional approach should be applied by African countries, implementing land reforms that will involve the local communities who own the land, stopping long-term leasing beyond 50 years, building capacity and creating awareness about land transactions of large magnitudes.

Furthermore, a report by Economic Commission for Africa [35] to show the linkages between land tenure and food security in Africa confirms that land plays an important role in the livelihoods of the majority of Africans. This asserts that food security and poverty reduction cannot be achieved unless issues of access to land, security of tenure and the capacity to use land productively and in a sustainable manner are addressed. This chapter suggests that land is central in promoting rural livelihoods in Africa because access to land and security of tenure are the main means through which food security and sustainable development can be realised because the livelihoods of over 70% of the population in Africa are mainly linked to land and natural resources exploitation.

This chapter studied the land policy generic model developed by Moyo [36], based upon Shivji et al. [37], derived from five analytical constructs of land management, namely, land distribution, land utilisation, land tenure security, land administration and land adjudication. It posits land tenure as one of the central factors determining food security and sustainable development.

In terms of land distribution, the major problem relates to unequal access to land according to race, gender, class and ethnic distinctions. The second analytical construct relates to land utilisation and how this has been economically and socially constructed. There is a tendency to view small farms as inefficient and large farms as highly efficient in terms of yields per unit of land [38], especially in settler countries. But such a perception has long since been debunked [39, 40] and multinational organisations (IMF and World Bank) have come to accept the efficiency of smallholder land use. The main question of analysis is how the regulation of land-use is consensual or coercive and whether it is free of discrimination. The third conceptual issue is how land tenure has been constructed and qualified in most African countries. The main question relates to how secure the tenure systems are and whether there is equity or not. It was argued that access to more productive land and control of natural resources by the poor offers the most stable form of security for poor households. In this case, livelihood security cannot be achieved without some form of redistribution of land held by wealthy classes (constituted of individuals and multinational companies). The assumption tends to be that enhancing access to land, security of tenure or sustainability of land resource use will ultimately enhance welfare, including food security.

http://dx.doi.org/10.5772/intechopen.79997

The fourth conceptual issue relates to how land is administered. The management regimes of land and natural resources differ due to the nature of the historical experiences. The main issues relate to the accessibility, transparency and accountability of the administrative systems. There has been a tendency towards too much administration, due to the different layers of the state, local and indigenous authority, particularly within customary tenure systems. The fifth analytical construct relates to systems of adjudicating land disputes. There are key questions with regard to how to resolve current and past land problems in situations where multiple tenure regimes exist. In most countries, the legal framework has been biased towards the market and the state.

The study also reviewed the model by Maxwell and Weibe [41] that illustrates a causal flow relationship between resources, production, income, consumption and nutritional status.

5. Conventional conceptual links between land and food

Below is an example of a conventional conceptual nexus between land and food: Resources → Production → Income → Consumption → Nutritional status (Source: Adopted from Maxwell and Wiebe [41].

But it was suggested that a simple linear model does not adequately capture the interrelationships between consumption and investment decisions, household endowments, production and exchange decisions, and household entitlements. A more comprehensive model illustrates a circular relationship between these four factors, which are further impacted upon by tenure institutions and asset markets and have outputs in terms of environmental impact, generation and redistribution of wealth. The most apparent qualitative link that was suggested was that increased security of tenure in productive resources enables more efficient and profitable agricultural production and hence greater access to food via both own production and trade.

Also, a report was given by the European Union on Development in 2011 that was driven mainly by concerns about the scarcity of food, energy and arable land and linked to these expectations of rising land values with issues relating to:

i. How the land deals come about, the quality of the contracts, and who will benefit and who will lose (i.e., governance and accountability issues with respect to decision-making on transactions and terms of the contracts).

ii. Whether there is recognition of all local rights (including informal and secondary rights) and adequate compensation.

iii. Whether the contracts override customary rights, resulting in smallholders, pastoralists and forest dwellers being driven from the land.

iv. The implications for food security/sovereignty, rural development and the future of smallholder farming.

v. Risks related to water availability, environmental degradation and loss of ecosystem services.

The report explained that it was not only transnational investors but also migrants who use their remittances to buy land in their country of origin [42] that were acquiring large areas of farmland. Although not a new phenomenon, studies indicated that this form of land acquisition is accelerating [43].

Also, LSLAs (large-scale land acquisitions) were also taking place for purposes other than food and biofuel or agricultural production [44, 45] that included nature conservation, parks and new initiatives under the United Nations Collaborative Programme on Reducing Emissions from Deforestation and Forest Degradation in Developing Countries (UN-REDD).

According to the report, the debate on global food security continues to stimulate the interest in large-scale farming, which is perceived as more productive and efficient. The argument runs as follows: given that by 2050, some 9 billion people will need to be fed and there is a need for a significant increase in production and productivity, which can be achieved only by replacing smallholder farming with industrial-type farming. Others contend, however, that smallholder farming can be just as efficient (e.g., [46]).

Moreso, the direct contribution of LSLAs to national food sovereignty depends on the type of crop grown and whether it is locally marketed at affordable prices. In Ethiopia, these large farms sell grain to WFP (World Food Programme), which uses them to feed those who left the famine areas and are now in camps. It is important to look at local as well as national food security. As mentioned, the first effect of these large-scale farms is often that local smallholders lose access to land and natural resources, which undermines their livelihoods, food security and ability to accumulate resources. These losses may be partly compensated for if new jobs are created that allow these people to earn sufficient to feed themselves and their families to a comparable level in terms of quantity and quality and also to pay for other services.

Fombe et al. [47] studied security of land ownership by women in Anglophone Cameroon. In most parts of Africa, land ownership is affected by traditional values, inheritance rights and government influence. This study illustrates that land acquisition and development by women constitute a problem because of traditional pressures and the law guiding the process of land certification. Aluko and Amidu (2006) however argue that the complexity of land tenure in Nigeria is the result of the co-existence of several systems; none of which is completely dominant. This legal pluralism causes a degree of uncertainty about land rights, particularly for vulnerable groups, like women. In most of Anglophone Cameroon, there are basically two ways in which women can acquire land either through family bond (users' rights) or through transactions (purchase, lease and rent). In the current context of land scarcity, population mobility, urbanisation and land reforms, the competitive demand for land has not only generated a diversity of struggles over land but has further complicated the prospects of women accessing land in a predominantly patriarchal setting (Lotsmart et al., 2010). In Anglophone Cameroon, the majority of women do not own land or have the right to inherit land and other property where statutory laws and customary practices co-exist.

The study's major aim was to ascertain the effectiveness of government's legal instrument (the Land Consultative Board) that regulates the purchase and ownership of land; random interviews were conducted with traditional and local administrative authorities as well as men and women landowners in five of six divisions in the North West and four of five in the South West Region. These two English-speaking regions were considered as Anglophone Cameroon because they were administered by British Cameroon during the colonial period.

A random but stratified survey method was adopted with 20% of the sampled population comprised males based on their traditional and societal roles, while 80% was addressed to rural and urban women in eight divisions.

Women's land rights can contribute to their empowerment and constitute a key to reducing poverty and developing a community. With secured land rights, crop production will increase; women will have better access to credit, develop self-esteem and will be able to develop their own skills (Mantobang, 2011). According to Sengupta (2000), empowerment is inextricably linked to rights language, whereby it constitutes the ability or opportunity of the poor to claim and exercise their rights.

Sustainable development should be inextricably linked to tenure rights. This concept is not mentioned in the Declaration of the Right to Development adopted by the UN General Assembly in 1986 (Marks, 2008). In as much as women have a right to development, so do they have a right to sustainable development, though this latter concept hinges more on rights and obligations of humans towards nature. Taking an ethical perspective on the concept of sustainable development, Marks (2008) looks at it as fairness at three levels; "intragenerational", "intergenerational" and "between humans and the environment", by enhancing people's moral consciousness and environmental protection, altering traditional ideologies and behaviours and promoting sustainable survival.

Women's lack of ownership and access to land results from some traditional practices that violate sustainable development. This is exemplified by those who uphold the values on the one hand and the environment on the other, with the victims (women) being at the centre. It follows therefore that depriving a woman of the right to inherit land for the simple fact that she is a woman puts her in a disadvantaged socio-economic and political position vis-à-vis the men. This only accentuates the feminization of poverty.

It was noted that there is need to exhume the barriers of government's legal instrument (the Land Consultative Board) that regulates the ownership of land and to revisit some traditional practices as regards land ownership that impact negatively on women in a changing and globalising world. A compromise approach was advocated for land acquisition that can transcend traditional barriers as well as render the process of land registration more realistic especially for women.

In the same vein, Odoemelam et al. [48] evaluated women access and rights to land and its implication on rural household food security in Abia State, Nigeria. The study established that it was not just the shortage of land that affects the output of agriculture but the structure of the land tenure, lack of proper land ownership and rights.

Multistage sampling procedure was used in the selection of the sample size with a total of 180 respondents. Data were generated through the use of Focus Group Discussion and participatory observation. Data generated were analysed through the use of simple descriptive statistics like frequency distribution and means.

Results showed that most of the women were in the economically active age and majority were married with a very high percentage having only up to primary school level of education. The level of income indicated that majority of the respondents 34% earned between (N160–180,000), while 16% earned between (N200–220,000) per annum. Methods of land inheritance showed that women acquired land for cultivation through majorly matrilineal ties and purchases, while other forms of acquisition such as renting, borrowing and through inheritance were minimal. In terms of access and use rights on lands, the study found about 30% of the respondents were allowed to cultivate only annual crops, 15% perennial crops, 11% were not allowed to harvest production from the trees, 13% were not allowed to retain land after long fallow and 8% complained about limited access to inputs.

On effect of access and land use rights on food security, the 25% respondents stated that due to the lack of access to land use right, their expertise on farming activities are underutilised leading to food insecurity, 15% said they lack commitment, 26% complained about their inability to get credit facilities from bank because they can not use their cultivating land to get collateral, unfertile plots 13% leading to low yield and nonmechanised farms, 24% due to scattered plots. The study concluded that ownership rights are critical to securing a sustainable livelihood and income, and the lack of rights is one of the main sources of economic insecurity. When men and women have equal access to land use rights, they have effective decision-making power in the household and food production will be enhanced to its full capacity to support the welfare of all family members and to promote food security in the society.

De Zoysa [1] examined the implications of large-scale land acquisition on small landholder's food security. This chapter analysed the implications of LSLA (large-scale land acquisition) on local population's food security and livelihoods using Ethiopia as a case study to measure the progression of vulnerability created by land reform policies that encourages such investments. Using the analytical framework of the Pressure and Release (PAR) model, this chapter argued that Ethiopia's land reform has systematically weakened small landholder's access to food and livelihood. The Pressure and Release (PAR) model was selected because the model allows one to understand that food insecurity as defined as the availability, access and utilisation of food does not occur due to a drought but more so a systematic breakdown of coping mechanisms and the lack of consideration paid to rectify root causes of food insecurity. It was recognised that all frameworks have limitations in analysing complex situations including the PAR, and it was treated with caution to ensure there is no oversimplifications.

The PAR originally developed by Blaikie et al. in 1994 demonstrates that disasters were not simply a random natural phenomenon but a result of development regression [49]. The PAR identified three factors that exacerbate vulnerabilities that include: (1) root causes; (2) dynamic pressures; (3) unsafe conditions.

The study was based on secondary data through the forms of both academic and grey literature.

In order to capture the impact of LSLA on affected populations, the case study needs to narrow the scope into a region of the country that faces the highest amount of LSLA such as the Gambela Region of Ethiopia.

Root Causes: Power: There was a lack of knowledge about land deals because they were rarely advised or consulted prior to the agreements. Even if they were consulted, their lack of bargaining power stunts their ability to negotiate fare rates of compensation. Blaikie et al. identified three causes that limit the access to power which are spatial, temporal and social.

Structures: Political and economic structures reproduce domination against the marginalised in order to maintain the control of power. In the case of the Gambela Region, the lack of secure land tenure was a structure that reinforced the deeper issue of an ethnic minority who had a lack of political power. **Resources:** Pastoral communities were increasingly vulnerable because they were denied access to grazing areas that provided access to livelihood resources. What can occur through this process is that these communities shred their cultural way of livelihood and become highly dependent on wages earned in farm labour to purchase food [50].

Dynamic Pressures: The notion of dynamic pressure exacerbates the root causes of tension entailed in limited access to power and resources. This was evident in the study area because of issues that dealt with the lack of press freedom and NGO advocacy. Blaikie et al. conveys that the root causes channelled by the dynamic pressure of the absence of press freedoms and limited scope of NGOs eventually force changes to the macro-economic and social conditions (2005).

Unsafe Conditions: Blaikie et al. expressed that root causes and dynamic pressures manifest into unsafe conditions in the physical and social environment making the affected populations more vulnerable to risks [51]. The physical environment of the land leased to investors steadily degraded through agricultural intensification because there is a lack of monitoring mechanisms and enforcement from the authority.

The major lessons learned from the Ethiopian case study was that the 'new phenomenon' of land acquisitions in the Gambela Region was actually a process that had historical roots dating to Ethiopia's imperial colonialism making it 'an old wine in a new bottle.' The local conditions of suppressing press freedoms and NGOs' operational space ensured that such oppressive actions continued in the future without critical review. Another lesson from the case study is that disasters are never a one-off event but manifests from embedded vulnerabilities. This chapter does not argue with the fact that a high level of investment is needed in Ethiopia, but the concern is that without rectifying the root causes of disasters that have plagued Ethiopia, large-scale land acquisition only exacerbates the progression of vulnerability.

6. Conclusion and recommendations

This study examined the issues around land acquisition and the implications on livelihood and food security. It was found that the land acquisition is bound by the Land Use Act 1978, which stipulates that all lands belong to the government. However, the bulk of the land transactions are carried out in informal markets under customary laws with poor or lack of documentation, especially in rural Nigeria. The land market in Nigeria has evolved over the years from the pre-colonial era when land titles were in custody of family and community heads through colonial era and post-colonial era. The current framework undermines food security as farmers typically lack the access to land and as such cannot scale their subsistence farming or even present land as collateral for formal loans. This study identified bureaucratic bottlenecks, high cost of registration of land title, weak land markets and policy inconsistency as the challenges of land acquisition and use in Nigeria. Based on the findings of the study, the following are recommended:

1. Government should reduce and make the processes to be completed easier in registering lands in Nigeria.
2. The Land Use Act 1978 should be amended to capture prevailing realities around customary laws and informal markets.
3. Government should reduce the cost of land registration in Nigeria.
4. Multilateral organisations and government should co-create and co-finance innovative interventions to improve activities in the land market.

Author details

Isaac B. Oluwatayo[1]*, Omowunmi Timothy[2] and Ayodeji O. Ojo[2]

*Address all correspondence to: isaacoluwatayo@yahoo.com

1 Department of Agricultural Economics, University of Limpopo, South Africa

2 Department of Agricultural Economics, University of Ibadan, Nigeria

References

[1] De Zoysa R. The Implications of Large Scale Land Acquisition on Small Landholder's Food Security. The Bartlett Development Planning Unit DPU Working Paper no. 156; 2013

[2] Adams M. Tenure security, livelihood and sustainable land use in southern Africa. In: Unpublished Paper Presented at the Conference on Land Reform and Poverty Alleviation in Southern Africa Convened by the Southern Africa Regional Poverty Netidok. Pretoria: Human Sciences Research Council; 2001

[3] National Population Commission. Nigeria's Population Hits 198 Million People. 2018 (January 15). Retrieved May 27, 2018, from National Population Commission: http://population.gov.ng/nigerias-population-hit-198m-people-npopc-chairman/

[4] Nuhu M. New development in land administration and surveying matters in Nigeria. In: Fig Commission 7 Conference on Cadastre and Land Management. Kuala Lumpur, Malayia: Fig Commission 7; 2009. pp. 1-17

[5] Omole F. Land use violations: Implications for sustainable development, the case study of the Federal Capital City Abuja, Nigeria. Current Research Journal of Social Sciences. 2009; **1**(1):31-37

[6] Udoekanem N, Adoga D, Onwumare V. Land ownership in Nigeria: Historical development, current issues and future expectations. Journal of Environment and Earth Science. 2014;**4**(2):182-187

[7] Babalola S, Abdulrahman A, Choon L, Van Oostorom P. Possibilities of Land Administration Domain Model (LADM) Implementation in Nigeria. In: Joint International Geoinformation Conference. Kuala Lumpur, Malaysia: ISPRS Annals of the Photogrammetry, Remote Sensing and Spatial Information Sciences; 2015. pp. 1-9

[8] Omuojine E. The land use act and the English doctrine of estate. Journal of the Nigerian Institute of Surveyors and Valuers. 1999;**22**(3):54-56

[9] Zoysa R. The Implications of Large Land Acquisition on Smallholder's Food Security. London: Development Planning Unit of University College London; 2015

[10] Chikaire JU, Anyoha N, Ani A, Atoma C. Land tenure reform: A vehicle for achieving agricultural transformation agenda in Nigeria. Merit Research Journal of Agricultural Science and Soil Sciences. 2014;**2**(9):114-122

[11] Durrand-Lasserve A, Selod H. Land markets and land delivery systems in rapidly expanding West African cities: The case of Bamako. In: Sixth Urban Research and Knowledge Symposium. Spain: World Bank; 2012. pp. 1-17

[12] Alarima C, Fabusoro E, Kolawole A, Uzoma K, Aromolaran A, Masunaga T, et al. Land rights and rental systems: Implications for management of conflicts related to land in Sawah-based rice production systems in Nigeria. African Study Monographs. 2012;**33**(3): 189-208

[13] Oloyede S, Ayedun C, Oni AS, Oluwatobi A. Land market challenges: The case of Ifo/Ota local government area of Ogun state, Nigeria. International Journal of Business and Social Science. 2014;**5**(12):208-214

[14] World Bank. 2017 (06, 30). Registering Property. Retrieved May 29, 2018, from World Bank website: http://www.doingbusiness.org/data/exploretopics/registering-property

[15] Dale P, Mahoney R. McLaren R. Land Markets and the Modern Economy. London: Royal Institution of Chartered Surveyors; 2002

[16] USAID. Property Rights and Resource Governance. Abuja: USAID; 2009

[17] Odoemelam L, Osahon E, Nechedo E. Effect of tenure security on livelihood activities of women farmers in Anambra state, Nigeria. Journal of Agriculture & Social Sciences. 2013; **13**(2):94-100

[18] National Bureau of Statistics; World Bank and Federal Ministry of Agriculture and Rural Development. LSMS-Integrated Surveys on Agriculture General Household Survey Panel 2015/2016 report. Abuja: NBS; 2016

[19] Eze CC, Konkwo SO, Orebiyi JS, Kadiri FA. Land tenure system, farm size, innovation and agricultural productivity in South-East Nigeria. In: Contributed Paper for the Agricultural Economics Society's 85th Annual Conference Held at University of Warwick; 2011

[20] Ojo SO. Effects of land acquisition for large scale farming on the performance of small scale farming in Nigeria. Journal of Human Ecology. 2008;**24**(1):35-40

[21] Olayide SO. Characteristics, problems and significance of small farmers. In: Olatide JAE, Card Bello-Osagie VE editors Nigerian Small Farmers: Problems and Prospects in Integrated Rural Development. University of Ibadan, Ibadan, Nigeria; 1980

[22] Ojo MO. The effectiveness of agricultural policies on Nigeria economic development. Rural Economy and Sociology, University of Ibadan, 1991;**8**:2

[23] Twene SK. Land Grabbing and rural livelihood sustainability: experiences from the bui dam construction in ghana. A Master of Philosophy Thesis submitted to the Department of Geography and Rural Development, Kwame Nkrumah University of Science and Technology for the award of Master of Philosophy Degree, 2016, pp. 1-147

[24] Hull S, Sehume T, Sothafile L. Land allocation, boundary demarcation and tenure security in tribal areas of South Africa. South African Journal of Geomatics. 2016;**5**(1):68-81. DOI: 10.4314/sajg.v5il.5

[25] Federal Ministry of Agriculture and Rural Development. The Agriculture Promotion Policy (2016–2020): Building on the Successes of the ATA, Closing Key Gaps. Abuja: FMARD; 2016

[26] Food and Agriculture Organisation of the United Nations. The Future of Food and Agriculture. Trends and Challenges. Rome: FAO; 2017

[27] Oladapo O, Olajide O. Land-income-nutrition nexus: Implication for food security of rural households in Nigeria. Journal of Agriculture and Sustainability. 2015;**8**(2):29-42

[28] Oluwatayo I, Ojo A. Awareness and adaptation to climate change among yam-based farmers in Oyo state. Nigeria. Journal of Developing Areas. 2016;**50**(2):97-108

[29] Onoja AO, Achike AI. Large-scale land acquisitions by foreign investors in West Africa: Learning points consilience. The Journal of Sustainable Development. 2015;**14**(2):173-188

[30] Sahel and West Africa Club (SWAC). Land, Investment and Development, FAO/SDC/IFAD Summary of Stakeholders Meeting held in Rome on 24; 2010

[31] FAO. Africa's Changing Landscape: Securing Land Access for the Rural Poor. FAO Regional Office for Africa; 2010. Retrieved on 25th August, 2011 from www.fao.org/africa. FAO: Rome; 2010

[32] Cotula L et al. Land Grab or Development Opportunity? London: IIED, FAO, IFAD; 2009

[33] Cotula L, Vermeulen S, Leonard R, Keeley J. Land Grab or Development Opportunity? Agricultural Investment and International Land Deals in Africa. London/Rome: IIED/FAO/IFAD; 2009. p. 2009

[34] Global Development. The Food Rush: Rising Demand in China and West Sparks African Land Grab. 2010. Retrieved on 29th August, 2011. Available from: http://www.guardianco.uk/global-development

[35] FAO. Compendium of agricultural-environmental indicators 1989-91 to 2000. In: Statistics Analysis Service Statistics Division Food and Agriculture Organization of the United Nations Rome; 2003

[36] Economic Commission for Africa. Land Tenure Systems and their Impacts on Food Security and Sustainable Development in Africa. ECA/SDD/05/09; 2004

[37] Moyo S. Land and Democracy in Zimbabwe. Monograph Series No. 7. Harare: SAPES Books; 1999

[38] Shivji IG, Moyo S, Gunby D, Ncube W. National Land Policy Frame-work. Draft discussion paper. Ministry of Lands and Agriculture: Harare; 1998

[39] Moyo S. The Land Question in Zimbabwe. Harare: SAPES Books; 1995

[40] Moyo S. The land question in Zimbabwe. In: Mandaza I, editor. Zimbabwe: The Political Economy of Transition. Dakar: CODESRIA; 1987

[41] Weiner D, Moyo S, Munslow B, O'Keefe P. Land use and agricultural productivity in Zimbabwe. The Journal of Modern African Studies. 1985;**23**(2)

[42] Maxwell D, Wiebe K. Land Tenure and Food Security: A Review of Concepts, Evidence and Methods, Research Paper No 129. Madison: University of Wisconsin; 1998

[43] Cotula L. Exploring the linkages between remittances and access to land: A conceptual framework. In: Cotula L, Toulmin C, editors. Till to Tiller: Linkages Between International Remittances and Access to Land in West Africa. London: IIED; 2004

[44] Hilhorst T, Nelen J, Traoré N. Agrarian Change Below the Radar Screen: Rising Farmland Acquisitions by Domestic Investors in West Africa: Results from a Survey in Benin, Burkina Faso and Niger. In: Paper presented at the International Conference on Global Land Grabbing; 6–8 April. Brighton: University of Sussex; 2011

[45] Zoomers A. Buy your own paradise. Over land, mobiliteit en ontwikkelingsbeleid. Faculteit Geowetenschappen, Universiteit Utrecht (oratie); 2008

[46] Zoomers A. Globalisation and the foreignisation of space: The seven processes driving the current global land grab. Journal of Peasant Studies. 2010;**37**(2):429-447

[47] Deininger K, Byerlee D, et al. Rising Global Interest in Farmland: Can it Yield Sustainable and Equitable Benefits. Washington, DC: World Bank; 2011

[48] Fombe LF, Sama-Lang IF, Fonjong L, Mbah-Fongkimeh A. Securing tenure for sustainable livelihoods: A case of women land ownership in Anglophone Cameroon. Ethics and Economics. 2013;**10**(2):74-86

[49] Aluko BT, Amidu A. Issues on Urban Growth: Customary Title and Informal Settlements. A paper presented at the 5th FIG Regional Conference, Accra, Ghana; March 8-11, 2006. pp. 1-13

[50] Lotsmart F, Sama-Lang I, Fombe FL. An assessment of the evolution of land tenure system in cameroon and its effect on women's land rights and food security. Perspectives on Global Development and Technology. 2010;**9**(1):154-169

[51] Mantohbang O.M. Women's Land rights and community development. Paper presented at workshop on Women and Land Right in Anglophone Cameroon by IDRC/UB Women and Land Rights Research Project, 22-23 October, 2010, pp. 1-15

[52] Odoemelam LE, Alamba C, Lekan-Akomolafe CN. Evaluation of women access and rights to land and its implications on rural household food security in selected rural communities, Abia state, Nigeria. Discourse Journal of Agriculture and Food Sciences. 2014;**2**(8):235-241

[53] Boano C. Disaster risk reduction in cities-session two. In: Development Planning Unit. London, UK: University College London; 2010

[54] Blaikie P et al. At Risk: Natural Hazards, People's Vulnerability and Disasters. 2nd ed. London: Taylor & Francis; 2005

Section 3

Land-Use Planning for Environmental, Economic and Social Development

Chapter 7

A Visual Quality Prediction Map for Michigan, USA: An Approach to Validate Spatial Content

Rüya Yilmaz, Chung Qing Liu and Jon Bryan Burley

Additional information is available at the end of the chapter

http://dx.doi.org/10.5772/intechopen.79490

Abstract

For a least the last half-century, scholars have been seeking methods to predict and assess the visual and environmental quality of the landscape. In these investigations, some scholars have been interested in applying predictors to create maps, representing visual and environmental quality. In our study, we employed a reliable environmental quality prediction equation that assesses environmental quality to create a validated visual quality map of Michigan containing a variance of 0.67, containing an overall p-value less than 0.0001, and p-values less than or equal to 0.05 for each predictor. Measures ranging in the mid-40s and 50s indicate a moderate level of environmental quality, while scores in the 80s through 110 indicate a very poor environmental quality. Through the Kendall's coefficient of concordance statistical test, we determined that the map is significantly reliable ($p \leq 0.005$) and conclude that constructing such a large area (250,493 km^2) is possible. This type of map can be employed to evaluate progress and decline in measuring the environmental quality/land-use change of extensive landscape areas.

Keywords: environmental psychology, landscape architecture, land-use planning, landscape planning

1. Introduction

For almost 50 years, investigators have been seeking quantitative methods to predict and assess the visual and environmental quality of the landscape. The literature on this subject is vast and continues as illustrated through recent investigations and contributions by psychologists, engineers, landscape architects, planners, and natural resource specialists [1–11]. One of the best summaries of the early efforts was described by Taylor et al. [12]. Despite the scientific advances, the application of equations and exploration of theories seemed somewhat impractical for many practitioners. In attempts to translate the research, two of the best practical summaries of the ideas revealed by the research were found in *The Experience of Nature: a Psychological Perspective* and in *With People in Mind: Design and Management of Everyday Nature* [13, 14]. Along with developing predictive equations, investigators were interested in constructing maps that predict visual quality. Brush and Shafer produced one of the earliest well known visual quality maps [15]. But this approach was not widely adopted. In the United States, a heuristic approach (without strong statistical evidence for the variables and coefficients but instead based upon normative theory) comprised of an index developed by Jones and Jones, becoming widely employed and over time refined [16, 17]. The success of this index may have hindered the development of more science-based equations, as stakeholders adopted this relatively easily understandable methodology and were unwilling to seek additional methods. The actual science supporting this normative theory approach has been relatively weak and not aggressively challenged by the scientific community. But to landscape architects not trained in the ways of statistical analysis, p-values, and variance explanations, the index seemed to make good logical sense. In some respects, there seemed to be a lull at the turn of the century concerning visual quality assessment, where a methodology to validate maps was not self-evident, and the production of equations to explain increased levels of the variance were at a standstill. The early attempts to predict visual quality were primarily focused upon artistic composition normative values such as foreground, mid-ground, background, geometrical harmony, and natural/urban components. By pursuing this set of esthetic and spatial variables, investigators were able to explain 30% to nearly 70% of the variance [18].

At the time, some scholars were frustrated with being unable to make much new headway. Some more recent investigators entered the subject area and began to explore the importance of other types of variables that were less esthetic in character, and more ecological, cultural, economic, and functional [19]. At a conference in 2005 in Switzerland (Our Shared Landscape 2005, Integrating ecological, socio-economic and esthetic aspects in landscape planning and management′ http://www.osl.group.shef.ac.uk, at Centro Stefano Franscini in Ascona, Switzerland), scholars were coming to the realization that respondents evaluated the landscape with more than just esthetic values. When these other potential predictors are added to the study, more of the variance was explained collectively and statistically. Stronger and more reliable predictive equations evaluating landscapes could be generated from respondents, explaining over 90% of the variance when esthetic, ecological, cultural, functional, and economic predictors are combined [20–22]. Along with the equations a series of theories to explain the results evolved [23]. Investigators noted that contents where humans infringed upon other humans (buildings, roads, people) were less preferred (even high quality architecture)—human disturbance theory; the benign environment (plants, waters, sky) were neutral in preference—natural preference theory; and temporal features (wildlife, momentary views of mountains, and flowers) were in preferred environments—temporal enhancement theory. Thus, theory and models were advancing together concerning evaluating and assessing the quality of the environment.

During this period, some scholars were beginning to explore the potential for globally universal predictive equations. Mo et al. recently reviewed much of this pertinent literature and discusses the perceptions of respondents in North America, France, Portugal, and PR China [24]. One finding suggests that Europeans and North Americans may have broad similar perceptions about landscape. Asians may have a different sensibility concerning environmental preference. Concurrently, on the digital visualization forefront Partin et al. studied the response of participants to computer generated images and reported that the perception of computer images was similar to the perception of photographs of landscapes [25]. In other words, it was possible for investigators to present computer-generated images to respondents and to obtain a similar response as if the respondents were examining photographs. He also demonstrated how a small study could be folded into a larger and widely studied set of images to obtain stable and reliable results. However, these equations and results are still formative and require duplication by others to refute them or support them. In addition, there are opportunities to explore the responses of various cultural groups and to refine these equations. While the literature on this topic is vast, there seems to be many more instances to add to the body of knowledge.

Also during this recent timeframe, attempts to explore mapping making potential were renewed. Lu et al. examined the Lower Muskegon watershed, located on the west side of the lower peninsula in Michigan [26]. He and his colleagues studied images of urban areas, farmland, wetlands, and forests and attempted to construct an environmental quality map of his study area. The results he obtained through statistical analysis revealed that the relationship between his predictions in the map and the real photographs are in concordance and at a reasonable (95%) confidence level. He concluded that visual/environmental quality could be mapped and reliably predicted in the Lower Muskegon Watershed. There was a strong relationship between the perception of environmental quality and land cover. Following Lu, Jin examined the southern portion of Michigan, an area much larger than the Lower Muskegon Watershed [27]. She reported similar results to Lu. She had demonstrated that it was possible to develop reliable maps for much larger areas, but still not at the scale of a province or state.

Figure 1. The location of Michigan, the study area in North America.

Lothian presents the fundamentals and an overview of various approaches to constructing visual quality maps and is a substantia update to the work of Taylor et al. [12, 28]

These recent studies provide a setting for our investigation. As an extension of Lu's et al. and Jin's research, we were interested in applying this approach to all of Michigan (**Figure 1**) [26, 27]. We wanted to make a validated map for all of Michigan.

2. Methodology

Michigan is located in the Great Lakes Region of the United States of America. It is the only state to consist of two peninsulas with the longest shoreline of any state of the lower 48 states. These two peninsulas are linked by the Mackinaw Bridge. The Straights of Mackinac separate the Upper Peninsula from the Lower Peninsula, whose shape looks like a mitten (**Figure 1**). The study area spans abundant agricultural lands in the south, hardwood forests in the middle portion, and mixed evergreen forests in the north. The study area also contains a large urban landscape comprised of metropolitan Detroit, plus numerous industrial sites, especially in the lower part of the state. The study area contains level glacial lake plains, hilly moraines, and ancient eroded mountain chains. The population of the state is only 10 million people (roughly the same population number as the Kingdom of Sweden but about half the size in overall land area).

The methodology used for this study was similar to that utilized by Lu et al., where land cover and visual/environmental quality covary [26]. A comprehensive explanation of this methodology can be found in Lu et al. [26]. Adhering closely to this method, images of various landscapes across the study area were collected and randomly sorted into two groups: one group to assist in making a prediction and another group to validate or refute the prediction. From the first group, scores for the images were generated by employing Eq. (1) developed by Burley [19]. The information presented by Burley is the formative paper in this line of work and investigators interested in understanding the fundamentals of this line of research are urged to examine this paper. Once the scores were obtained, they were applied to similar land-uses to form a map predicting environmental quality.

$$\begin{aligned} Y = {} & 68.30 - (1.878 * HEALTH) - (0.131 * X1) - (0.064 * X6) + (0.020 * X9) \\ & + (0.036 * X10) + (0.129 * X15) - (0.129 * X19) - (0.006 * X32) \\ & + (0.00003 * X34) + (0.032 * X52) + (0.0008 * X1 * X1) + (0.00006 * X6 * X6) \\ & - (0.0003 * X15 * X15) + (0.0002 * X19 * X19) - (0.0009 * X2 * X14) \\ & - (0.00003 * X52 * X52) - (0.0000001 * X52 * X34) \end{aligned} \tag{1}$$

where

HEALTH = environmental quality index (**Table 1**)

X1 = perimeter of immediate vegetation

X2 = perimeter of intermediate non-vegetation

X3 = perimeter of distant vegetation

X4 = area of intermediate vegetation

X6 = area of distant non-vegetation

X7 = area of pavement

X8 = area of building

X9 = area of vehicle

X10 = area of humans

X11 = area of smoke

X14 = area of wildflowers in foreground

X15 = area of utilities

X16 = area of boats

X17 = area of dead foreground vegetation

X19 = area of wildlife

X30 = open landscapes = X2 + X4 + (2 × (X3 + X6))

X31 = closed landscapes = X2 + X4 + (2 × (X1 + X17))

X32 = openness = X30 − X31

X34 = mystery = X30 × X1 × X7/1140

X52 = noosphericness = X7 + X8 + X9 + X15 + X16

Next, the second group of images was compared to predictions made by the map through the use of Kendall's Concordance, a statistical technique that examines and tests for significant agreement/similarity [29, 30]. If the scores statistically agree, it is possible to create a reliable visual quality prediction map. This step in the methodology used here is explained in great detail by Jin [31]. Investigators interested in applying this methodology are advised to obtain copies of Lu et al. and Jin for a complete explanation [26, 31].

The test statistics were determined by applying Eqs. (2) and (3) to the data. The results are based upon rankings of treatment scores across rows. In this case, the rows are pairs of images between two treatments: (1) the predicted score for a randomly chosen site in the study area and (2) the actual score from a photograph taken at that location. There are 30 rows (pairs of scores) for this study (n = 30). The treatments are the columns (m = 2). The rankings are summed and squared, to compute Kendall's W value (Eq. 2). $(Rj)^2$ is the sum of the squares of the rankings for a column in computing the Kendall's W value [29, 30].

$$W = 12\sum_{j=1}^{n} (Rj)^2 - [3\,m^2\,n\,(n+1)^2]/[m^2\,n(n^2-1)] \tag{2}$$

Kendall's W value is a number ranging between 0 and 1. When W is near 0, there is no strong overall trend of agreement among the respondents. If W is near 1, then the responses could be

Health index	
Variable	**Score**
A. Purifies Air	+1 0–1
B. Purifies Water	+1 0–1
C. Builds Soil Resources	+1 0–1
D. Promotes Human Cultural Diversity	+1 0–1
E. Preserves Natural Resources	+1 0–1
F. Limits Use of Fossil Fuels	+1 0–1
G. Minimizes Radioactive Contamination	+1 0–1
H. Promotes Biological Diversity	+1 0–1
I. Provides Food	+1 0–1
J. Ameliorates Wind	+1 0–1
K. Prevents Soil Erosion	+1 0–1
L. Provides Shade	+1 0–1
M. Presents Pleasant Smells	+1 0–1
N. Presents Pleasant Sounds	+1 0–1
O. Does not Contribute to Global Warming	+1 0–1
P. Contributes to the World Economy	+1 0–1
Q. Accommodates Recycling	+1 0–1
R. Accommodates Multiple Use	+1 0–1
S. Accommodates Low Maintenance	+1 0–1
T. Visually Pleasing	+1 0–1
Total score	

Table 1. Variables for the environmental quality/health index in Eq. (1).

regarded as close to unanimous in their agreement. The W test statistic approximates a Chi-square distribution with n−1 degrees of freedom (Eq. (3)). If computed values for Chi-square (results of (Eq. (3)) are greater than significant values in a Chi-square table for n–1 degrees of freedom (in this case 29 = 30–1), then there is a high level of agreement/concordance—the predicted scores and the actual scores are in agreement.

$$X^2 = m(n-1)W \tag{3}$$

3. Results

The sample of images gathered in the investigation includes forested lands (**Figure 2**), agricultural lands (**Figure 3**), residential environments (**Figure 4**) (known as urban savanna), downtown-like environments (**Figure 5**) (known as cliff detritus), industrial sites (**Figure 6**), and open water (**Figure 7**) [32].

Figure 2. Image of sample number 67 of a forested landscape in the upper peninsula of Michigan (visual score of 38.548).

Figure 3. Image of sample number 80 of a farmland landscape in north-east of the lower peninsula of Michigan (visual score of 44.09).

Figure 4. An image of a residential landscape (urban savanna with visual quality score of 46.587), sample number 12 in the northwest of the lower peninsula.

Figure 5. An image of sample number 73 of a downtown environment (cliff detritus with a visual score of 82.766).

Figure 6. An image of sample number 77 of an industrial environmental (with a visual score of 78.778).

Figure 7. An image of sample number 46 of a primarily open water environment of Lake superior (with a visual score of 41.498).

http://dx.doi.org/10.5772/intechopen.79490

Table 2 presents the rankings of the images from the study. The predicted ranks are scores generated from the expected a land-use scores and applying the expected score to a land-use map of the study area. The actual scores are values taken and measured from random sites in the study area. Kendall's Concordance analysis revealed a Chi-square score of 54.267. The

Property	Predicted ranking	Mean expected score	Actual score	Set 2 ranking
Residential 1	18	63.06312	73.69797	21
	18	63.06312	55.94164	15
	18	63.06312	46.58693	8
	18	63.06312	70.73992	20
	18	63.06312	68.34915	18
Downtown	23	81.56982	69.28333	19
	23	81.56982	84.30117	24
	23	81.56982	86.72108	27
	23	81.56982	82.76631	23
	23	81.56982	84.77721	25
Farmland	13	55.92384	59.12320	17
	13	55.92384	57.74200	16
	13	55.92384	45.25820	7
	13	55.92384	44.09459	6
	13	55.92384	53.27078	13
Industrial	28	90.91938	86.56068	26
	28	90.91938	94.63968	28
	28	90.91938	96.94017	29
	28	90.91938	78.77838	22
	28	90.91938	97.67801	30
Forested	3	42.67716	54.40120	14
	3	42.67716	46.98920	9
	3	42.67716	36.70580	2
	3	42.67716	36.70580	2
	3	42.67716	38.58380	4
Water	8	44.99797	52.89468	12
	8	44.99797	47.78978	11
	8	44.99797	35.62803	1
	8	44.99797	47.27917	10
	8	44.99797	41.39820	5

Table 2. Comparison of ranks between the average expected score and actual site photographs.

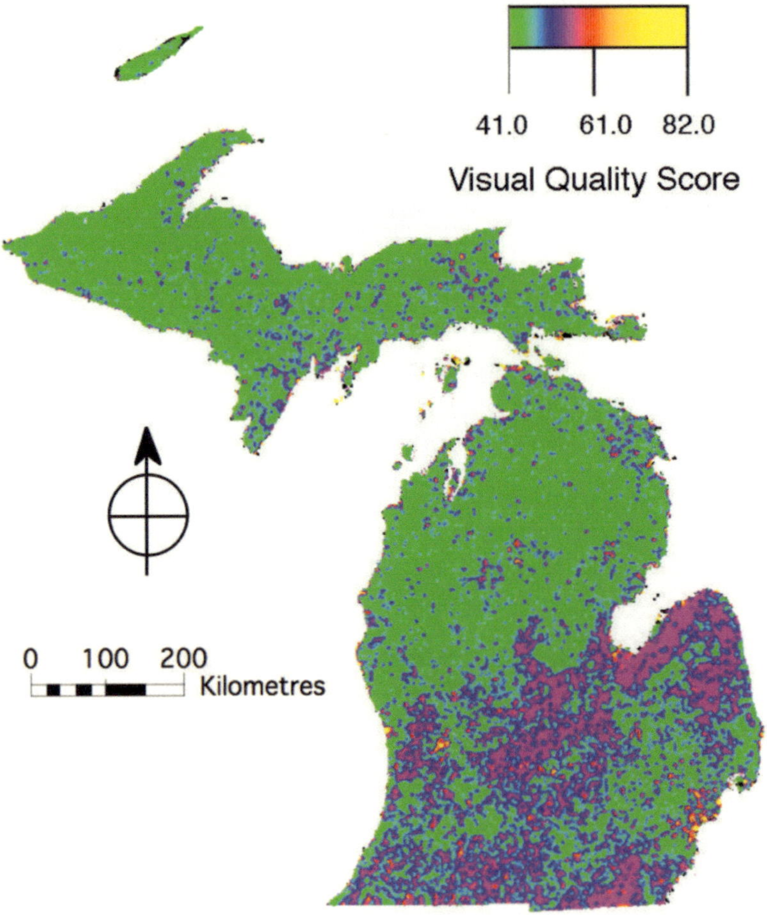

Figure 8. A map of the predicted environmental/visual quality of the Michigan landscape. Green areas are environments with good visual quality. Blue, magenta, red, and orange indicate successively less preferred environments. Yellow areas are the least preferred.

number found in a standard table for such a Chi-square is 52.336 (a 99.5% confidence level ($p \leq 0.005$) for 29 degrees of freedom. Since 54.267 is larger than 52.336, the predicted scores and the actual scores are in agreement at a 99.5% confidence level ($p \leq 0.005$). These results suggest that it is possible to construct an environmental/visual quality map of Michigan that is relatively statistically reliable (**Figure 8**). In other words it is possible to predict the visual quality of any site in Michigan by knowing the land-use and to accurately predict the expected visual quality correctly 199 times in 200 attempts for any one attempt.

4. Discussion

4.1. Understanding the map

To interpret the scores, Burley notes that scores in the 30s indicate highly preferred environments [23]. From randomly selected sites across the study area, no landscape scored in this category. The study revealed numerous landscapes with scores in the 40s and 50s. Such landscapes are often modestly preferred environments. Scores in the 70s are less preferred and scores near or above 100 are not preferred [23]. Across the state, least preferred environments were rarely encountered, primarily in the southeastern portion of the state. The 95% confidence interval for any scores is ±5 points [33]. Thus, it takes a separation of 10 points for any pairs of images to be notably different as perceived by respondents.

4.2. Applications of the map

The average expected score (the sum of the products between the number of grid cells by their expected score, divided by the total number of grid cells) for the whole state is approximately 47.4. This score suggests that collectively the whole state of Michigan may not be the most beautiful of all environments; on the other hand, the environmental score is quite respectable and viewed by respondents as at least somewhat scenic.

To understand the context of the average expected score in Michigan, it is useful to explore the land-uses, published perceptions, recreational activities, and agriculture within the state. In an article published on the 29th of June, 2015, the Detroit Free Press reported that Thrillist who ranked all of the states in America, placed Michigan as the top state [34]. While the results of the list do not definitively demonstrate that Michigan is at the top, it does indicate that the environment of Michigan merits consideration as a noteworthy place in regard to visual quality and may be similar in score to numerous rural mixed agrarian and woodland environments around the world, such as in Poland, Romaina, or Hubei Province, PR China. It may be reasonable to consider the extensive shorelines, vast expanse of national forest lands (3 national forests), state forest lands, national parks, 99 state parks, a national lakeshore, wildlife refuges, woodlots, and agrarian landscapes assist in maintaining a relatively preferred environment [35]. According to the map (**Figure 8**), the impact of the urban and industrial development is not widespread in the state and has not yet affected the state with large megalopolis expanses. Without attempting to be promotional, the state remains a big fishing (salmon, trout, walleye, northern, pan-fish), hunting (black bear, elk, white-tailed deer, and wild turkey), recreation state (camping, boating, hiking, cross country trails, snowshoeing trails, biking trails, snowmobiling trails, and horseback-riding—actually thousands of miles of trails) and in the Upper Peninsula there are 150 waterfalls. Michigan has also been rated as one of the top places in the world for watching sunrises on earth [36]. Such features are promoted in Michigan's 'Pure Michigan' tourist campaign [37]. The state has the most diverse agricultural economy after California [36]. Michigan produces cherries, apples, blueberries, many vegetable crops, nursery plants, surgarbeets, a strong wine producing industry, potatoes, dairy, cattle, hogs, chickens, turkeys, timber, hardwoods for flooring and furniture, and paper pulp, as well as the staple corn, soybeans, and winter wheat. While the state is associated with the famed mid-west rust-belt and the failures of Detroit, the overall impact upon the state's extensive forested lands and agrarian landscape is relatively minimal [38]. In addition, visitors to the Detroit metropolitan area often are surprised with the activity and prosperous nature of the Detroit metropolitan area. The Detroit metropolitan area is a distributed urban environment with no dominant central district, the opposite of many other large cities which are more concentrated in the core such as a city like Shanghai, P.R. of China. The state is a major constituent of the third coast, the longest coastline associated with the United States and Canada (excluding the Arctic) [39]. The results of this study reflect the impression that the state is still predominantly a rural environment and is modestly beautiful in a 'low-key' manner. In other words, the average expected score and the existing landscape condition are mutually similar in expectations.

The map resulting for this study is an image representing the spatial perceptions of the respondents concerning the quality of the landscape across the state. As the land is managed and developed, the scores can be recomputed to estimate the perceived changes (both improvements and degradation) for the state. With additional research, the map could be considered a metric of the state's general perceived environmental quality. Across the globe, people are concerned about the impacts of human's transforming the environment; yet, comparatively, the compiled total environmental quality has not appreciatively changed since the arrival of Europeans, Africans, and Asians (a score predicted to be around 43 between the years 1816 and 1858 to a score of 47.4 in the year 2001, which is not significantly different in perceived quality) [40]. **Figures 9–11** are images from the landscape with scores near the current expected mean for much of the state. Notice none of the images are spectacular; however, none of the images age dismal either.

Figure 9. An image of a modest rural residential setting collected in the study area with a score of 46.6.

Figure 10. An image of a typical agricultural landscape found in much of Michigan. This image produced a score of 44.1.

Figure 11. An image of a great lakes coastline, generating a score of 47.8.

The results of this study and related studies indicate that land cover type is a strong predictor of visual/environmental quality. In planning and design circles, the nuances of the built environment are heavily debated with small details carefully examined by experts. Yet respondents evaluate these cover types relatively uniformly. In other words the refined changes and differences observed by experts and taught in planning and design schools are not necessarily observed/detected by the public. For experts the differences between communities can be quite distinct. As Charles Jencks indicated, there are at least two levels of cognition: the expert's and the public's [41]. For example, architects debate the merits of buildings; however, the research suggests that a poor building in a warehouse district or a noted piece of architecture such as Frank Lloyd Wright's Falling water house are perceived as simply structures and the less of them comprising the view, the more the environment is preferred [23].

Similarly, landscape architects debate the merits of various plant material and landscape settings; yet, the public sees a noble red pine tree (*Pinus resinosa* Sol. Ex Aiton.) or a weedy boxelder tree (*Acer negundo* L.) as basically the same thing—a tree. It is only when the plant has ornamental flowers, is there an increased appreciation during the period of flowering. Within any cover type, there is variation (scores can range up or down), but the expected mean within the cover type is quite consistent when measured repeatedly across time, across locations, and among respondent groups in America and Europe. In other words, the most important characteristic that the planner or designer makes concerning visual/environmental quality may be in determining the cover type for a parcel of land, not the refined details of a design. In the future, designers may develop improved techniques to mask human intrusions and abundantly incorporate preferred features such as wildlife, flowers, and views of distant landscapes, creating strong variation in the range of scores possible for a given cover type. Then the covariation of land cover type with visual/environmental quality may no longer hold true.

The relationships between the perception of environmental quality and cultural settings are explored in the book From Eye to Heart: Exterior Spaces Explored and Explained [42]. This book begins with discussing expectation values concerning the environment. Then the book examines some of the many interpretations for planning and designing across the globe and through time. The book concludes with the relationship between environmental science and the built/managed landscape. This book provides some greater and broader context to the visual mapping study presented in this investigation.

4.3. Limitations

Additional research should be conducted to refute, validate, or refine the findings presented here. However, it may be surprising that with only 60 images from across the state, the selected images facilitate a significant statistical result and make a reliable map for a whole state. In some studies, weakly developed experiments with large data sets and numerous observations may produce unremarkable results. The results presented in this paper follow a philosophy that relies upon methodologies that have yielded significant results (such as employing Q-sort techniques as opposed to Likert scales), an exploration of reliable predictors, and non-parametric statistical procedures. We would like to believe that well-grounded, focused, simple experiments often yield results that some expensive and elaborate studies fail to yield. The senior investigator in this study had spent over 30 years, carefully plotting each step, conducting the next logical step/experiment before proceeding. So each study is simple (for example the statistics were done on a spreadsheet), yet yielding meaningful results. This philosophical approach has served the team well and we encourage others to have this level of insight and commitment when formulating environmental quality studies. Investigators sometimes believe that technology, huge sample sizes, and big money make impressive research.

The results produced in the map are dependent upon the quality of the land cover type map. Many cover maps concentrate upon the great variation of vegetation and naturalistic cover types. Urban and suburban cover types are rarely produced with the same level of sensitivity. Brady et al. at the University of Waterloo developed an excellent example concerning the classification of human disturbed areas, based upon morphological and ecological features [32]. Yet this level of description has not yet permeated many land cover maps. We believe that classification systems that adopt ideas embedded in work by investigators such as Brady et al. will produce higher quality maps in urban areas than the maps that are currently produced [32].

In the landscape there are several types of cover types that exist in the landscape such as large sand dunes, mud flats, and bare rock that may be beyond the predicative capabilities of this study. These more rare landscape types were neither studied in the prediction models nor in creating the map presented in this study. Cover types such as these would need to be studied in detail to make a more comprehensive and complete map. This is computed to be a substantial area of land, approximately 1700 km^2. Yet this area is only about 0.68% of the land of Michigan.

The resolution of the map is relatively coarse (2.207 km^2 of land). Large maps with finer resolution will yield more refined results. A grid cell of that size will certainly have variation in it. The score represents the mean expected value within the cell.

5. Conclusions

Predictive, respondent based models have been constructed to measure environmental and visual quality. This work is based upon over 50 years of research by investigators in the social, recreational, and planning and design disciplines/profession. The attributes of the landscape can be measured to form reliable maps of environmental/visual quality, providing a metric to assess landscapes, including urban landscapes. We were able to produce such a metric map for Michigan. We believe our approach allows investigators to evaluate these visions and assess, measure, and quantify environmental perceptions. Furthermore, we believe the methods are reproducible, allowing investigators around the world to produce similar maps of additional areas.

Author details

Rüya Yilmaz[1], Chung Qing Liu[2] and Jon Bryan Burley[3*]

*Address all correspondence to: burleyj@msu.edu

1 Namk Kemal University, Tekirdag, Turkey

2 Jiangxi Agricultural University, Nanchang, PR China

3 Michigan State University, USA

References

[1] Martin B, Ortega E, Otero I, Arce RM. Landscape character assessment with GIS using map-based indicators and photographs in the relationship between landscape and roads. Journal of Environmental Management. 2016;**180**:324-334

[2] Bratman GN, Hamilton JP, Hahn KS, Daily GC, Gross JJ. Nature experience reduces rumination and subgenual prefrontal cortex activation. Proceedings of the National Academy of Sciences. 2015;**112**(28):8567-8572

[3] Dupont L, Antrop M, Van Eetvelde V. Does landscape related expertise influence the visual perception of landscape photographs? Implications for participatory landscape planning and management. Landscape and Urban Planning. 2015;**141**:68-77

[4] Filova L, Vojar J, Svobodova K, Sklenicka P. The effect of landscape type and landscape elements on public visual preferences: Ways to use knowledge in the context of landscape planning. Environmental Planning and Management. 2015;**58**(11):2037-2055

[5] Meagher BR, Marsh KL. Testing an ecological account of spaciousness in real and virtual environments. Environment and Behavior. 2015;**47**(7):782-815

[6] Seresinhe CI, Preis T, Moat HS. Quantifying the impact of scenic environments on health. Scientific Reports. 2015;**5**:16899. DOI: 10.1038/srep16899

[7] Wagtendonk AJ, Vermaat JE. Visual perception of cluttering in landscapes: Developing a low resolution GIS-evaluation method. Landscape and Urban Planning. 2014;**124**:85-92

[8] Kuper R. Here and gone: The visual effects of seasonal changes in plant and vegetative characteristics on landscape preference criteria. Landscape Journal. 2013;**32**(1):65-78

[9] Scarfò F, Mercurio R, del Peso C. Assessing visual impacts of forest operations on a landscape in the Serre Regional Park of southern Italy. Landscape & Ecological Engineering. 2013;**9**:1-10

[10] Rechtman O. Visual perception of agricultural cultivated landscapes: Key components as predictors for landscape preferences. Landscape Research. 2013;**38**(3):273-294

[11] Yao Y, Zhu X, Xu Y, Yang H, Wu X, Li Y, Zhang Y. Assessing the visual quality of green landscape in rural residential areas: The case of Guangzhou, China. Environmental Monitoring and Assessment. 2012;**184**(2):951-967

[12] Taylor JG, Zube EH, Sell JL. Landscape assessment and perception research methods. In: Bechtel RB, Marans RW, Michelson W, editors. Methods in Environmental and Behavioral Research. New York, New York: Van Nostrand Reinhold Company; 1987. pp. 361-393

[13] Kaplan R, Kaplan S. The Experience of Nature: A Psychological Perspective. Cambridge: Cambridge University Press; 1989

[14] Kaplan R, Kaplan S, Ryan R. With People in Mind: Design and Management of Everyday Nature. Washington, DC: Island Press; 1998

[15] Brush RO, Shafer EL. Application of a landscape-preference model to land management. In: Zube EH, Brush RO, Fabos JG, editors. Landscape Assessment: Values, Perceptions, and Resources. Stroudsburg, Pennsylvania: Dowden, Hutchinson and Ross, Inc; 1975. pp. 168-182

[16] Jones and Jones. Esthetics and Visual Resource Management for Highways. Washington, DC: U.S. Department of Transportation, Federal Highway Administration, National Highway Institute and Office of Environmental Policy Seminar Notes; 1979

[17] Jones and Jones. Visual Impact Assessment for Highway Projects. Washington, DC: U.S. Department of Transportation, Federal Highway Administration; 1981

[18] Kaplan R, Kaplan S, Brown T. Environmental preference: A comparison of four domains of predictors. Environment and Behavior. 1989;**21**(5):509-530

[19] Burley JB. Visual and ecological environmental quality model for transportation planning and design. Transportation Research Record. 1997;**1549**:54-60

[20] Burley JB, Yilmaz R. Visual quality preference: The Smyser index variables. International Journal of Energy and Environment. 2014;**8**:147-153

[21] Yilmaz R, Burley JB. Dissecting the Smyser index in visual quality. Modern landscape architecture. In: Proceedings of the 6th WSEAS International Conference on Landscape Architecture (LA'13); November 17-19; Nanjing, China. 2013. pp. 32-36

[22] Burley JB, Deyoung G, Partin S, Rokos J. Reinventing Detroit: Grayfields—New metrics in evaluating urban environments. Challenges. 2011;**2**:45-54

[23] Burley JB. The science of design: Green vegetation and flowering plants do make a difference: Quantifying visual quality. The Michigan Landscape. 2006;**49**(8):27-30

[24] Mo F, Le Cléach G, Sales M, Deyoung G, Burley JB. Visual and environmental quality perception and preference in the People's Republic of China, France, and Portugal. International Journal of Energy and Environment. 2011;**4**(5):549-556

[25] Partin S, Burley JB, Schutzki R, Crawford P. Concordance between photographs and computer generated 3D models in a Michigan highway transportation setting. In: Buhmann W, Ervin S, Pietsch M, editors. Peer Reviewed Proceedings of Digital Landscape Architecture 2012 at Anhalt University of Applied Science. Berlin: Wichmann; 2012. pp. 482-471

[26] Lu D, Burley JB, Crawford P, Schutzki R, Loures L. Chapter 7: Quantitative methods in environmental and visual quality mapping and assessment: A Muskegon, Michigan watershed case study with urban planning implication. In: Burian J, editor. Advances in Spatial Planning. Rijeka, Croatia: InTech; 2012. pp. 127-142

[27] Jin Y, Burley JB, Machemer P, Crawford P, Xu H, Wu Z, Loures L. The Corbusier dream and frank Lloyd Wright vision: Cliff detritus vs. urban savanna. In: Ergen M, editor. Urban Agglomeration. Rijeka, Croatia: Intech; 2018, 2018. pp. 211-230

[28] Lothian A. The Science of Scenery: How We See Scenic Beauty, What It Is, Why We Love It, and How to Measure and Map It. North Charleston, South Carolina: CreateSpace Independent Publishing Platform; 2017

[29] Daniel WW. Applied Nonparametric Statistics. Boston, MA: Houghton Mifflin Company; 1978

[30] Kendall M. The problem of m rankings. The Annals of Mathematical Statistics. 1939;**10**(3):275-287

[31] Jin Y. Validating a Visual Quality Prediction Map of Southern Michigan [Thesis]. East Lansing, Michigan: Michigan State University; 2012

[32] Brady RF, Tobias T, Eagles PFJ, Ohrner R, Micak J, Veale B, Dorney RS. A typology for the urban ecosystem and its relationship to larger biogeographical landscape units. Urban Ecology. 1979;**4**:11-28

[33] Burley JB. A quantitative method to assess aesthetic/environmental quality for spatial surface mine planning and design. WSEAS Transactions on Environment and Development. 2006;**5**(2):524-529

[34] Munzullo B. Confirmed: Michigan Is America's #1 State [Internet]. Detroit Free Press; 2015. http://www.freep.com/story/news/local/michigan/2015/06/28/michigan-best-state--america/29434215/. [Accessed: 1 July 2015]

[35] Sommers LM. Atlas of Michigan. Easting Lansing: Michigan State University Press; 1977

[36] Tuschman SA. Good Morning! The 11 Best Sunrises on Earth [Internet]. 2015. Yahoo travel. https://www.yahoo.com/travel/good-morning-the-11-best-sunrises-on-earth-12-2252814962.html. [Accessed: 1 July 2015]

[37] State of Michigan. Pure Michigan [Internet]. 2015. http://www.michigan.org. [Accessed: 1 July 2015]

[38] LeDuff C. Detroit: An American Autopsy. New York, New York: The Penguin Press; 2013

[39] Ibanez D, Lyster C, Waldheim C, White M. Third Coast Atlas: Prelude to a Plan. New York, New York: Actar Publishers; 2017. 344 p

[40] Corner PJ, Albert DA, Wells HA, Hart BL, Price DL, Moore JR, Kashian DM, Schuen DW. Michigan's Native Landscape as Interpreted from the General Land Office Surveys 1816-1856. Lansing, Michigan: Michigan Natural Features Inventory; 1995

[41] Jencks C. The Language of Post-Modern Architecture. New York, New York: Rizzoli; 1977

[42] Burley JB, Machemer T. From Eye to Heart: Exterior Spaces Explored and Explained. San Diego, California: Cognella Academic Publishing; 2016. 601 p

Chapter 8

Assessing Land-Use Changes in European Territories: A Retrospective Study from 1990 to 2012

José Manuel Naranjo Gómez, Luis Carlos Loures,
Rui Alexandre Castanho, José Cabezas Fernández,
Luis Fernández-Pozo,
Sérgio António Neves Lousada and Patrícia Escórcio

Additional information is available at the end of the chapter

http://dx.doi.org/10.5772/intechopen.78258

Abstract

The need to understand what land use is has motivated the development of programmes that aims to identify it and quantify it—CORINE Land Cover (CLC) in 1985. From this official and open geodatabase—through the using of geographic information system (GIS) tools—the amount of area established for each land use has been identified in all the 28 member states of the EU. This mostly corresponds to agricultural and forestry uses. Between 1990 and 2012, it was possible to determine countries with variable land use models such as Finland, Latvia, Portugal and Spain—the rest of the states presenting stable land use models. Additionally, some countries are characterized by the predominance of one or two land uses. Contextually, the proposal aims to develop a retrospective study regarding the land-use changes in the EU territories from 1990 to 2012, through the available tools such as CLC.

Keywords: land uses, CORINE Land Cover (CLC), European territories, geographic information system (GIS) tools, planning

1. Introduction

The increasing need for comprehensive and reliable information about land cover, land uses and their dynamics and patterns has catalysed the development of several sets of global land cover data, derived from Earth observation by satellites [1]. Such development was motivated by different initiatives and programmes national and international. In fact, the variety of mapping standards reflects the wide scope of interests and programmes [2].

Precisely, the land use coverage maps are data extremely useful, as evidenced by its widespread use and interdisciplinarity that they provide. These maps enable us to obtain information on the occupation of the land—biophysics coverage on the surface of the Earth [3]. For this reason, their use is essential for the study and modelling of territorial dynamics [4].

Among the available products Global Land Cover (GLC2000) should be highlighted; it had a global coverage by the year 2000 [5]. Europe stresses on Pan-European Land Use and Land Cover Monitoring (PELCOM), created from images of the year 1996, with a resolution of 1 km [4]. However, in Europe, at national and regional levels, it has included Coordination of Information on the Environment (CORINE) maps [4].

In this regard, in Europe, a special effort to monitor the change of land cover in a standardized manner has been carried out. The so-called inventory of CORINE Land Cover (CLC or 'Corine') has been created from satellite images. This common database used by a large number of organizations in Europe and co-funded by the European Commission and the member states has been processed by the European Environment Agency (EEA) considering the different land use covers—through the guidelines of the System of Environmental and Economic Accounts (SEEA) for 'land and ecosystem'. Thus, the database is now the core element for integration of the information system of the EEA [6]. In fact, the CORINE project containing the use coverage of European Union (EU) is seen as a relevant complement for the knowledge regarding major changes in land cover [7].

Although traditionally the CLC has been generated from the photo interpretation of satellite images, nevertheless, in some countries, such as Germany, Austria, Finland, Ireland, Iceland, Norway, United Kingdom, Sweden and Switzerland (mainly since 2006), the map is obtained from generalization techniques of national maps with greater detail [8]. In other cases—Slovakia, Hungary and Poland—CLC is used to obtain further details, scale 1:50000 maps, with a minimum map unit (MMU) from 4 and a legend adapted to the specific geographic features of the territory [8]. The same techniques have been used to obtain land use data prior to 1990 CLC [7].

Therefore, there are different ways of producing CLC. Still, countries like Germany or Ireland have changed its methodology in the production of CORINE land use maps—as for the photo interpretation for the general use. A similar scenario occurred in the Netherlands, once the government decided to produce the CLC independently [8, 9].

However, from CORINE, land use maps remain a tool of major relevance that enables one to analyse soil applications—regardless of the problems arising at the administrative and technical level. According to the directive INSPIRE 2007-2-EC [10], CLC is one of the most outstanding harmonized European data sets and CLC even has achieved a semantic and technical standardization, considering that the CLC is a set of reference data in common use for European scale assessments since it uses a generic land cover class definition throughout Europe [11, 12].

http://dx.doi.org/10.5772/intechopen.78258

Indeed, other sources of information, so far, have only compatibility and comparability enclosed between different maps sources of land cover and its legends' theme—since they exist as independent datasets [1]. Usually, the heterogeneity in land cover maps result from different methods and underlying patterns—several layouts, syntactic issues, schematic heterogeneity and semantic aspects [13]. Different mapping methodologies are difficult to separate land use changes, once those changes are the results of a different used approach in creating the map [1].

Nevertheless, land use data sets are crucial in exploring socio-economic, political, cultural and environmental factors that influence land use decisions [14–16].

In this regard, the changing landscape of European territories is the subject of several studies and researches—pointing to significant change [17–19]. However, despite the pace of change of land uses in the European panorama, there is only a limited research that analyses the patterns of change in the use of the land on a pan-European level. Most of the existing research related to land-use change patterns have consisted of case studies from specific regions or local areas [20–24].

The harmonized CLC data have been used for the analysis of multiple disciplines, such as environmental [11] in social and economic analyses [25], transportation management [26] and demographic studies [27].

On the one hand, local case studies provide evidence on change catalysts in land use in a more detailed local context. Still, they are often verified in particular contexts, actors, processes, resolutions or scales [28]. Also, European land use change studies can lead to a more global view, whereas the analysis of the land use changes and the associated factors can be generalized and even their methodology can be transferred between different scenarios [29].

The European landscape has a wide variety of regional features and a well-defined dynamic structure—where agriculture is one of the most dominant land uses [30]. The agricultural land use covers more than 35% of the European territories—almost ten times more than the urban land use [31, 32]. Nevertheless, this is not the only type of soil that is changing in Europe.

The overall objective of the present study is to perform a retrospective analysis of the European land use changes. Contextually, it will determine the extension that EU state members dedicated to land uses, according to CLC. So, specific objectives can be summarized: (a) identify countries, where there is land use which is widespread and dominant over the remaining land uses and (b) if the surface extension dedicated to land uses has been constant or variable between 1990, 2000, 2006 and 2012.

2. Material and methods

To carry out the study, firstly, data have been collected such as official information that is detailed with sufficient precision and accuracy to characterize each of the countries part of the EU, according to their land use in 2018. It was decided to analyse the EU for its economic relevance and also according to the significant expansion of territory on a global scale.

Regarding information sources, the EEA provides the CLC, through the Copernicus Global Land Service[1]. This inventory was initiated in 1985 although the first 'visible' results date from 1990, and updates have occurred in 2000, 2006 and 2012 (**Table 1**) [9]. Another two main goals, of the CLC programme, are: (a) providing quantitative coverage of the soil—consistent and comparable data across Europe for stakeholders in European environmental policy and (2) developing a digital land cover database covering the EU Member States and other European and North African sovereign states [1].

	CLC1990	**CLC2000**	**CLC2006**	**CLC2012**
Satellite data	Landsat-5 MSS/TM	Landsat-7 ETM	SPOT-4/5 and IRS P6 LISS III	IRS P6 LISS III and rapid eye
	Single date	Single date	Dual date	Dual date
Time consistency	1986–1998	2000 +/− 1 year	2006 +/− 1 year	2011–2012
Geometric accuracy, satellite data	≤ 50 m	≤ 25 m	≤ 25 m	≤ 25 m
Min. mapping unit/ width	25 ha/ 100 m	25 ha/100 m	25 ha/100 m	25 ha/100 m
Geometric accuracy, CLC	100 m	Better than 100 m	Better than 100 m	Better than 100 m
Thematic accuracy, CLC	≥ 85% (Probably not achieved)	≥ 85% (Achieved)	≥ 85% (Not checked)	≥ 85%
Change mapping (CLCC)	Not implemented	Boundary displacement min. 100 m;	Boundary displacement min.100 m;	Boundary displacement min.100 m;
		Change area for existing polygons ≥5 ha; for isolated changes ≥25 ha	All changes ≥5 ha are to be mapped	All changes ≥5 ha are to be mapped
Thematic accuracy, CLCC	—	Not checked	≥ 85% (Achieved)	≥ 85%
Production time	10 years	4 years	3 years	2 years
Documentation	Incomplete metadata	Standard metadata	Standard metadata	Standard metadata
Access to the data (CLC, CLCC)	Unclear dissemination policy	Dissemination policy agreed from the start	Free access for all users	Free access for all users
Number of countries involved	26 (27 with late implementation)	30 (35 with late implementation)	38	39

Table 1. Evolution of Land Cover CORINE [33].

[1]The CORINE programme was established in 1985 by the European Commission at: http://land.copernicus.eu/global.

http://dx.doi.org/10.5772/intechopen.78258

Additionally, it has a spatial resolution of 100 m to linear phenomena. Also, different land uses have been classified using three levels of details—from the first with a higher degree of aggregation, the third party with the greatest degree of detail and therefore more disaggregated. The third comprises a total of 44 classes allowing one to characterize the land uses of each country (**Table 2**).

Regarding the CLC spatial coverage, additionally to the 28 EU member states, it also covers Albania, Bosnia and Herzegovina, Iceland, Kosovo, Liechtenstein, Macedonia, Monte Negro, Norway, Serbia, Switzerland and Turkey. Nevertheless, for this large set of countries, information only has been available in CLC 2000, 2006 and 2012 updates—and in the 1990 version countries not belonging to the EU was not included.

Then, GIS tools (ArcGIS) along with management tools (Microsoft Access) have been used. Considering that CLC updates generate a map of land use changes, only changes larger than 5 hectares, the first map corresponds to the changes between 1990 and 2000. With the first map, and combining with other intersections features, it has been possible to generate two new maps: (a) reference data and (b) a review of the previous map. According to [33]: '*the study of the territorial changes should be studied from the change maps and not from the intersection of the CORINE maps for the years of reference, given that the cross-tabulation of various maps can produce technical changes not real, arising from variations in production methodology*'.

Regarding the methodological framework, the objective was to obtain the representative land use through polygons and their corresponding alphanumeric information for Europe in 1990, 2000, 2006, and 2012.

The graphical information layer consists of polygonal graphics entities, each of 44 kinds of reported soil applications. Also, the alphanumeric information contains information fields associating an identifier - a code for the use of the soil for level 3 (**Table 2**); the area of the polygon is measured in hectares as well as the length of the surface of each of the polygons is also calculated.

Considering the aim of the study, it has been necessary to count the number of hectares of land use classified by CLC for each of the countries—aiming to achieve that this was also represented by polygonal entities of each of the EU countries and administrative boundaries. This layer of information has a scale of 1: 1,000,000 being the graphic equivalent to 200 m, and the coordinate reference system is European Terrestrial Reference System 1989 (ETRS89), the same used for CLC for flooring applications. The origin is the centre of the mass of the Earth, including oceans and atmosphere. In addition, the z-axis is parallel to the direction of the pole Conventional International Origin (CIO). The x-axis intersects the Greenwich Meridian origin, and the origin plane is perpendicular to the z-axis.

Using GIS tools, a file representing the administrative boundaries of each of the 28 EU Member States has been generated throughout territorial polygons that have been processed.

After, have been overlapping polygons previously obtained for CLC land uses representing all polygons with EU land uses. This new layer inherits the thematic attributes of the layer on the CLC land uses.

LEVEL 1	Nomenclature definition	LEVEL 2	Nomenclature definition	LEVEL 3	Nomenclature definition
1	Artificial surfaces	11	Urban fabric	111	Continuous urban fabric: most of the land is covered by buildings, roads and artificially surfaced area cover almost all the ground. Non-linear areas of vegetation and bare soil are exceptional.
				112	Discontinuous urban fabric: most of the land is covered by structures, buildings, roads and artificially surfaced areas associated with vegetated areas and bare soil, which occupy discontinuous but significant surfaces
		12	Industrial, commercial and transport units	121	Industrial or commercial units: artificially surfaced areas (with concrete, asphalt, tarmacadam, or stabilized, e.g. beaten earth) devoid of
					vegetation, occupy most of the area in question, which also contains buildings and/or vegetated areas.
				122	Road and rail networks associated land: motorways, railways, including associated installations (stations, platforms, embankments). Minimum width
					to include: 1 m.
				123	Port areas: infrastructure of port areas, including quays, dockyards and marinas.
				124	Airports: airport installations like runways, buildings and associated land.
		13	Mine, dump and construction sites	131	Mineral extraction sites: areas with open-pit extraction of industrial minerals (sandpits, quarries) or other minerals (opencast mines). Includes flooded gravel pits, except for river-bed extraction.
				132	Dump sites: landfill or mine dump sites, industrial or public.
				133	Construction sites: spaces under construction development, soil or bedrock excavations, earthworks.
		14	Artificial, non-agricultural vegetated areas	141	Green urban areas: areas with vegetation within urban fabric. Includes parks and cemeteries with vegetation.
				142	Spot and leisure facilities: camping grounds, sports grounds, leisure parks, golf courses, racecourses, etc. Includes formal parks not surrounded by urban zones.

LEVEL 1	Nomenclature definition	LEVEL 2	Nomenclature definition	LEVEL 3	Nomenclature definition
2	Agricultural areas	21	Arable land	211	Non-irrigated arable land: cereals, legumes, fodder crops, root crops and fallow land. Includes flower and tree (nurseries) cultivation and vegetables, whether open field, under plastic or glass (includes market gardening). Includes aromatic, medicinal and culinary plants. Excludes permanent pastures.
				212	Permanently irrigated land: crops irrigated permanently and periodically, using a permanent infrastructure (irrigation channels, drainage network). Most of these crops could not be cultivated without an artificial water supply. Does not include sporadically irrigated land.
				213	Rice fields: land developed for rice cultivation. Flat surfaces with irrigation channels. Surfaces regularly flooded.
		22	Permanent crops	221	Vineyards: areas planted with vines.
				222	Fruit trees and berry plantations: parcels planted with fruit trees or shrubs: single or mixed fruit species, fruit trees associated with permanently grassed surfaces. Includes chestnut and walnut groves.
				223	Olive groves: areas planted with olive trees, including mixed occurrence of olive trees and vines on the same parcel.
		23	Pastures	231	Pastures: dense, predominantly graminoid grass cover, of floral composition, not under a rotation system. Mainly used for grazing, but the fodder may be harvested mechanically. Includes areas with hedges (bocage).
		24	Heterogeneous agricultural areas	241	Annual crops associated with permanent crops: non-permanent crops (arable lands or pasture) associated with permanent crops on the same parcel.
				242	Complex cultivation: juxtaposition of small parcels of diverse annual crops, pasture and/or permanent crops.
				243	Land principally occupied by agriculture: areas principally occupied by agriculture, interspersed with significant natural areas.
				244	Agro-forestry areas: annual crops or grazing land under the wooded cover of forestry species.

LEVEL 1	Nomenclature definition	LEVEL 2	Nomenclature definition	LEVEL 3	Nomenclature definition
3	Forest and semi-natural areas	31	Forests	311	Broad-leaved forest: vegetation formation composed principally of trees, including shrub and bush understories, where broadleaved species predominate.
				312	Coniferous forest: vegetation formation composed principally of trees, including shrub and bush understories, where coniferous species predominate.
				313	Mixed forest: vegetation formation composed principally of trees, including shrub and bush understories, where broadleaved and coniferous species co-dominate.
		32	Scrub and/ or herbaceous vegetation associations	321	Natural grassland: low productivity grassland. Often situated in areas of rough uneven ground. Frequently includes rocky areas, briars, and heathland.
				322	Moors and heathland: vegetation with low and closed cover, dominated by bushes, shrubs and herbaceous plants (heath, briars, broom, gorse, laburnum, etc.).
				323	*Sclerophyllous* vegetation: bushy sclerophyllous vegetation. Includes maquis and garrigue. Maquis: a dense vegetation association composed of numerous shrubs associated with siliceous soils in the Mediterranean environment. Garrigue: discontinuous bushy associations of Mediterranean calcareous plateaus. Generally composed of kermes oak, arbutus, lavender, thyme, cistus, etc. May include a few isolated trees.
				324	Transitional woodland-shrub: bushy or herbaceous vegetation with scattered trees. Can represent either woodland degradation or forest regeneration/ colonization.
		33	Open spaces with little or no vegetation	331	Beaches, dunes, and sand plains: beaches, dunes and expanses of sand or pebbles in coastal or continental, including beds of stream channels with torrential regime.
				332	Bare rock: scree, cliffs, rocks and outcrops.
				333	Sparsely vegetated areas: includes steppes, tundra and badlands. Scattered high-attitude vegetation.
				334	Burnt areas: areas affected by recent fires, still mainly black.
				335	Glaciers and perpetual snow: land covered by glaciers or permanent snowfields.

LEVEL 1	Nomenclature definition	LEVEL 2	Nomenclature definition	LEVEL 3	Nomenclature definition
4	Wetlands	41	Inland wetlands	411	Inland marshes: low-lying land usually flooded in winter, and more or less saturated by water all year round.
				412	Peat bogs: peatland consisting mainly of decomposed moss and vegetable matter. May or may not be exploited.
		42	Maritime wetlands	421	Salt marshes: vegetated low-lying areas, above the high-tide line, susceptible to flooding by sea water. Often in the process of filling in, gradually being colonized by halophilic plants.
				422	Salines: salt-pans, active or in process of. Sections of salt marsh exploited for the production of salt by evaporation. They are clearly distinguishable from the rest of the marsh by their segmentation and embankment systems.
				423	Intertidal flats: generally unvegetated expanses of mud, sand or rock lying between high and low water-marks. On contour on maps.
5	Water bodies	51	Inland waters	511	Water courses: natural or artificial water-courses serving as water drainage channels. Includes canals. Minimum width to include: 100 m.
				512	Water bodies: natural or artificial stretches of water.
		52	Marine waters	521	Coastal lagoons: unvegetated stretches of salt or brackish waters separated from the sea by a tongue of land or other similar topography. These water bodies can be connected with the sea at limited points, either permanently or for parts of the year only.
				522	Estuaries: the mouth of a river within which the tide ebbs and flows.
				523	Sea and ocean: zone seaward of the lowest tide limit.

Table 2. CLC nomenclature [33].

To avoid the appearance of *slivers*, in the layers overlapping, that is, a country's boundary, CLC flooring applications, a margin of tolerance (distance) between two lines was set in order that two similar lines are considered as a single. In the present chapter, more graphic tolerances correspond to 200 meters of the layer corresponding to countries' boundaries.

Once geo-database was obtained for EU territories, and considering the CLC land uses for the years 1990, 2000, 2006, and 2012, the overlay process was performed four times for each of the countries. Taking into account that four countries in the EU 28 Members had no registered CLC land uses for the year 1990, 432 geodatabases were obtained in total.

Subsequently, all these geo-database alphanumeric information were analysed by country and by year basis, using the Microsoft Access database. For each of these geo-databases there was a table of alphanumeric information, applying a query that is based on the Standard Query Language (SQL). In this regard, the surface of EU Member States has been summarized through CLC land use (**Table 2**). Relating the number of hectares of each country allocated to particular land use (**Table 2**), it was possible to characterize the EU countries according to land uses and determine what changed according to hectares' numbers dedicated to different land uses in the years 1990, 2000, 2006 and 2012. Also, this synthetic methodology has been based on actual and open-access EU data—possible to replicate in future years/periods.

3. Results and discussion

The results come from the analysis of the land uses for each of the European countries in the years 1990, 2000, 2006 and 2012. The results will be exposed through the graphs, tables and thematic cartography. This typology of results allows to extract the most relevant information and to characterize each of the European countries on the basis of the 44 uses of the soil determined by CLC—through an easy read.

According to the latitude, EU Member States have been classified into three groups: (i) further to the North—'North EU group countries'; (ii) further to the South—'South EU group countries'; (iii) countries that occupy an intermediate position—'Central EU group countries' (**Figure 1**). Also, the obtained surfaces can be observed in **Table 3**.

Initially, the 'North EU group countries' have been analysed—Estonia, Finland, Latvia, Lithuania and Sweden (**Table 4** and **Figure 2**).

Estonia seems to be a country dominated by two land uses—mixed forest (313) and coniferous forest (312), corresponding to the forest and semi-natural areas. The following higher percentage of land use corresponds to non-irrigated arable land (211). Therefore, if there was a greater exploitation of agricultural resources, there would be an increase in food production. In fact, the abovementioned land uses present an expansion; however, it does not differ significantly, considering the extension of the rest of the land uses—which are from 5–10%, corresponding to transitional woodland-shrub, broad-leaved forest, pastures and land occupied by agriculture (324, 311, 231, 243).

Finland is a predominantly forest country, characterized by two major land uses: coniferous forest (312) and mixed forests (313). Surprisingly, between 2000 and 2006, the extension occupied by those land uses was approximately similar; nevertheless, in 2012, coniferous forest cover increased. Therefore, it seems that the use of the coniferous forest land has increased in detriment of the mixed forest. The third land use with major relevance in Finland is transitional woodland shrub (324). However, this land use has decreased in 2012, until reaching an area similar to water bodies' land use (512).

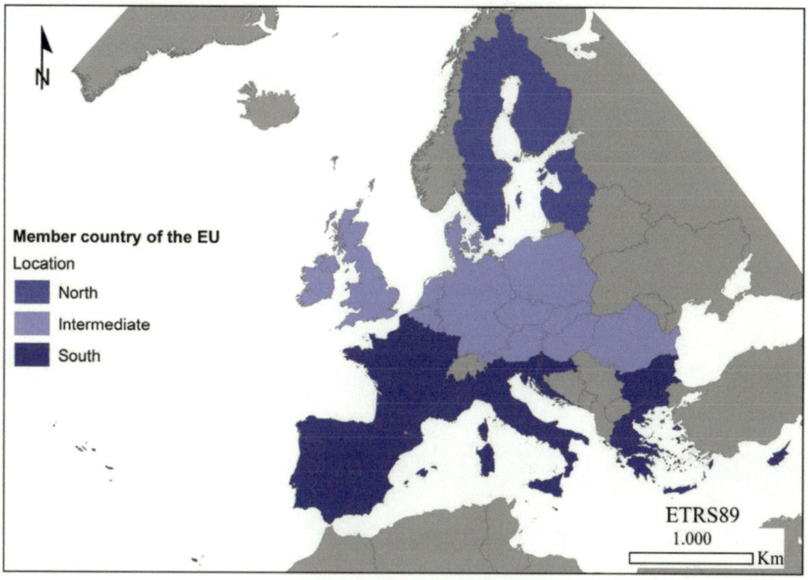

Figure 1. EU Member States (authors).

Latvia does not seem to highlight by a specific land use as all the land uses in 2012 comprised 0–16%. The major land uses are vineyards (211), mixed forests (313), transitional woodland shrub (324), coniferous forests (312) and pastures (231).

Countryt	Area (hectares)	Country	Area (hectares)
Austria	8,728,000	Italy	31,300,000
Belgium	3,086,000	Latvia	6,914,000
Bulgaria	12,620,000	Lithuania	6,950,000
Croatia	5,977,000	Luxembourg	2,631,000
Cyprus	1,215,000	Malta	33,180
Czech Republic	8,228,000	Netherlands	3,766,000
Denmark	4,379,000	Poland	33,010,000
Estonia	4,834,000	Portugal	9,267,000
Finland	35,320,000	Romania	26,690,000
France	55,190,000	Slovakia	5,240,000
Germany	36,540,000	Slovenia	2,119,000
Greece	14,970,000	Spain	50,660,000
Hungary	9,969,000	Sweden	46,000,000
Ireland	7,013,000	United Kingdom	24,490,000

Table 3. Surface of EU Member States (authors).

	Estonia				Finland			Latvia				Lithuania				Sweden		
LEVEL 3	1990	2000	2006	2012	2000	2006	2012	1990	2000	2006	2012	1990	2000	2006	2012	2000	2006	2012
111	0.01	0.01	0.01	0.01	0.00	0.00	0.00	0.01	0.01	0.01	0.01	0.00	0.00	0.01	0.01	0.01	0.01	0.01
112	1.08	1.12	1.25	1.27	1.05	1.07	0.96	0.80	0.82	1.17	1.20	2.26	2.26	2.29	2.33	0.89	0.90	0.91
121	0.41	0.40	0.37	0.39	0.15	0.15	0.19	0.23	0.26	0.36	0.37	0.58	0.57	0.51	0.51	0.14	0.15	0.15
122	0.07	0.07	0.05	0.05	0.01	0.01	0.00	0.03	0.02	0.03	0.03	0.09	0.09	0.08	0.08	0.05	0.06	0.06
123	0.01	0.01	0.01	0.02	0.01	0.01	0.01	0.01	0.01	0.03	0.03	0.01	0.01	0.01	0.01	0.00	0.01	0.01
124	0.05	0.05	0.05	0.05	0.02	0.03	0.03	0.03	0.03	0.03	0.03	0.04	0.05	0.05	0.05	0.03	0.03	0.03
131	0.15	0.13	0.13	0.17	0.06	0.06	0.07	0.05	0.05	0.06	0.07	0.09	0.09	0.07	0.06	0.02	0.03	0.03
132	0.08	0.08	0.08	0.07	0.01	0.01	0.02	0.00	0.00	0.00	0.01	0.01	0.01	0.02	0.01	0.02	0.02	0.02
133	0.00	0.00	0.01	0.01	0.00	0.00	0.00	0.00	0.00	0.02	0.01	0.04	0.03	0.04	0.02	0.00	0.01	0.00
141	0.05	0.05	0.05	0.05	0.02	0.02	0.02	0.13	0.10	0.10	0.10	0.12	0.12	0.12	0.12	0.07	0.07	0.07
142	0.03	0.03	0.04	0.05	0.04	0.05	0.04	0.01	0.01	0.14	0.14	0.04	0.04	0.04	0.05	0.10	0.12	0.12
211	14.59	14.62	15.15	15.11	4.76	5.09	4.58	14.07	15.41	16.34	16.68	33.63	34.17	32.79	32.86	6.67	6.66	6.65
212	0.00	0.00	0.00	0.00	0.00	0.00	0.00	0.00	0.00	0.00	0.00	0.00	0.00	0.00	0.00	0.00	0.00	0.00
213	0.00	0.00	0.00	0.00	0.00	0.00	0.00	0.00	0.00	0.00	0.00	0.00	0.00	0.00	0.00	0.00	0.00	0.00
221	0.00	0.00	0.00	0.00	0.00	0.00	0.00	0.00	0.00	0.00	0.00	0.00	0.00	0.00	0.00	0.27	0.26	0.26
222	0.04	0.04	0.03	0.03	0.00	0.00	0.00	0.06	0.05	0.06	0.06	0.15	0.14	0.13	0.11	1.25	1.24	1.24
223	0.00	0.00	0.00	0.00	0.00	0.00	0.00	0.00	0.00	0.00	0.00	0.00	0.00	0.00	0.00	0.00	0.00	0.00
231	6.14	5.68	6.78	6.71	0.01	0.01	0.01	14.43	13.21	11.91	11.48	7.55	6.54	6.18	5.93	4.42	4.45	4.46
241	0.00	0.00	0.00	0.00	0.00	0.00	0.00	0.00	0.00	0.00	0.00	0.00	0.00	0.00	0.00	47.95	46.37	49.74
242	3.53	3.91	3.27	3.24	0.00	0.00	0.00	8.54	8.40	8.19	8.21	12.27	12.72	13.48	13.45	3.63	3.68	4.52
243	8.20	8.20	6.56	6.58	3.93	3.65	3.96	6.76	6.75	5.21	5.22	8.08	7.91	7.85	7.79	0.43	0.43	0.43
244	0.00	0.00	0.00	0.00	0.00	0.00	0.00	0.00	0.00	0.00	0.00	0.00	0.00	0.00	0.00	6.13	6.14	6.14

	Estonia				Finland			Latvia				Lithuania				Sweden		
311	9.56	9.57	8.22	8.40	2.20	2.23	2.03	8.94	8.71	8.38	7.96	6.45	6.47	6.65	6.65	9.82	11.25	7.01
312	18.55	18.09	17.99	18.00	29.51	28.47	41.70	15.26	14.61	14.15	12.98	11.58	11.20	11.06	11.00	0.83	0.84	0.84
313	18.80	18.43	19.23	20.04	26.19	27.39	17.87	19.59	18.52	17.64	16.33	11.55	11.23	11.44	11.42	1.52	1.52	1.52
321	0.82	0.80	0.71	0.71	0.01	0.01	0.05	0.10	0.08	0.12	0.13	0.01	0.02	0.02	0.02	0.06	0.06	0.06
322	0.31	0.31	0.20	0.20	1.24	1.23	2.07	0.00	0.00	0.00	0.00	0.05	0.05	0.05	0.05	0.13	0.14	0.14
323	0.00	0.00	0.00	0.00	0.00	0.00	0.00	0.00	0.00	0.00	0.00	0.00	0.00	0.00	0.00	6.29	6.32	6.32
324	7.93	8.85	9.98	9.03	14.14	13.91	9.61	6.48	8.46	11.27	14.19	2.51	3.40	4.20	4.53	0.27	0.27	0.27
331	0.10	0.07	0.07	0.07	0.00	0.00	0.00	0.06	0.06	0.03	0.03	0.04	0.03	0.02	0.02	8.06	8.06	8.06
332	0.00	0.00	0.00	0.00	0.01	0.01	0.42	0.00	0.00	0.00	0.00	0.00	0.00	0.00	0.00	0.00	0.00	0.00
333	0.05	0.02	0.01	0.01	0.32	0.32	0.13	0.01	0.01	0.06	0.06	0.01	0.01	0.01	0.01	0.00	0.00	0.00
334	0.00	0.00	0.00	0.00	0.00	0.00	0.00	0.00	0.00	0.00	0.00	0.00	0.00	0.00	0.00	0.00	0.00	0.00
335	0.00	0.00	0.00	0.00	0.00	0.00	0.00	0.00	0.00	0.00	0.00	0.00	0.00	0.00	0.00	0.57	0.57	0.57
411	1.59	1.60	1.71	1.71	0.08	0.08	0.10	0.38	0.36	0.30	0.30	0.28	0.28	0.30	0.30	0.03	0.00	0.00
412	2.69	2.71	2.90	2.90	6.58	6.52	6.30	2.02	2.08	2.26	2.27	0.60	0.60	0.62	0.62	0.01	0.01	0.01
421	0.01	0.01	0.01	0.01	0.02	0.03	0.03	0.00	0.00	0.00	0.00	0.00	0.00	0.00	0.00	0.00	0.00	0.00
422	0.00	0.00	0.00	0.00	0.00	0.00	0.00	0.00	0.00	0.00	0.00	0.00	0.00	0.00	0.00	0.00	0.00	0.00
423	0.00	0.00	0.00	0.00	0.00	0.00	0.00	0.00	0.00	0.00	0.00	0.00	0.00	0.00	0.00	0.00	0.00	0.00
511	0.07	0.07	0.07	0.07	0.20	0.21	0.18	0.23	0.23	0.25	0.25	0.28	0.27	0.28	0.28	0.00	0.00	0.00
512	4.53	4.53	4.55	4.55	9.14	9.14	9.28	1.62	1.61	1.74	1.75	1.63	1.64	1.65	1.66	0.01	0.01	0.01
521	0.03	0.03	0.03	0.03	0.00	0.00	0.00	0.00	0.00	0.00	0.00	0.02	0.02	0.02	0.02	0.02	0.02	0.02
522	0.00	0.00	0.00	0.00	0.00	0.00	0.00	0.00	0.00	0.00	0.00	0.00	0.00	0.00	0.00	0.30	0.30	0.30
523	0.48	0.48	0.49	0.49	0.28	0.28	0.32	0.11	0.11	0.12	0.12	0.03	0.03	0.03	0.03	0.00	0.00	0.00

Table 4. Percentage of land use from 1990 to 2012 for Estonia, Finland, Latvia, Lithuania and Sweden (authors).

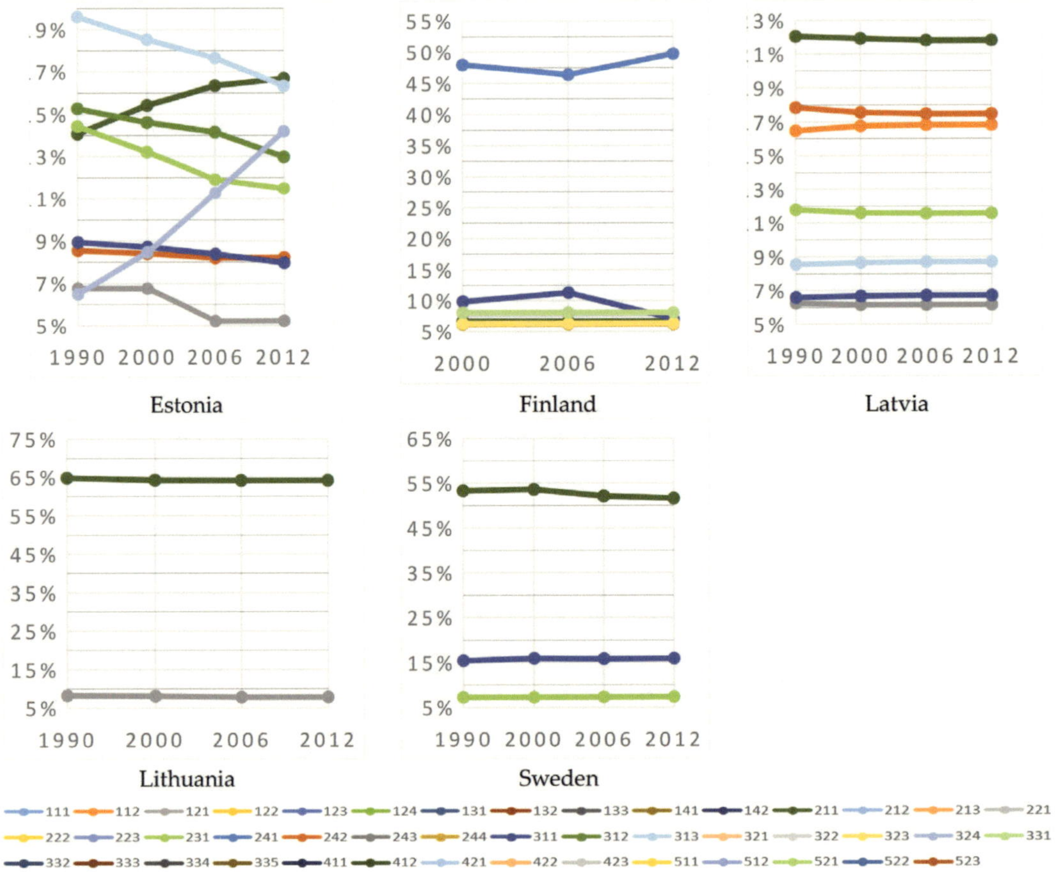

Figure 2. Trend of land uses higher than 5% for the North EU group countries (authors).

Lithuania stands out as an eminent agricultural country—once approximately one-third of the land comprised vineyards (211). Additionally, the area designated for vineyards tends to be fairly constant. Even this percentage is far superior to the second major land use, corresponding to complex cultivation (242). The following land uses with the highest percentage correspond to the forest and semi-natural areas, mixed forests (313) and coniferous forests (312).

Sweden stands out as a prominent agricultural country with approximately half of the territory earmarked for annual crops associated with permanent crops (241). Additionally, this trend over the analysed period seems to increase. Thus, it is possible that such values will increase even further in future. However, the second major land use in Sweden should also be considered, corresponding to forestry use, which is broad-leaved forest (331) (**Tables 5–7** and **Figure 3**).

Through the analysis of the developed graphics for the Central EU countries, it is possible to verify that the trend of variation of the land uses in countries such as Austria, Belgium, Czech Republic, Denmark, Germany, Hungary, Ireland, Netherland, Poland, Romania and Slovakia is low or very low. So, constant and stable land use models predominate in this area.

	Austria				Belgium				Czech Republic				Denmark				Germany			
CODE	1990	2000	2006	2012	1990	2000	2006	2012	1990	2000	2006	2012	1990	2000	2006	2012	1990	2000	2006	2012
111	0.09	0.10	0.12	0.12	0.16	0.17	0.17	0.17	0.02	0.02	0.02	0.02	0.14	0.14	0.14	0.14	0.06	0.06	0.07	0.04
112	3.64	4.10	4.47	4.49	16.47	16.75	16.82	16.82	4.54	4.73	4.80	4.85	4.27	4.37	4.59	4.59	5.94	6.20	6.42	6.93
121	0.11	0.19	0.33	0.34	1.32	1.67	1.79	1.79	0.66	0.73	0.76	0.80	0.50	0.58	0.68	0.68	0.69	0.86	0.93	1.38
122	0.03	0.04	0.05	0.06	0.30	0.35	0.34	0.34	0.06	0.07	0.08	0.09	0.02	0.02	0.02	0.02	0.05	0.05	0.05	0.06
123	0.00	0.00	0.00	0.00	0.17	0.20	0.23	0.23	0.00	0.00	0.00	0.00	0.08	0.08	0.08	0.08	0.03	0.03	0.03	0.02
124	0.04	0.04	0.04	0.04	0.18	0.17	0.17	0.17	0.07	0.07	0.07	0.07	0.16	0.17	0.17	0.17	0.13	0.13	0.13	0.11
131	0.07	0.10	0.11	0.11	0.25	0.23	0.24	0.24	0.23	0.21	0.21	0.21	0.08	0.12	0.09	0.09	0.33	0.29	0.28	0.20
132	0.00	0.00	0.01	0.01	0.05	0.04	0.04	0.04	0.20	0.14	0.12	0.10	0.00	0.01	0.01	0.01	0.05	0.05	0.05	0.04
133	0.01	0.00	0.00	0.00	0.10	0.05	0.08	0.08	0.03	0.01	0.03	0.01	0.00	0.02	0.00	0.00	0.02	0.02	0.01	0.01
141	0.03	0.04	0.06	0.06	0.14	0.15	0.15	0.15	0.08	0.08	0.08	0.08	0.25	0.26	0.24	0.24	0.12	0.12	0.12	0.20
142	0.02	0.15	0.30	0.32	0.62	0.66	0.66	0.66	0.15	0.18	0.20	0.22	1.18	1.29	1.44	1.44	0.22	0.27	0.30	0.45
211	13.16	13.90	15.37	15.35	22.06	21.94	21.82	21.82	45.03	39.07	37.90	36.75	64.79	64.21	64.12	64.12	38.96	38.21	37.81	37.93
212	0.00	0.00	0.00	0.00	0.00	0.00	0.00	0.00	0.00	0.00	0.00	0.00	0.00	0.00	0.00	0.00	0.00	0.00	0.00	0.00
213	0.00	0.00	0.00	0.00	0.00	0.00	0.00	0.00	0.00	0.00	0.00	0.00	0.00	0.00	0.00	0.00	0.00	0.00	0.00	0.00
221	0.83	0.84	0.79	0.79	0.00	0.00	0.00	0.00	0.14	0.16	0.20	0.21	0.00	0.00	0.00	0.00	0.36	0.36	0.36	0.35
222	0.00	0.00	0.01	0.01	0.25	0.27	0.26	0.26	0.42	0.38	0.40	0.37	0.01	0.01	0.09	0.09	0.42	0.34	0.34	0.42
223	0.00	0.00	0.00	0.00	0.00	0.00	0.00	0.00	0.00	0.00	0.00	0.00	0.00	0.00	0.00	0.00	0.00	0.00	0.00	0.00
231	9.86	8.91	8.26	8.26	11.78	11.59	11.57	11.57	3.21	8.17	9.11	10.08	1.27	1.28	1.32	1.32	12.38	12.66	12.26	17.98
241	0.00	0.00	0.00	0.00	0.00	0.00	0.00	0.00	0.00	0.00	0.00	0.00	0.00	0.00	0.00	0.00	0.00	0.00	0.00	0.00
242	7.54	6.80	4.49	4.49	17.84	17.55	17.47	17.47	0.55	0.62	0.61	0.60	2.49	2.48	2.15	2.15	5.79	5.76	6.09	0.19
243	1.41	1.98	3.07	3.07	6.26	6.16	6.16	6.16	8.53	8.95	8.97	9.01	8.26	8.13	7.86	7.86	2.44	2.43	2.53	0.25
244	0.00	0.00	0.00	0.00	0.00	0.00	0.00	0.00	0.00	0.00	0.00	0.00	0.00	0.00	0.00	0.00	0.00	0.00	0.00	0.00

	Austria				Belgium				Czech Republic				Denmark				Germany			
311	4.10	4.89	5.26	5.25	6.59	6.67	6.70	6.70	3.17	3.52	3.53	3.60	1.63	1.62	1.68	1.68	6.69	6.70	6.74	9.72
312	25.52	26.25	26.87	26.69	4.60	4.63	4.48	4.48	21.01	21.59	21.86	21.73	4.61	4.16	3.93	3.93	15.83	15.73	15.70	16.56
313	15.19	13.39	12.12	12.11	8.55	8.64	8.70	8.70	7.41	7.77	7.82	8.03	3.16	3.04	3.16	3.16	6.57	6.61	6.69	4.07
321	6.48	7.14	7.22	7.26	0.03	0.03	0.03	0.03	0.51	0.35	0.33	0.32	0.55	0.54	0.51	0.51	0.55	0.49	0.47	0.42
322	3.26	2.92	2.63	2.63	0.57	0.55	0.52	0.52	0.03	0.02	0.02	0.02	1.11	1.14	1.16	1.16	0.16	0.16	0.15	0.27
323	0.00	0.00	0.00	0.00	0.00	0.00	0.00	0.00	0.00	0.00	0.00	0.00	0.00	0.00	0.00	0.00	0.00	0.00	0.00	0.00
324	0.09	0.09	0.25	0.40	0.74	0.54	0.60	0.60	3.14	2.30	2.02	1.93	1.07	1.92	2.07	2.07	0.40	0.59	0.59	0.63
331	0.00	0.00	0.00	0.00	0.04	0.04	0.04	0.04	0.00	0.00	0.00	0.00	0.18	0.19	0.22	0.22	0.03	0.02	0.02	0.03
332	3.33	2.90	3.00	3.02	0.00	0.00	0.00	0.00	0.00	0.00	0.00	0.00	0.00	0.00	0.00	0.00	0.05	0.05	0.05	0.03
333	3.40	3.59	3.62	3.62	0.00	0.01	0.00	0.00	0.00	0.00	0.00	0.00	0.00	0.00	0.00	0.00	0.12	0.13	0.09	0.03
334	0.00	0.00	0.00	0.00	0.00	0.00	0.00	0.00	0.00	0.00	0.00	0.00	0.00	0.00	0.00	0.00	0.00	0.00	0.00	0.00
335	0.65	0.52	0.43	0.41	0.00	0.00	0.00	0.00	0.00	0.00	0.00	0.00	0.00	0.00	0.00	0.00	0.00	0.00	0.00	0.00
411	0.26	0.25	0.24	0.24	0.14	0.09	0.10	0.10	0.07	0.08	0.08	0.08	0.68	0.68	0.66	0.66	0.14	0.14	0.14	0.10
412	0.04	0.03	0.02	0.02	0.16	0.16	0.16	0.16	0.05	0.05	0.06	0.06	0.59	0.57	0.56	0.56	0.25	0.25	0.25	0.21
421	0.00	0.00	0.00	0.00	0.01	0.01	0.02	0.02	0.00	0.00	0.00	0.00	0.45	0.49	0.53	0.53	0.03	0.04	0.04	0.05
422	0.00	0.00	0.00	0.00	0.00	0.00	0.00	0.00	0.00	0.00	0.00	0.00	0.00	0.00	0.00	0.00	0.00	0.00	0.00	0.00
423	0.00	0.00	0.00	0.00	0.01	0.01	0.01	0.01	0.00	0.00	0.00	0.00	0.07	0.08	0.08	0.08	0.05	0.05	0.05	0.04
511	0.27	0.27	0.27	0.27	0.16	0.19	0.19	0.19	0.06	0.06	0.06	0.06	0.00	0.00	0.00	0.00	0.21	0.21	0.21	0.21
512	0.54	0.56	0.57	0.57	0.30	0.33	0.34	0.34	0.62	0.65	0.66	0.67	0.83	0.85	0.88	0.88	0.83	0.88	0.93	0.91
521	0.00	0.00	0.00	0.00	0.00	0.00	0.00	0.00	0.00	0.00	0.00	0.00	0.20	0.20	0.22	0.22	0.03	0.03	0.03	0.04
522	0.00	0.00	0.00	0.00	0.13	0.13	0.14	0.14	0.00	0.00	0.00	0.00	0.00	0.00	0.00	0.00	0.05	0.05	0.05	0.07
523	0.00	0.00	0.00	0.00	0.02	0.02	0.02	0.02	0.00	0.00	0.00	0.00	1.37	1.36	1.34	1.34	0.04	0.04	0.04	0.04

Table 5. Percentage of land use from 1990 to 2012 for Austria, Belgium, Czech Republic, Denmark and Germany (authors).

	Hungary				Ireland				Luxembourg				The Netherlands				Poland			
CODE	1990	2000	2006	2012	1990	2000	2006	2012	1990	2000	2006	2012	1990	2000	2006	2012	1990	2000	2006	2012
111	0.03	0.02	0.02	0.02	0.07	0.04	0.04	0.04	0.28	0.28	0.27	0.27	0.00	0.00	0.00	0.00	0.03	0.03	0.03	0.03
112	4.44	4.58	4.66	4.70	0.99	1.31	1.53	1.56	6.36	6.74	7.38	7.44	6.79	7.98	8.59	8.89	2.47	3.13	4.54	4.57
121	0.51	0.56	0.62	0.66	0.05	0.11	0.17	0.18	0.71	0.82	1.16	1.23	1.02	1.64	2.00	2.18	0.31	0.33	0.37	0.39
122	0.04	0.05	0.06	0.10	0.00	0.03	0.07	0.09	0.12	0.14	0.20	0.19	0.14	0.15	0.22	0.25	0.04	0.04	0.05	0.08
123	0.00	0.00	0.00	0.00	0.01	0.01	0.01	0.01	0.01	0.01	0.01	0.01	0.27	0.33	0.34	0.37	0.01	0.01	0.01	0.01
124	0.06	0.07	0.08	0.08	0.03	0.03	0.03	0.03	0.13	0.14	0.17	0.17	0.17	0.16	0.18	0.19	0.07	0.06	0.06	0.07
131	0.06	0.07	0.10	0.10	0.08	0.11	0.13	0.13	0.09	0.12	0.11	0.12	0.04	0.07	0.11	0.12	0.10	0.12	0.12	0.16
132	0.05	0.06	0.06	0.06	0.00	0.01	0.02	0.02	0.23	0.25	0.15	0.14	0.01	0.02	0.06	0.06	0.04	0.04	0.04	0.04
133	0.01	0.02	0.07	0.03	0.01	0.04	0.05	0.00	0.01	0.01	0.00	0.10	0.35	0.34	0.48	0.38	0.02	0.01	0.02	0.06
141	0.06	0.06	0.06	0.06	0.05	0.05	0.05	0.05	0.08	0.10	0.12	0.12	0.27	0.30	0.37	0.39	0.09	0.05	0.05	0.05
142	0.33	0.37	0.37	0.37	0.12	0.23	0.29	0.30	0.03	0.12	0.17	0.18	0.82	1.10	1.30	1.38	0.11	0.17	0.19	0.19
211	53.29	53.56	52.07	51.59	5.73	7.71	5.18	4.87	8.74	8.62	12.37	12.28	21.14	20.61	20.10	19.77	44.83	44.67	43.77	43.61
212	0.00	0.00	0.00	0.00	0.00	0.00	0.00	0.00	0.00	0.00	0.00	0.00	0.00	0.00	0.00	0.00	0.00	0.00	0.00	0.00
213	0.16	0.13	0.12	0.09	0.00	0.00	0.00	0.00	0.00	0.00	0.00	0.00	0.00	0.00	0.00	0.00	0.00	0.00	0.00	0.00
221	1.63	1.54	1.26	1.20	0.00	0.00	0.00	0.00	0.60	0.57	0.61	0.59	0.00	0.00	0.00	0.00	0.00	0.00	0.00	0.00
222	0.69	0.79	0.90	0.75	0.00	0.00	0.00	0.00	0.00	0.00	0.05	0.05	0.19	0.21	0.20	0.19	0.29	0.41	0.51	0.52
223	0.00	0.00	0.00	0.00	0.00	0.00	0.00	0.00	0.00	0.00	0.00	0.00	0.00	0.00	0.00	0.00	0.00	0.00	0.00	0.00
231	7.31	7.28	7.35	7.39	54.23	51.13	54.65	54.90	11.91	11.78	14.58	14.55	30.41	28.63	27.38	27.09	8.87	8.70	8.88	8.80
241	0.00	0.00	0.00	0.00	0.00	0.00	0.00	0.00	0.58	0.52	0.03	0.03	0.00	0.00	0.00	0.00	0.00	0.00	0.00	0.00
242	3.45	2.68	3.06	3.11	1.62	2.04	0.89	0.84	23.83	23.67	17.92	17.80	15.26	14.79	14.43	14.22	5.58	4.54	2.76	2.76
243	1.79	1.63	1.78	1.78	6.04	6.28	6.92	6.92	10.09	9.96	8.44	8.44	2.92	2.92	3.13	3.11	4.90	4.64	4.09	4.07
244	0.00	0.00	0.00	0.00	0.00	0.00	0.00	0.00	0.00	0.00	0.00	0.00	0.00	0.00	0.00	0.00	0.00	0.00	0.00	0.00

	Hungary				Ireland				Luxembourg				The Netherlands				Poland			
311	15.41	15.90	15.81	15.88	0.43	0.41	0.41	0.41	24.56	24.29	24.39	24.46	1.33	1.55	1.62	1.62	4.71	4.82	4.90	4.92
312	1.04	1.08	1.02	0.98	3.55	3.34	3.93	3.92	4.87	4.57	4.54	4.50	4.35	4.32	4.29	4.25	17.77	17.86	17.94	18.04
313	1.62	1.68	1.66	1.63	0.33	0.41	0.99	0.99	6.20	5.95	6.83	6.81	2.50	2.52	2.52	2.53	7.09	7.39	7.79	7.89
321	2.42	2.45	2.45	2.45	1.31	1.25	0.61	0.62	0.07	0.07	0.00	0.00	0.68	0.84	1.11	1.25	0.14	0.12	0.11	0.11
322	0.00	0.00	0.00	0.00	0.82	0.77	1.31	1.31	0.00	0.00	0.00	0.00	1.00	1.01	1.07	1.13	0.01	0.01	0.01	0.01
323	0.00	0.00	0.00	0.00	0.00	0.00	0.00	0.00	0.00	0.00	0.00	0.00	0.00	0.00	0.00	0.00	0.00	0.00	0.00	0.00
324	2.61	2.61	3.60	4.12	3.05	5.20	4.00	4.13	0.19	0.95	0.20	0.20	0.02	0.04	0.04	0.04	0.59	0.96	1.81	1.70
331	0.00	0.00	0.00	0.00	0.13	0.09	0.11	0.11	0.00	0.00	0.00	0.00	0.37	0.32	0.28	0.28	0.02	0.01	0.01	0.01
332	0.00	0.00	0.00	0.00	0.24	0.20	0.22	0.22	0.00	0.00	0.00	0.00	0.00	0.00	0.00	0.00	0.01	0.01	0.01	0.01
333	0.03	0.03	0.03	0.03	0.28	0.29	0.76	0.76	0.00	0.00	0.00	0.00	0.00	0.00	0.00	0.00	0.08	0.03	0.03	0.03
334	0.00	0.00	0.00	0.00	0.00	0.00	0.00	0.00	0.00	0.00	0.00	0.00	0.00	0.00	0.00	0.00	0.01	0.00	0.00	0.00
335	0.00	0.00	0.00	0.00	0.00	0.00	0.00	0.00	0.00	0.00	0.00	0.00	0.00	0.00	0.00	0.00	0.00	0.00	0.00	0.00
411	0.98	0.82	0.83	0.83	0.26	0.23	0.27	0.27	0.00	0.00	0.01	0.01	0.79	0.90	0.90	0.97	0.34	0.31	0.32	0.32
412	0.13	0.10	0.10	0.10	17.60	15.84	14.60	14.56	0.00	0.00	0.00	0.00	0.20	0.21	0.21	0.22	0.03	0.03	0.03	0.03
421	0.00	0.00	0.00	0.00	0.03	0.04	0.05	0.05	0.00	0.00	0.00	0.00	0.21	0.22	0.22	0.22	0.00	0.00	0.00	0.00
422	0.00	0.00	0.00	0.00	0.00	0.00	0.00	0.00	0.00	0.00	0.00	0.00	0.00	0.00	0.00	0.00	0.00	0.00	0.00	0.00
423	0.00	0.00	0.00	0.00	0.22	0.26	0.23	0.23	0.00	0.00	0.00	0.00	0.26	0.21	0.21	0.21	0.00	0.00	0.00	0.00
511	0.51	0.50	0.50	0.50	0.13	0.11	0.10	0.10	0.09	0.09	0.08	0.07	1.24	1.24	1.24	1.24	0.24	0.24	0.24	0.24
512	1.32	1.37	1.39	1.40	1.75	1.68	1.61	1.61	0.23	0.23	0.22	0.22	7.07	7.16	7.20	7.23	1.18	1.22	1.25	1.26
521	0.00	0.00	0.00	0.00	0.01	0.01	0.02	0.02	0.00	0.00	0.00	0.00	0.00	0.00	0.00	0.00	0.01	0.01	0.01	0.01
522	0.00	0.00	0.00	0.00	0.08	0.08	0.10	0.10	0.00	0.00	0.00	0.00	0.03	0.03	0.03	0.03	0.00	0.00	0.00	0.00
523	0.00	0.00	0.00	0.00	0.72	0.65	0.64	0.64	0.00	0.00	0.00	0.00	0.17	0.17	0.17	0.18	0.03	0.03	0.03	0.03

Table 6. Percentage of land use from 1990 to 2012 for Hungary, Ireland, Luxembourg, the Netherlands and Poland (authors).

	Romania				Slovakia				United Kingdom		
CODE	1990	2000	2006	2012	1990	2000	2006	2012	2000	2006	2012
111	0.05	0.05	0.04	0.04	0.02	0.02	0.02	0.02	0.12	0.13	0.13
112	5.39	5.40	4.59	4.59	4.54	4.43	4.68	4.72	4.95	5.28	5.31
121	0.57	0.58	0.43	0.43	0.56	0.56	0.61	0.63	0.57	0.79	0.82
122	0.03	0.03	0.02	0.02	0.03	0.05	0.05	0.09	0.03	0.05	0.05
123	0.01	0.01	0.01	0.01	0.01	0.00	0.00	0.01	0.04	0.05	0.05
124	0.01	0.01	0.02	0.02	0.05	0.05	0.04	0.04	0.18	0.20	0.20
131	0.09	0.10	0.11	0.11	0.07	0.06	0.07	0.08	0.22	0.28	0.29
132	0.03	0.03	0.02	0.02	0.03	0.03	0.03	0.03	0.03	0.03	0.03
133	0.01	0.01	0.01	0.01	0.10	0.01	0.03	0.04	0.02	0.03	0.03
141	0.03	0.03	0.02	0.02	0.02	0.03	0.03	0.03	0.24	0.27	0.27
142	0.03	0.03	0.02	0.02	0.20	0.17	0.20	0.22	0.93	1.13	1.13
211	34.04	34.19	36.48	36.48	34.12	34.27	32.98	32.88	24.79	27.18	27.16
212	0.00	0.00	0.00	0.00	0.00	0.00	0.00	0.00	0.00	0.00	0.00
213	0.15	0.03	0.13	0.13	0.00	0.00	0.00	0.00	0.00	0.00	0.00
221	1.78	1.72	1.32	1.32	0.57	0.49	0.47	0.46	0.00	0.00	0.00
222	1.60	1.56	1.52	1.52	0.27	0.23	0.25	0.24	0.07	0.04	0.04
223	0.00	0.00	0.00	0.00	0.00	0.00	0.00	0.00	0.00	0.00	0.00
231	10.63	10.59	10.39	10.39	6.51	5.58	5.31	5.27	27.34	28.43	28.40
241	0.00	0.00	0.00	0.00	0.00	0.00	0.00	0.00	0.00	0.00	0.00
242	3.50	3.55	3.32	3.32	0.50	1.24	1.25	1.25	3.79	0.13	0.13
243	4.96	4.98	4.00	4.00	8.17	6.60	7.32	7.31	2.24	0.44	0.44
244	0.00	0.00	0.00	0.00	0.00	0.00	0.00	0.00	0.00	0.00	0.00
311	20.20	20.40	20.65	20.65	21.26	21.97	22.14	22.10	2.70	2.15	2.16
312	4.85	4.72	5.41	5.41	10.97	10.81	10.52	9.81	5.17	5.45	5.13
313	4.19	4.21	4.22	4.22	7.26	8.09	8.80	8.92	0.21	1.11	1.10
321	1.46	1.46	2.48	2.48	0.66	0.59	0.57	0.57	7.96	5.78	5.79
322	0.31	0.31	0.31	0.31	0.28	0.29	0.31	0.31	11.83	7.37	7.38
323	0.00	0.00	0.00	0.00	0.00	0.00	0.00	0.00	0.00	0.00	0.00
324	2.64	2.55	1.36	1.36	2.99	3.50	3.36	4.02	0.78	1.08	1.34
331	0.10	0.10	0.05	0.05	0.00	0.00	0.00	0.00	0.08	0.14	0.14
332	0.03	0.03	0.03	0.03	0.13	0.12	0.12	0.12	0.25	0.08	0.08
333	0.08	0.08	0.02	0.02	0.12	0.11	0.10	0.10	1.39	1.05	1.05

	Romania				Slovakia				United Kingdom		
334	0.00	0.00	0.00	0.00	0.00	0.00	0.00	0.00	0.00	0.01	0.01
335	0.00	0.00	0.00	0.00	0.00	0.00	0.00	0.00	0.00	0.00	0.00
411	1.60	1.60	1.29	1.29	0.12	0.06	0.08	0.08	0.07	0.06	0.06
412	0.00	0.00	0.00	0.00	0.00	0.00	0.00	0.00	2.09	9.31	9.31
421	0.00	0.00	0.03	0.03	0.00	0.00	0.00	0.00	0.13	0.15	0.15
422	0.00	0.00	0.00	0.00	0.00	0.00	0.00	0.00	0.00	0.00	0.00
423	0.00	0.00	0.00	0.00	0.00	0.00	0.00	0.00	0.18	0.25	0.25
511	0.68	0.68	0.74	0.74	0.15	0.21	0.22	0.22	0.02	0.02	0.02
512	0.67	0.67	0.66	0.66	0.31	0.43	0.44	0.44	0.89	0.91	0.91
521	0.28	0.28	0.28	0.28	0.00	0.00	0.00	0.00	0.00	0.00	0.00
522	0.00	0.00	0.00	0.00	0.00	0.00	0.00	0.00	0.12	0.09	0.09
523	0.02	0.02	0.02	0.02	0.00	0.00	0.00	0.00	0.56	0.54	0.54

Table 7. Percentage of land use from 1990 to 2012 for Romania, Slovakia and the United Kingdom (authors).

Focusing on Austria, the country shows that the land use for the coniferous forest (312) predominates above others. In fact, it occupies more than one-quarter of the Austrian territory—so, the country is considered forest. The following major land uses correspond to non-irrigated arable land and mixed forests (211 and 313).

The most representative land use in Belgium corresponds to non-irrigated arable land (211). The second most widespread use corresponds to complex cultivation (242). However, it practically occupies the same extension to discontinuous urban fabric (111), equivalent to most of the land covered by structures, buildings, roads and artificially surfaced areas associated with vegetated areas and bare soil, occupying discontinuous but significant surfaces. Therefore, even though it can be said that this country is eminent in agriculture, there is also the development of associated structures indicating the degree of development of the country. Also, this model seems consolidated and not variable in future years—once the lines that describe land uses are mostly horizontal and parallel.

Although the area destined to non-irrigated arable land (211) has been descending in Czech Republic, its extension is far above other land uses, occupying more than one-third of the country. The second relevant land use corresponds to the coniferous forest (312) occupying almost one-fifth of Czech Republic surface. It also should be highlighted that the third major land use corresponding to pastures (231) has increased significantly in 2000.

Denmark presents a surface of approximately 65% occupied by non-irrigated arable land (211). The country's agricultural character seems such that it will not change in the next few years—once the line that determines the percentage of land use (211) remains horizontal.

Germany and Hungary have repeated the model of land uses with a high predominance of non-irrigated arable land (211). Non-irrigated arable land in Germany occupies approximately

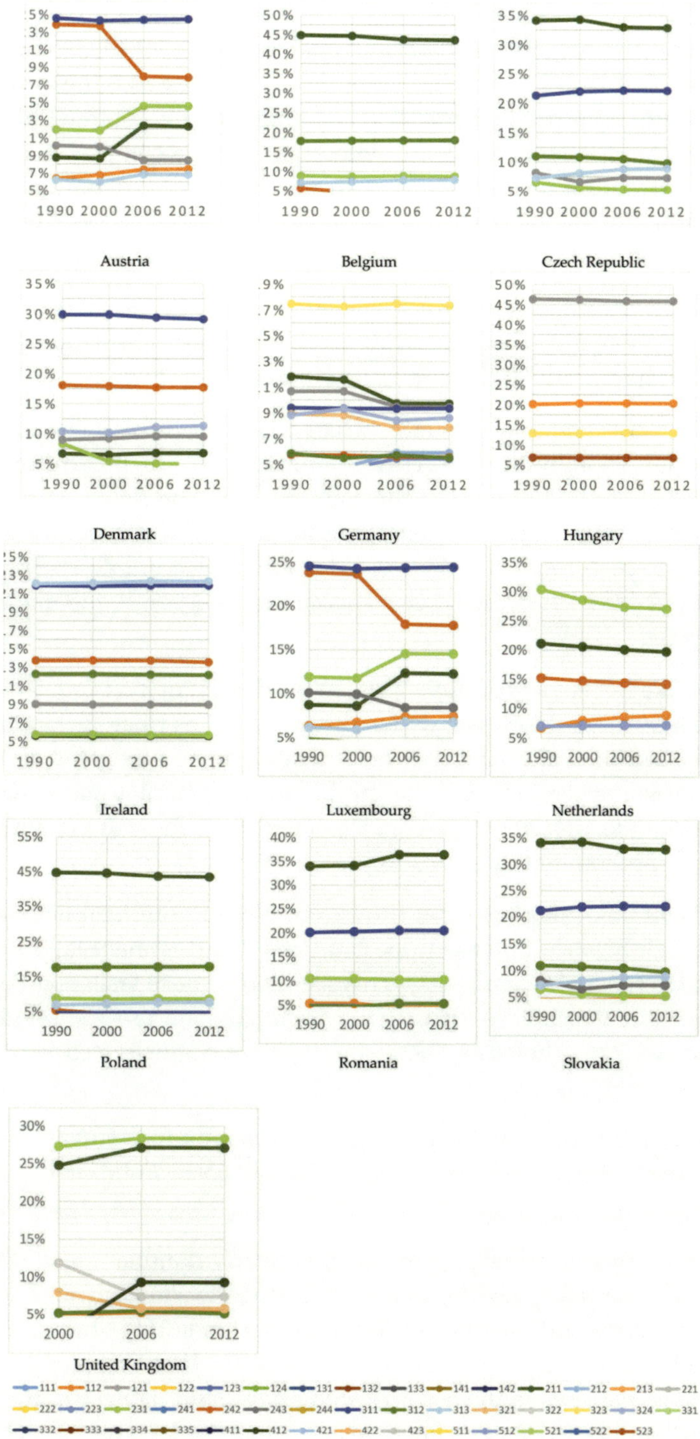

Figure 3. Trend of land uses higher than 5% for the Central EU group countries (authors).

52% of the territory and in Hungary approximately 40% of the territory. In Germany, land use stands out for coniferous forests (312) and pastures (231), which increased substantially between 2006 and 2012. In 2012, some land uses clearly increase as is the case of broad-leaved forests (311), discontinuous urban fabric (112) and others; on the contrary, mixed forests (313) descend, some of them suddenly becoming almost non-existent complex cultivation (242), and land principally is occupied by agriculture, with significant areas of natural vegetation (243). Conversely, in Hungary, the model is very steady and is similar to what occurs in Denmark due to the great dominance of single land use—non-irrigated arable land (211).

Regarding Ireland a clear dominance of single land use is also possible to verify—pastures (231), occupying more than half of the territory surface. In fact, this land use is much higher than the second most relevant land use in Ireland, peat bogs (412).

Luxembourg is a clear example of a country where the opposite happens, noting a very variable land use model. Though there is a clear dominant land use, broad-leaved forests (311), such use occupies about one-quarter of the country area. The second most relevant use is complex cultivation (242), which has greatly declined since 2000. There are also other two significant land use: pastures (231) and non-irrigated arable land (211). Therefore, it seems that the agricultural production model is changing and as a result, the model of land use is changing as a whole.

The Netherlands like Ireland has pastures (231) as the dominant land use, occupying approximately 27% of the territory. However, its dominance is not as clear as in other cases such as Ireland—once the second major land use corresponding to non-irrigated arable land (211)—and occupies approximately 20%, and the third land use complex cultivation (242) reaches approximately 15% of territorial occupation. However, these three dominant land uses imply that this country is predominantly agricultural.

Poland, Romania and Slovakia are other three examples of dominant land use and also the remaining uses slightly vary. In these three countries, the dominant land use clearly corresponds to non-irrigated arable land (211). This scenario is more visible in Poland where the land use is above 40%, which is also clearly the dominant land use scenario in Romania and Slovakia, both above 30%. Therefore, these countries are characterized by agricultural land uses.

In the Polish case, the second major land use corresponds to coniferous forests (312). In Romania and Slovakia, the second most relevant land use corresponds to broad-leaved forests (311).

Interestingly, a country where there is not only one clearly dominant land use but two is the United Kingdom. Although something similar happened in Finland, none of the two dominant land uses—pastures (231) and non-irrigated arable land (211)—has descended to please each other throughout the analysed years. Possibly, this effect would occur if natural resources are explored, that is, pastures in the non-irrigated arable land. However, the tendency notes great stability and uniformity. So, it is possible to say that the land use model varied between 2000 and 2006 and has been more stable in the 2006–2012 period. In fact, between 2000 and 2006, a tremendous increase of peat bogs (412) has occurred; as well as the significant decline in moors and heathland (322), complex cultivation (242) and natural grasslands (321) (**Tables 8** and **9** and **Figure 4**).

Curiously, all South European countries with an exception for Portugal have shown a well-defined land use model in which there is one or two dominant land uses that determine the country's land use pattern.

In the Bulgarian case, the land use is denominated by non-irrigated arable land (211) and occupies approximately one-third of the country's territory— which is clearly superior to the second major land use in Bulgaria, broad-leaved forests (311). Thus, a consistent land use model is identified in the Bulgarian territory and is possible that it will remain in the coming years.

A similar scenario occurs in Croatia, where there is clearly a dominant land use, the broad-leaved forest (311), prevailing over others and occupying approximately 30% of the country. There is also a second land use with relevance, corresponding to complex cultivation (242)—occupying approximately 17%. The situation is similar to Bulgaria but with some disparities in the period from 1990 to 2000 where a significant variability in these land uses is observed; the situation has stabilized from 2000 and in fact is similar to what occurred in the United Kingdom.

Once, in the case of Cyprus, there was a dominance of non-irrigated land (211), which occupied about one-quarter of the country. However, different from what occurs in Bulgaria and Croatia, there is also a single secondary major land use, but Cyprus presents two land uses side by side that virtually occupies the same surface extension: *sclerophyllous* vegetation (323) and coniferous forests (312).

The same that has been seen in Cyprus is verified in France, where the dominant land use is above the 25%, the non-irrigated arable land (211). Additionally, two land uses exist with major relevance: broad-leaved forests (311) and pastures (231). The rest of the uses are in percentages lower than 11% while remaining stable over the analysed years.

A considerable amount of land uses have been developed in Greece. Here, it should be highlighted that the predominant land use is *sclerophyllous* vegetation (323), occupying below 18% of the territory. Thus, Greece presents a great diversity of land uses. Also, it's possible to notice that the land uses whose percentage of extension is between 3% and 12% have suffered the vast majority of variability between the years 2000 and 2006. Such changes contrast with the constancy shown in 1990 and then in 2012. Land use, where a decrease has been identified between 2000 and 2006, corresponds to arable land (211)—land mostly occupied by agriculture, with significant areas of natural vegetation (243)—transitional Woodland shrub (324) and natural grassland (321). On the contrary, land uses that have increased are permanently irrigated land (212), olive groves (223) and mixed forests (313).

Once again, in Italy, a predominant land use is also found—arable soil land (211), occupying more than one-quarter of the Italian territory. A second predominant land use—but in much lower amount, occupying approximately 18%—is the former broad-leaved forest (311) and finally, the rest of the soils due to the supremacy of the first use of the soil is virtually stagnant.

	Bulgaria				Croatia				Cyprus			France				Greece			
CODE	1990	2000	2006	2012	1990	2000	2006	2012	2000	2006	2012	1990	2000	2006	2012	1990	2000	2006	2012
111	0.01	0.01	0.01	0.01	0.01	0.01	0.01	0.01	0.06	0.06	0.06	0.08	0.08	0.08	0.08	0.12	0.12	0.18	0.18
112	3.68	3.68	3.48	3.48	2.32	2.38	2.56	2.57	4.41	5.06	5.25	3.48	3.70	3.99	4.07	1.16	1.21	1.57	1.59
121	0.69	0.69	0.68	0.69	0.19	0.20	0.21	0.24	1.36	1.50	1.55	0.50	0.58	0.64	0.69	0.21	0.26	0.36	0.38
122	0.04	0.04	0.04	0.04	0.02	0.02	0.14	0.18	0.03	0.07	0.07	0.05	0.07	0.09	0.09	0.01	0.07	0.12	0.15
123	0.01	0.01	0.01	0.01	0.01	0.01	0.01	0.01	0.04	0.04	0.04	0.02	0.02	0.02	0.02	0.01	0.01	0.01	0.01
124	0.03	0.03	0.05	0.05	0.04	0.04	0.04	0.04	0.27	0.27	0.28	0.08	0.08	0.08	0.08	0.06	0.07	0.09	0.09
131	0.25	0.27	0.30	0.31	0.06	0.08	0.08	0.09	0.30	0.28	0.27	0.14	0.14	0.15	0.16	0.13	0.20	0.23	0.24
132	0.04	0.04	0.03	0.03	0.01	0.01	0.01	0.01	0.03	0.03	0.03	0.02	0.01	0.01	0.01	0.00	0.00	0.00	0.00
133	0.00	0.00	0.01	0.01	0.01	0.01	0.03	0.03	0.13	0.24	0.15	0.03	0.02	0.02	0.02	0.05	0.09	0.08	0.08
141	0.04	0.04	0.04	0.04	0.03	0.03	0.03	0.03	0.12	0.11	0.11	0.04	0.04	0.04	0.04	0.01	0.01	0.01	0.01
142	0.08	0.08	0.10	0.11	0.08	0.08	0.10	0.11	0.39	0.56	0.66	0.16	0.19	0.20	0.21	0.05	0.05	0.09	0.10
211	34.98	35.20	34.65	34.64	6.68	6.53	6.80	6.82	25.77	24.58	24.49	27.91	28.08	27.44	27.40	11.80	11.58	9.75	9.73
212	0.00	0.00	0.00	0.00	0.17	0.17	0.19	0.19	2.06	2.73	2.73	0.00	0.01	0.00	0.00	4.63	4.75	5.92	5.91
213	0.21	0.08	0.17	0.17	0.00	0.00	0.00	0.00	0.00	0.00	0.00	0.06	0.07	0.08	0.08	0.19	0.22	0.28	0.28
221	1.36	1.32	1.12	1.15	0.50	0.51	0.47	0.49	1.52	1.51	1.51	2.24	2.09	1.94	1.94	0.69	0.65	0.62	0.62
222	0.66	0.58	0.46	0.45	0.17	0.17	0.14	0.14	1.72	1.79	1.77	0.34	0.32	0.34	0.34	0.91	0.91	0.96	0.96
223	0.00	0.00	0.00	0.00	0.30	0.33	0.36	0.36	0.70	0.77	0.77	0.02	0.02	0.02	0.02	4.56	4.58	5.45	5.44
231	3.74	3.72	3.59	3.55	8.34	5.39	5.02	4.98	0.13	0.10	0.10	16.03	15.87	15.30	15.25	0.55	0.53	0.78	0.75
241	0.00	0.00	0.00	0.00	0.00	0.00	0.00	0.00	3.56	3.46	3.45	0.00	0.00	0.01	0.01	0.02	0.02	0.02	0.02
242	1.78	1.79	2.33	2.33	18.06	17.91	17.72	17.73	7.76	7.62	7.55	10.78	10.71	10.45	10.40	5.74	5.71	5.60	5.58
243	8.94	8.94	9.48	9.48	9.00	9.23	9.60	9.57	4.39	5.03	5.01	2.68	2.72	2.86	2.86	10.66	10.67	9.49	9.47
244	0.00	0.00	0.00	0.00	0.00	0.00	0.00	0.00	0.00	0.00	0.00	0.00	0.00	0.00	0.00	0.00	0.00	0.00	0.00

	Bulgaria				Croatia				Cyprus			France				Greece			
311	21.00	21.07	20.73	20.70	29.87	29.87	29.40	29.15	0.08	0.07	0.07	16.18	16.04	16.97	17.07	9.38	9.34	9.32	9.35
312	4.89	4.86	4.90	4.87	1.75	1.79	1.75	1.72	16.49	16.42	16.38	6.76	6.48	6.21	5.96	5.84	5.49	5.73	5.52
313	5.49	5.53	5.86	5.84	4.77	4.72	4.77	4.75	0.04	0.04	0.04	3.43	3.55	3.51	3.52	3.10	3.08	4.20	4.10
321	3.59	3.54	3.65	3.65	1.33	4.38	4.39	4.40	3.18	2.80	2.80	2.46	2.27	2.23	2.23	8.90	8.81	7.87	7.87
322	0.29	0.29	0.24	0.24	0.12	0.07	0.05	0.05	0.00	0.00	0.00	0.82	0.70	0.73	0.73	0.40	0.41	0.38	0.38
323	0.00	0.01	0.01	0.01	2.29	2.41	1.85	1.80	17.01	16.62	16.44	0.87	1.04	1.04	1.05	17.45	17.26	17.48	17.35
324	6.69	6.68	6.53	6.62	10.36	10.16	11.13	11.36	3.20	4.33	4.33	1.95	2.23	2.39	2.52	8.80	9.31	8.42	8.61
331	0.02	0.02	0.02	0.02	0.01	0.00	0.00	0.00	0.55	0.49	0.49	0.06	0.06	0.06	0.06	0.21	0.22	0.20	0.20
332	0.10	0.10	0.11	0.11	0.21	0.19	0.06	0.06	0.24	0.12	0.12	0.78	0.75	0.75	0.75	0.12	0.11	0.21	0.22
333	0.38	0.38	0.35	0.35	1.07	1.01	0.84	0.85	1.30	1.33	1.30	0.76	0.78	0.78	0.78	1.38	1.39	1.67	1.80
334	0.02	0.01	0.00	0.00	0.03	0.06	0.00	0.04	1.22	0.02	0.22	0.05	0.02	0.02	0.00	0.05	0.06	0.07	0.14
335	0.00	0.00	0.00	0.00	0.00	0.00	0.00	0.00	0.00	0.00	0.00	0.07	0.07	0.05	0.05	0.00	0.00	0.00	0.00
411	0.08	0.09	0.08	0.08	0.31	0.33	0.33	0.34	0.06	0.05	0.05	0.13	0.14	0.34	0.34	0.18	0.18	0.17	0.17
412	0.01	0.01	0.01	0.01	0.00	0.00	0.00	0.00	0.00	0.00	0.00	0.01	0.01	0.01	0.01	0.00	0.00	0.00	0.00
421	0.00	0.00	0.00	0.00	0.01	0.01	0.01	0.01	0.21	0.21	0.21	0.11	0.13	0.13	0.13	0.22	0.22	0.21	0.21
422	0.01	0.01	0.01	0.01	0.01	0.01	0.01	0.01	0.00	0.00	0.00	0.05	0.02	0.02	0.02	0.04	0.04	0.03	0.03
423	0.00	0.00	0.00	0.00	0.00	0.00	0.00	0.00	0.00	0.00	0.00	0.07	0.07	0.06	0.06	0.00	0.00	0.00	0.00
511	0.27	0.27	0.31	0.31	0.43	0.44	0.44	0.44	0.00	0.00	0.00	0.21	0.21	0.22	0.22	0.17	0.17	0.17	0.17
512	0.57	0.57	0.59	0.59	0.53	0.53	0.53	0.52	0.17	0.22	0.23	0.36	0.39	0.38	0.39	0.65	0.65	0.69	0.73
521	0.00	0.00	0.01	0.01	0.00	0.00	0.00	0.00	0.00	0.00	0.00	0.13	0.13	0.12	0.12	0.07	0.07	0.08	0.08
522	0.00	0.00	0.00	0.00	0.00	0.00	0.00	0.00	0.00	0.00	0.00	0.03	0.03	0.08	0.08	0.00	0.00	0.01	0.01
523	0.04	0.04	0.04	0.04	0.91	0.91	0.91	0.91	1.50	1.49	1.49	0.06	0.06	0.13	0.13	1.46	1.47	1.46	1.46

Table 8. Percentage of land use in 1990, 2000, 2006, and 2012 for Bulgaria, Croatia, Cyprus, France and Greece (authors).

	Italy				Malta				Portugal				Slovenia				Spain			
CODE	1990	2000	2006	2012	1990	2000	2006	2012	1990	2000	2006	2012	1990	2000	2006	2012	1990	2000	2006	2012
111	0.48	0.47	0.46	0.46	1.20	1.20	1.20	1.20	0.13	0.13	0.13	0.13	0.01	0.01	0.01	0.01	0.49	0.52	0.57	0.42
112	2.89	3.07	3.31	3.34	20.10	20.34	20.36	20.36	1.43	2.34	2.65	2.67	2.05	2.05	2.07	2.07	0.44	0.54	0.67	0.97
121	0.64	0.75	0.89	0.94	2.12	2.35	2.35	2.35	0.17	0.33	0.40	0.43	0.32	0.33	0.33	0.33	0.15	0.24	0.29	0.48
122	0.04	0.04	0.05	0.05	0.00	0.00	0.00	0.00	0.00	0.02	0.08	0.11	0.06	0.09	0.09	0.29	0.01	0.01	0.04	0.08
123	0.03	0.03	0.03	0.03	0.63	0.67	0.67	0.67	0.01	0.01	0.02	0.02	0.01	0.01	0.01	0.01	0.01	0.01	0.02	0.02
124	0.07	0.07	0.07	0.08	1.19	1.19	1.19	1.19	0.05	0.06	0.06	0.06	0.03	0.03	0.03	0.04	0.03	0.03	0.04	0.04
131	0.14	0.16	0.16	0.17	1.10	1.04	1.04	1.04	0.07	0.14	0.16	0.16	0.06	0.06	0.06	0.06	0.09	0.13	0.15	0.16
132	0.01	0.01	0.01	0.01	0.05	0.11	0.13	0.20	0.00	0.01	0.01	0.02	0.02	0.02	0.02	0.02	0.01	0.01	0.01	0.03
133	0.02	0.02	0.02	0.01	0.04	0.00	0.00	0.00	0.02	0.06	0.07	0.05	0.03	0.01	0.03	0.03	0.03	0.06	0.14	0.17
141	0.03	0.03	0.04	0.04	0.58	0.58	0.58	0.58	0.02	0.02	0.03	0.03	0.01	0.02	0.02	0.02	0.01	0.01	0.01	0.04
142	0.05	0.07	0.08	0.09	0.69	0.69	0.69	0.69	0.05	0.10	0.13	0.15	0.06	0.06	0.08	0.07	0.02	0.04	0.06	0.09
211	26.74	26.88	26.82	26.78	0.48	0.39	0.39	0.39	12.73	11.16	9.80	9.36	5.54	5.51	5.55	5.55	20.49	19.76	19.09	19.68
212	0.14	0.14	0.14	0.14	0.00	0.00	0.00	0.00	1.04	1.10	0.87	0.88	0.01	0.00	0.00	0.00	1.57	1.67	1.75	2.26
213	0.92	0.95	0.98	0.98	0.00	0.00	0.00	0.00	4.86	4.41	3.57	3.57	0.00	0.00	0.00	0.00	0.29	0.28	0.28	0.06
221	1.77	1.75	1.91	1.91	0.18	0.18	0.18	0.18	7.11	6.81	6.96	6.96	0.77	0.76	0.76	0.76	7.62	7.63	7.60	3.66
222	1.32	1.33	1.40	1.39	0.00	0.00	0.00	0.00	8.40	8.19	8.91	8.90	0.18	0.18	0.18	0.17	4.90	4.89	4.90	2.80
223	4.17	4.02	4.00	3.99	0.00	0.00	0.00	0.00	6.19	6.85	6.85	6.82	0.00	0.00	0.00	0.00	4.72	4.84	4.90	4.86
231	1.51	1.42	1.42	1.42	0.00	0.00	0.00	0.00	12.89	12.62	10.87	11.56	5.73	5.74	5.69	5.69	7.47	7.51	7.39	10.04
241	1.30	1.26	0.68	0.68	0.00	0.00	0.00	0.00	8.57	7.84	5.61	5.37	0.00	0.00	0.00	0.00	7.98	7.87	7.61	9.08
242	7.30	7.17	7.27	7.23	3.26	3.26	3.26	3.26	6.09	6.06	5.59	5.60	13.70	13.72	13.72	13.54	2.86	2.96	2.95	2.76
243	6.59	6.79	7.03	7.02	46.35	46.18	45.88	45.88	2.39	2.19	1.32	1.32	8.98	8.94	8.95	8.95	5.33	5.16	5.22	7.76
244	0.62	0.58	0.57	0.57	0.00	0.00	0.00	0.00	4.29	3.41	3.95	3.98	0.00	0.00	0.00	0.00	1.89	1.84	1.85	3.88

	Italy				Malta				Portugal				Slovenia				Spain			
311	18.16	18.20	18.42	18.41	0.00	0.00	0.00	0.00	9.21	11.24	16.63	16.29	21.86	21.82	21.88	21.88	8.77	8.82	9.16	4.03
312	4.38	4.27	4.30	4.28	0.21	0.21	0.21	0.21	0.48	0.26	0.04	0.04	12.24	12.25	12.19	12.16	0.41	0.41	0.47	0.56
313	3.42	3.63	3.66	3.65	0.45	0.45	0.45	0.45	0.89	1.14	0.95	0.95	22.08	22.14	22.35	22.34	1.82	1.82	1.83	1.47
321	4.81	4.88	4.58	4.58	0.00	0.00	0.00	0.00	0.00	0.00	0.00	0.00	1.03	1.02	1.02	1.02	0.00	0.00	0.00	0.00
322	0.91	0.48	0.50	0.50	0.00	0.00	0.00	0.00	0.01	0.01	0.02	0.02	1.12	1.12	1.12	1.12	0.11	0.11	0.11	0.09
323	3.14	3.28	3.31	3.31	12.85	12.75	13.01	12.94	0.06	0.06	0.06	0.06	0.00	0.00	0.00	0.00	0.00	0.00	0.00	0.00
324	3.35	3.55	3.45	3.46	0.00	0.00	0.00	0.00	0.23	0.23	0.21	0.21	2.11	2.15	1.90	1.90	0.09	0.09	0.10	0.09
331	0.27	0.25	0.25	0.25	0.00	0.00	0.00	0.00	0.32	0.39	0.66	0.68	0.03	0.03	0.03	0.03	0.40	0.48	0.49	0.51
332	1.58	1.47	1.43	1.43	0.00	0.00	0.00	0.00	0.60	0.59	0.46	0.50	0.84	0.84	0.83	0.83	0.20	0.26	0.28	0.27
333	1.60	1.38	1.19	1.19	1.63	1.52	1.52	1.52	2.25	2.43	2.30	2.32	0.53	0.52	0.52	0.52	1.65	1.63	1.64	2.12
334	0.01	0.03	0.01	0.03	0.00	0.00	0.00	0.00	3.02	2.85	3.07	3.48	0.02	0.00	0.00	0.00	3.41	3.54	3.65	4.47
335	0.18	0.15	0.14	0.14	0.00	0.00	0.00	0.00	1.20	1.07	1.50	1.48	0.00	0.00	0.00	0.00	1.30	1.28	1.27	1.72
411	0.05	0.05	0.06	0.06	0.00	0.00	0.00	0.00	2.48	2.45	2.38	2.38	0.13	0.12	0.12	0.12	10.85	10.54	10.25	9.80
412	0.00	0.00	0.00	0.00	0.00	0.00	0.00	0.00	0.12	0.11	0.10	0.10	0.00	0.00	0.00	0.00	0.09	0.09	0.09	0.07
421	0.08	0.07	0.08	0.08	0.00	0.00	0.00	0.00	0.49	0.32	0.40	0.21	0.00	0.01	0.01	0.01	0.16	0.15	0.11	0.05
422	0.03	0.03	0.03	0.03	0.08	0.08	0.08	0.08	0.20	0.19	0.18	0.18	0.02	0.02	0.02	0.02	0.06	0.06	0.06	0.07
423	0.00	0.00	0.00	0.00	0.00	0.00	0.00	0.00	0.08	0.07	0.07	0.07	0.00	0.00	0.00	0.00	0.04	0.04	0.04	0.04
511	0.16	0.16	0.15	0.15	0.00	0.00	0.00	0.00	0.01	0.02	0.03	0.03	0.26	0.25	0.25	0.25	0.01	0.01	0.01	0.00
512	0.56	0.56	0.57	0.58	0.00	0.00	0.00	0.00	0.09	0.09	0.09	0.09	0.14	0.13	0.14	0.14	0.02	0.02	0.04	0.02
521	0.15	0.15	0.15	0.15	0.00	0.00	0.00	0.00	0.09	0.09	0.09	0.09	0.00	0.00	0.00	0.00	0.02	0.02	0.03	0.02
522	0.00	0.00	0.00	0.00	0.00	0.00	0.00	0.00	0.31	0.31	0.31	0.31	0.00	0.00	0.00	0.00	0.18	0.18	0.51	0.48
523	0.41	0.41	0.41	0.41	6.81	6.79	6.79	6.79	1.34	2.21	2.45	2.38	0.02	0.02	0.02	0.02	4.01	4.43	4.31	4.81

Table 9. Percentage of land use in 1990, 2000, 2006, and 2012 Italy, Malta, Portugal, Slovenia and Spain (authors).

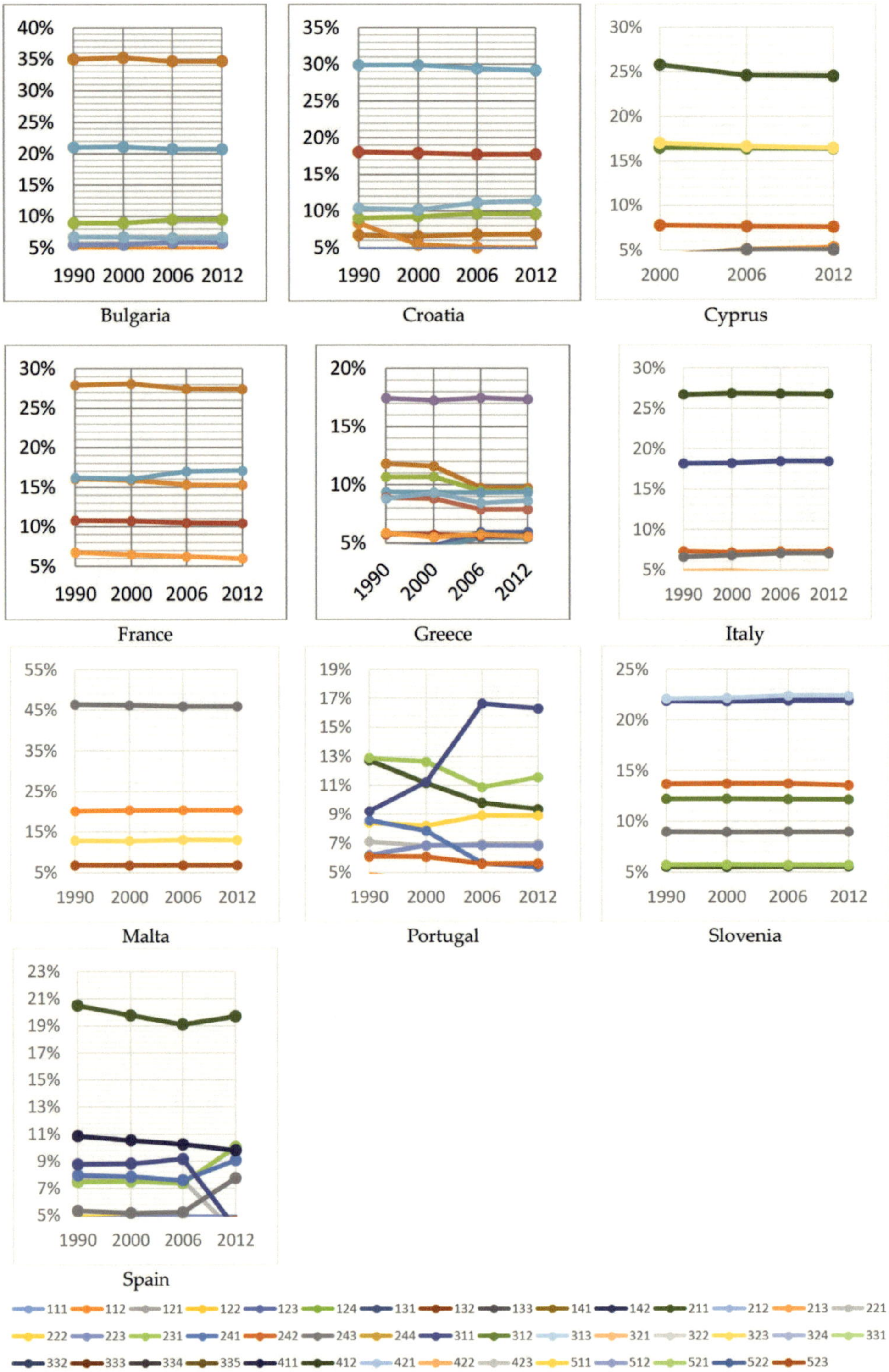

Figure 4. Trend of the land uses of percentage higher than 5% for the South EU group countries (authors).

Also, a low variability pattern of land use is seen in Malta. Nevertheless, the most relevant land use is occupied by agriculture, with significant areas of natural vegetation (243)—representing approximately half of the extension of the country. Additionally, the second major land use is the discontinuous urban fabric (112).

A pattern of land use that breaks with the shown tendency of conservative models over the analysed years is found in the Portuguese case. In Portugal, the extension of the different uses of the soil has varied considerably. The increase in the broad-leaved forest (311) from 9.21% in 1990 to 16.29% in 2012 should be highlighted. It is also noteworthy that the land use pastures (231) and non-irrigated arable land (211) have both decreased. In fact, this last one (non-irrigated arable land) presents similar values to vineyards (221) in 2012. However, if there is variability in the extent of the land uses, what occurred in other countries like Austria, Luxembourg or the United Kingdom should be taken into account, between 2006 and 2012, which seems to play a critical role in the decrease of data variability.

Slovenia is another example of highly stable and consolidated land use patterns, once all tendency lines are horizontal. In this case, two land uses co-exist, mixed forests (313) and broad-leaved forests (311), both over 20%. The combination of these two land uses—40% of the territory—establishes a forest character for Slovenia. As an example, the fourth important use of the land corresponds to the coniferous forest (312) and the third to the complex cultivation (242).

In context, Spain does not escape from the predominance of a single land use pattern, the non-irrigated arable land (211), which occupies approximately one-fifth of the Spanish mainland. Additionally, the remaining land use covers an area below 11%. Regarding the surface extension variability for each land use, although there was a trend of low variability between 1990 and 2006, between 2006 and 2012, this trend broke with high variability. In this sense, increases in the land use include pastures (231), annual crops associated with permanent crops (241), land principally occupied by agriculture with significant areas of natural vegetation (243) and burnt areas (334) and in a lower level of increase comes the land use agro-forestry areas (244), permanently irrigated land (212), glaciers and perpetual snow (335), discontinuous urban fabric (112); in terms of decreasing more dramatically, the land uses include broad-leaved forests (311), vineyards (221), fruit and three berry plantations (222) and less-pronounced inland marshes (411). Therefore, it seems that this model in the future can present great variability and probably will need time to be able to stabilize.

After the analysis of the EU territories, the major land uses are represented on a map (**Figure 5**). The map enables us to verify that most of the land use corresponds to agricultural and forestry, the two being the most predominant agricultural uses. Even within agricultural use, it is possible to notice that the majority corresponds to non-irrigated land (211). Therefore, it can be argued that EU territories are characterized by agricultural and forest uses—mostly intended for agricultural-use non-irrigated land.

Also, in countries located in the North of Europe, their land uses are both agricultural and forestry. In Central EU territories, under the use of agricultural land, the non-irrigated land is the predominant one (211). This is similar to what happens in the EU South territories. However, in this area, the predominance of agricultural use is not so dominant, alternating in some countries the majority use to forestry use.

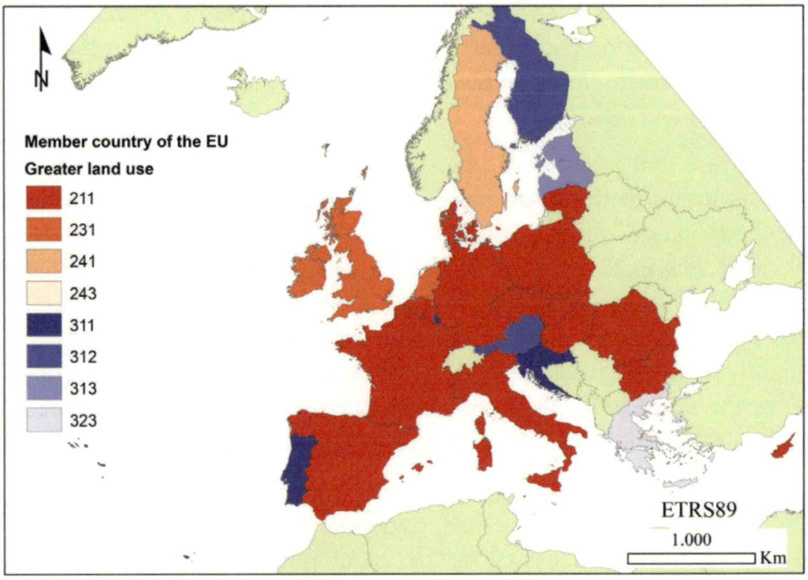

Figure 5. Major land uses in EU territories (authors).

4. Final remarks

The synthetic methodology analysis shown to characterize each of the EU Member States according to the area dedicated to different land uses—defining land use patterns, models and dynamics. Also, this typology of study is possible to replicate using the official and open-access tools mentioned above. In fact, through CLC and its available data, the analysis can be expanded for 2012 and onwards.

In this regard, the performed analyses provide valuable results and knowledge for the decision-makers, in territorial governance and land use planning, which can influence directly and indirectly the socio-economic aspects, such as the environmental paradigm.

Precisely, different trends regarding the presence of certain typologies of land uses in the EU territories between the periods of 1990, 2000, 2006 and 2012 determine that the majority use in Europe is the agrarian use, followed by the forest, in which the majority is the non-irrigated land. Also, it is possible to verify the high variability in land use pattern of some countries—as the case of Finland, Latvia, Portugal and Spain. The rest of the countries present are deeply consolidated models determined by the scarce variation trend of land use.

It is also possible to verify as the land use in some countries is not very varied, since one or more land uses very prominently predominate over others. This is the case of countries like Finland, Lithuania, Sweden, Austria, Czech Republic, Denmark, Germany, Hungary, Ireland, Netherlands, Poland, Romania, Slovakia, Bulgaria, Croatia, Cyprus, France, Greece, Italy, Malta, Slovenia and Spain. Therefore, in these countries it is not easy to observe quick changes on the land use model and pattern. As a result, if for some reason in some of the abovementioned countries it is deemed appropriate to change the land use, it is necessary to change major land uses, to achieve higher variability.

Author details

José Manuel Naranjo Gómez[1,2*], Luis Carlos Loures[2,3,4], Rui Alexandre Castanho[2,5,6,7], José Cabezas Fernández[2,5], Luis Fernández-Pozo[5], Sérgio António Neves Lousada[8] and Patrícia Escórcio[8]

*Address all correspondence to: jnaranjo@unex.es

1 Polytechnic School, University of Extremadura, Caceres, Spain

2 VALORIZA – Research Centre for Endogenous Resource Valorization, Portalegre, Portugal

3 Polytechnic Institute of Portalegre, Portugal and Research Centre for Spatial and Organizational Dynamics (CIEO), University of Algarve, Algarve, Portugal

4 Research Centre for Tourism, Sustainability and Well-being (CinTurs), University of Algarve, Algarve, Portugal

5 Environmental Resources Analysis Research Group (ARAM), University of Extremadura, Badajoz, Spain

6 Department of Landscape, Environment and Planning, School of Science and Technology, University of Évora, Évora, Portugal

7 ICAAM—Institute for Agrarian and Environmental Sciences, University of Évora, Évora, Portugal

8 Department of Civil Engineering and Geology, Faculty of Exact Sciences and Engineering, University of Madeira, Funchal, Portugal

References

[1] Neumann K, Herold M, Hartley A, Schmullius C. Comparative assessment of CORINE2000 and GLC2000: Spatial analysis of land cover data for Europe. International Journal of Applied Earth Observation and Geoinformation. 2007;**9**(4):425-37. DOI: 10.1016/j.jag.2007.02.004

[2] Herold M, Woodcock CE, Di Gregorio A, Mayaux P, Belward AS, Latham J, et al. A joint initiative for harmonization and validation of land cover datasets. IEEE Transactions on Geoscience and Remote Sensing. 2006;**44**(7):1719-1727. DOI: 10.1109/TGRS.2006.871219

[3] Delgado J, editor. Armonización de las bases de datos de ocupación del suelo y su importancia en la evaluación de parámetros e indicadores medioambientales. In: Congreso Nacional de Medio Ambiente Cumbre Del Desarrollo Sostenible. 2008

[4] García-Álvarez D, Camacho Olmedo MT. Changes in the methodology used in the production of the Spanish CORINE: Uncertainty analysis of the new maps. International Journal of Applied Earth Observation and Geoinformation. 2017;**63**:55-67. DOI: 10.1016/j.jag.2017.07.001

[5] Bartholomé E, Belward AS. GLC2000: A new approach to global land cover mapping from earth observation data. International Journal of Remote Sensing. 2005;**26**(9):1959-1977. DOI: 10.1080/01431160412331291297

[6] Weber J-L. Implementation of land and ecosystem accounts at the European Environment Agency. Ecological Economics. 2007;**61**(4):695-707. DOI: 10.1016/j.ecolecon.2006.05.023

[7] Feranec J, Jaffrain G, Soukup T, Hazeu G. Determining changes and flows in European landscapes 1990-2000 using CORINE land cover data. Applied Geography. 2010;**30**(1):19-35. DOI: 10.1016/j.apgeog.2009.07.003

[8] Hazeu G, Büttner G, Arozarena A, Valcárcel N, Feranec J, Smith G. Detailed CLC data: Member states with CLC level 4/level 5 and (semi-) automated solutions. European Landscape Dynamics CORINE Land Cover Data. 2016:275-304. DOI: 10.1201/9781315372860-27

[9] Büttner G. CORINE land cover and land cover change products. In: Land Use and Land Cover Mapping in Europe. Springer; 2014. pp. 55-74. DOI: 10.1007/978-94-007-7969-3

[10] European Union. Directive 2007/2/EC of the European Parliament and of the Council of 14 March 2007 Establishing an Infrastructure for Spatial Information in the European Community (INSPIRE). Published in the official Journal on the 25th April. 2007

[11] Modugno S, Balzter H, Cole B, Borrelli P. Mapping regional patterns of large forest fires in Wildland–urban Interface areas in Europe. Journal of Environmental Management. 2016;**172**:112-126. DOI: 10.1016/j.jenvman.2016.02.013

[12] Pereira MG, Aranha J, Amraoui M. Land cover fire proneness in Europe. Forest Systems. 2014;**23**(3):598-610. DOI: 10.5424/fs/2014233-06115

[13] Bishr Y. Overcoming the semantic and other barriers to GIS interoperability. International Journal of Geographical Information Science. 1998;**12**(4):299-314. DOI: 10.1080/136588198241806

[14] Pinto-Correia T, Kristensen L. Linking research to practice: The landscape as the basis for integrating social and ecological perspectives of the rural. Landscape and Urban Planning. 2013;**120**:248-256. DOI: 10.1016/j.landurbplan.2013.07.005

[15] Hennig EI, Schwick C, Soukup T, Orlitová E, Kienast F, Jaeger JAG. Multi-scale analysis of urban sprawl in Europe: Towards a European de-sprawling strategy. Land Use Policy. 2015;**49**:483-498. DOI: 10.1016/j.landusepol.2015.08.001

[16] Salvati L, Zambon I, Chelli FM, Serra P. Do spatial patterns of urbanization and land consumption reflect different socioeconomic contexts in Europe? Science of The Total Environment. 2018;**625**:722-730. DOI: 10.1016/j.scitotenv.2017.12.341

[17] Verburg PH, van Berkel DB, van Doorn AM, van Eupen M, van den Heiligenberg HARM. Trajectories of land use change in Europe: A model-based exploration of rural futures. Landscape Ecology. 2010;**25**(2):217-232. DOI: 10.1007/s10980-009-9347-7

[18] van Vliet J, de, Groot HLF, Rietveld P, Verburg PH. Manifestations and underlying drivers of agricultural land use change in Europe. Landscape and Urban Planning. 2015;**133**:24-36. DOI: 10.1016/j.landurbplan.2014.09.001

[19] Levers C, Müller D, Erb K, Haberl H, Jepsen MR, Metzger MJ, et al. Archetypical patterns and trajectories of land systems in Europe. Regional Environmental Change. 2015 December 24. DOI: 10.1007/s10113-015-0907-x

[20] Mottet A, Ladet S, Coqué N, Gibon A. Agricultural land-use change and its drivers in mountain landscapes: A case study in the Pyrenees. Agriculture, Ecosystems and Environment. 2006;**114**(2):296-310. DOI: 10.1016/j.agee.2005.11.017

[21] Sluiter R, de, Jong SM. Spatial patterns of Mediterranean land abandonment and related land cover transitions. Landscape Ecology. 2007 April 01;**22**(4):559-576. DOI: 10.1007/s10980-006-9049-3

[22] Hersperger AM, Bürgi M. Going beyond landscape change description: Quantifying the importance of driving forces of landscape change in a Central Europe case study. Land Use Policy. 2009;**26**(3):640-648. DOI: 10.1016/j.landusepol.2008.08.015

[23] Bieling C, Plieninger T, Schaich H. Patterns and causes of land change: Empirical results and conceptual considerations derived from a case study in the Swabian Alb, Germany. Land Use Policy. 2013;**35**:192-203. DOI: 10.1016/j.landusepol.2013.05.012

[24] Primdahl J. Agricultural landscape sustainability under pressure: Policy developments and landscape change. Landscape Research. 2014;**39**(2):123-140. DOI: 10.1080/01426397.2014.891726

[25] Stuczynski T, Siebielec G, Korzeniowska-Puculek R, Koza P, Pudelko R, Lopatka A, et al. Geographical location and key sensitivity issues of post-industrial regions in Europe. Environmental Monitoring and Assessment. 2009;**151**(1-4):77-91. DOI: 10.1007/s10661-008-0251-4

[26] Suau-Sanchez P, Burghouwt G, Pallares-Barbera M. An appraisal of the CORINE land cover database in airport catchment area analysis using a GIS approach. Journal of Air Transport Management. 2013;**34**:12-16. DOI: 10.1016/j.jairtraman.2013.07.004

[27] Goerlich FJ, Cantarino I. A population density grid for Spain. International Journal of Geographical Information Science. 2013;**27**(12):2247-2263. DOI: 10.1080/13658816.2013.799283

[28] Bürgi M, Hersperger AM, Schneeberger N. Driving forces of landscape change — Current and new directions. Landscape Ecology. 2004 December 01;**19**(8):857-868. DOI: 10.1007/s10980-004-0245-8

[29] Rindfuss RR, Entwisle B, Walsh SJ, Mena CF, Erlien CM, Gray CL. Frontier land use change: Synthesis, challenges, and next steps. Annals of the Association of American Geographers. 2007;**97**(4):739-754. DOI: 10.1111/j.1467-8306.2007.00580.x

[30] Ustaoglu E, Williams B. Determinants of urban expansion and agricultural land conversion in 25 EU countries. Environmental Management. 2017;**60**(4):717-746. DOI: 10.1007/s00267-017-0908-2

[31] Food and Agricultural Organization. The agri-environmental statistics of the Food and Agriculture Organisation of the United Nations. United Nations; 2014

[32] Martin J. The European Environment: State and Outlook 2010: Synthesis: European Environment Agency; 2010

[33] European Commission [Internet]. 2014. Available from: http://www.copernicus.eu/ [Accessed: 2018-03-09]

Chapter 9

Land Use Change Monitoring and Modelling using GIS and Remote Sensing Data for Watershed Scale in Thailand

Patchareeya Chaikaew

Additional information is available at the end of the chapter

http://dx.doi.org/10.5772/intechopen.79167

Abstract

Landsat 7 Enhanced Thematic Mapper (ETM), Landsat 8 Operational Land Imager (OLI) and Thermal Infrared Sensor (TIRS) images obtained in 1991, 2005 and 2014 with maps, and field survey data were used to classify land use and land cover (LULC) changes over 23 years and predict soil erosion risk locations in the Khlong Kui watershed (73,700 ha), Prachuap Khiri Khan province, Thailand. Classified images together with soil features, slope and rainfall data were used to identify potential risk areas of soil erosion. Based on field check data, the overall classification accuracy was accessed from random samples that resulted as 80% for 1991, 83% for 2005 and 86% for 2014. The study discovered that rice field and rangeland increased by 1.12 and 2.81%, respectively, deciduous forest, and on the other hand, it decreased by 8.28%. GIS analysis identified the potential risk areas of soil erosion as 46,431 ha (0.63%) at very high risk.

Keywords: landsat, DEM, land use land cover, watershed, remote sensing, GIS, soil erosion

1. Introduction

Since the 1990s, global, regional and local studies of land use and land cover changes (LULCC) have greatly developed, thanks to advances in earth observation and monitor methods including remote sensing and GIS techniques. The matter of land use changes has been measured in many international and interdisciplinary researches such as remote sensing, environment and biogeography [1, 2].

In Southeast Asia, including Thailand, deforestation has been occurring during the last 15 years because of an increase in agricultural crops [3]. Land use land cover change in Prachuap Khiri Khan province was reported by the Office of Agriculture Economics (OAE) in 2014 that deforestation has been occurring 6.96% while agriculture and other land use increase 34.97 and 45.44% respectively [4].

Recently, remote sensing is widely applied for monitoring changes and dynamics in land use and land cover (LULC) observation and its impact to the environment. It offers a variety of benefits in LULC study and an opportunity to assess remote area such as tropical forest, high mountains, update land and terrain information and explore historical LULC. To offer more efficiency in identifying land cover changes, remote sensing is often combined with Geographic Information System (GIS) technique. GIS technology refers to for analyzing and managing spatial and temporal data associated with their features [5]. Both technologies provide capability to collect land use characteristics and changes by integrating existing remotely sensed data and relevant environments such as tropical forests, urban areas and coastal zone and different land transformations such as deforestation, urban development and desertification [2, 6–8]. This study shows environmental problems such as deforestation and soil erosion in Thailand caused by human activities. The results of this study could support local governments, local residents and farmers to focus on environmental problems in their regions. The erosion risk map can be used as the potential disaster information to establish field experiments plots for warning the risk area of soil erosion.

2. Data and methods

2.1. Study area

The Khlong Kui watershed is a large watershed in the Southwestern Thailand and is located between 11° 58′16″N and 12°15′50″N and between 99°31′56″E and 99°58′30″E as mapped in **Figure 1**. The entire area of the watershed covers approximately 73,700 ha (460,625 rai) in Kui Buri district, Prachuap Khiri Khan province, Thailand. Khlong Kui watershed, with the main river of the watershed, named the Kui Buri River, is surrounded by three main watersheds as (1) Pran Buri; (2) Khlong Khao Daeng and (3) Khlong Saphan Yai of the Prachuap Khiri Khan coast basin, the major river basin in Thailand.

Topography: Khlong Kui watershed includes high mountain range (max. 958 m) on the West, hilly and rolling land, plain and floodplain to the cost on the East as presented in **Figure 1**. High mountain ranges, the major landscape of the Khlong Kui watershed, are mostly in the upper watershed and are mostly covered by forest. Forests in these areas are strictly conserved as water sources. Plains, which cover a small part of the watershed, are used for crop, orchard, and vegetable cultivations. Floodplains, the second large landscape, surround the main rivers and are mostly located in the lower watershed. These areas are generally used for rice cultivations. Deforestation and soil erosion are the major environmental problems in the watershed. These problems are more prominent in the mountain ranges and hilly and rolling lands.

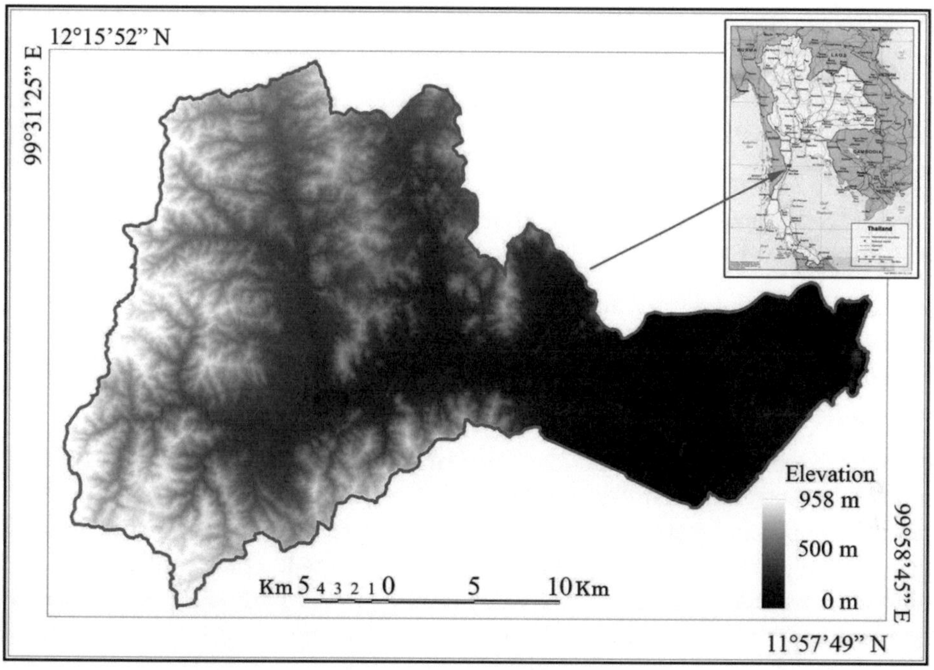

Figure 1. The study area in Khlong Kui watershed (ASTER GDEM is a product of METI and NASA).

Climate: The study area has a tropical savanna climate with drying season from January to May and raining season from June to December. The annual rainfall 30-year average is 1153 mm as the highest in November and the annual average temperature is 31.4°C as the highest in April [9]. Due to the highest rainfall in November, soil erosion and land slide might be occurred in the area where nonvegetation and bared land with high slope are the types of land use in Khlong Khui watershed.

2.2. Data use

In this study, Landsat 7 Enhanced Thematic Mapper (ETM) images in 1991, 2005 and Landsat 8 Operational Land Imager (OLI) and Thermal Infrared Sensor (TIRS) image in 2014 were used for land use and land cover (LULC) classification of the Khlong Kui watershed. A digital form of the watershed boundary was utilized. Field survey, topographic maps and LULC thematic maps were used for classification accuracy assessment. Digital Elevation Model (DEM), soil series digital maps and rainfall data were used as ancillary data to identify potential risk areas of soil erosion.

Landsat 7 ETM acquired on 02 Dec. 1991 and 17 Feb. 2005 were provided Global Land Cover Facility at Maryland University was available at http://glcf.umd.edu/data/landsat/

Landsat 8 OLI-TIRS image dated 02 Feb. 2014 was available at: http://earthexplorer.usgs.gov. The study area was covered by Landsat images with path 129/row 52. The multispectral bands contain spatial resolution at 30 × 30 m and the panchromatic band has spatial resolution at 15 × 15 m.

The Khlong Kui watershed boundary was GIS vector in ESRI Shape file that was derived from the Forest and Watershed Management Project in 2005 of the Royal Forest Department.

Topographic maps were acquired in 1995 from the Land Development Department, Thailand with a scale of 1:50,000.

LULC thematic maps were shape files for Prachuap Khiri Khan province that were created by the Land Development Department, Thailand with a scale of 1:50,000 (surveyed between 2000 and 2002).

Digital Elevation Model was Advanced Spaceborne Thermal Emission and Reflection Radiometer (ASTER) Global Digital Elevation Model (GDEM) 30 x 30 m that was provided by Japan Space Systems, Earth Remote Sensing Division, available at http://gdem.ersdac.jspace systems.or.jp/.

Rainfall data were composed from The Tropical Rainfall Measuring Mission (TRMM) is a joint mission between National Aeronautics and Space Administration (NASA) and the Japan Aerospace Exploration Agency (JAXA) designed to monitor and study tropical rainfall. The rainfall measuring instruments on the TRMM satellite include the Precipitation Radar (PR), an electronic scanning radar operating at 13.8 GHz; TRMM Microwave Image (TMI), a nine-channel passive microwave radiometer; and Visible and Infrared Scanner (VIRS), a five-channel visible/infrared radiometer. The purpose updated algorithm is to produce the best-estimate precipitation rate (in mm/h) and root-mean-square (RMS) precipitation-error estimates from TRMM and other data sources [10]. Vertical hydrometeor profiles and surface rainfall means are computed monthly with the grid size as 0.5°×0.5°. The amounts of annual rainfall in the study area and its six categories with higher amount of rainfall were ranked with the higher scores (**Table 1**).

Factor	**Ranking scores**					
	1	**2**	**3**	**4**	**5**	**6**
Slope (%)	≤2.0	2–5	5–10	10–15	15–30	>30
Land use/land cover	Water bodies, urban and built-up land, and wetland	Deciduous forest	Evergreen forest	Rice field	Orchard	Cropland
Parent material	Very high resistant to water erosion (water bodies, rock land, igneous rock formations, more diorite, andesite, and basalt)	High resistant to water erosion (alluvial deposits of plains)	Moderate resistant to water erosion (various rock and metamorphic formations, quartzite, slate, phyllite, some andesite, and some shale)	Slight low resistant to water erosion (combination of metamorphic and sedimentary rock formations, quartzite, slate, phyllite, more sandstone and shale)	Low resistant to water erosion (sedimentary rock formations, more shale and limestone)	Very low resistant to water erosion (badland, residuum and colluviums form sandstone and old alluvium, rock mountainous and eroded land)
Rainfall (mm)	≤1000	1000–1150	1150–1300	1300–1450	1450–1600	>1600

Table 1. Factors ranking used in the model of risk assessment of soil erosion.

Soil series maps of six provinces that were collected during field work between 1999 and 2002 were created by Land Development Department, Thailand in the shape file format. They came with soil series' soil materials properties in the Excel format. The attributes of soil series' soil materials properties were in the Excel format which was standardized with type of lithology prepared by FAO (in 2006). The soil materials were graded into six classes based on their resistant to water as provided in **Table 1**.

Slope is shown as the percentage of slope gradient that was calculated from Triangulated Irregular Networks (TIN) come from Digital Elevation Model (DEM) by using the Spatial Analyst Surface. The slope gradient structures were classified into six classes in accordance with the slope gradient classes [11] and the slope classes for water erosion [12]. The classes were classified from 1 to 6 as presented in **Table 1**.

Land use/land cover in 2014 classified from Landsat 8 OLI-TIRS images that were reclassified into six LULC types based on the crop management factor values provided by Land Development Department, Thailand (in 2000). The ranking scores of LULC are described in **Table 1**.

2.3. Methods

This study was accomplished using three major procedures: image classification and analysis, modeling LULC changes in 23 years (during 1991–2014) and identification of potential risk areas of soil erosion in the Khlong Kui watershed described as shown in **Figure 2**.

2.3.1. Satellite image geometric correction

The geometric correction process geometrically converts the image coordinates from (x, y) into Universal Transverse Mercator (UTM) Zone 47P map projection coordinate by using eight ground control points (GCPs). For the purpose of and use change and soil erosion analysis, all the satellite image and maps must be registered in the same pixel size and map projection with precise overlaying together. The second-order polynomial transformation and cubic convolution are used for image registration. In this study, the GCPs have been collected during 10–20 April, 2014 by using GPS-GLONASS L1 receiver brand ASTECH model Promark 100.

2.3.2. Satellite image enhancement

With respect original multispectral data set, the color distortion of pan-sharpening technique is significant limitation as shown in **Figure 3**. The statistics analysis was used to evaluate the digital value and characteristic of original data before pan-sharpening transform with enhanced data after pan-sharpening transform.

2.3.3. Image classification and analysis

The Landsat satellite images described in the previous section were used to investigate LULC in the Khlong Kui watershed, Thailand in 1991, 2005 and 2014. The images were analyzed with the image processing software GEOMATICA Ver. 2013, a widely used image processing software package, which is often used to perform LULC classification of remotely sensed data.

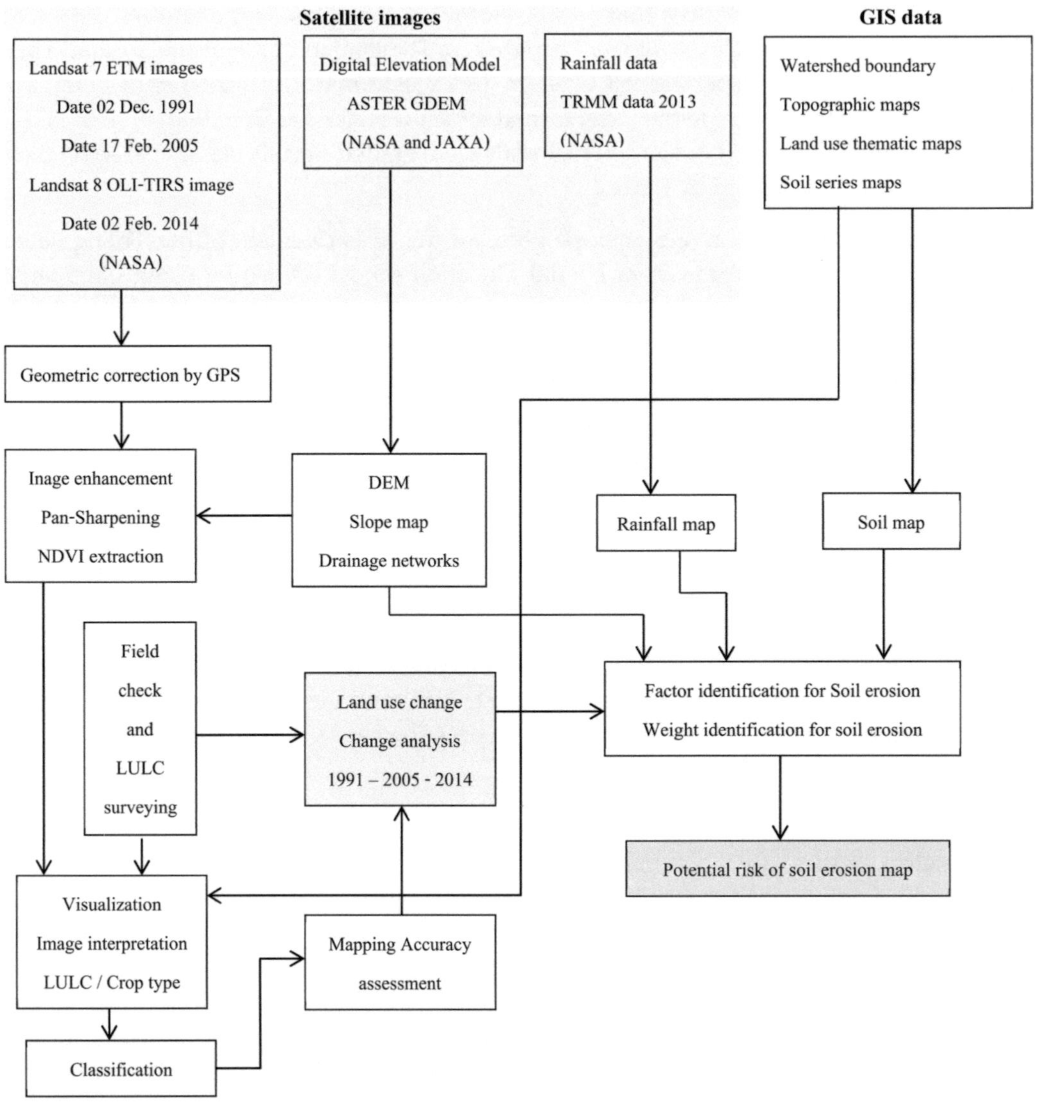

Figure 2. Methods for land use land cover classification and soil erosion risk mapping.

2.3.4. Image classification for land use and land cover map

Digital image classification is the process of recognizing pixels, which are given in multispectral bands of a satellite image. The process generates clusters of pixels with similar digital values into the same informational categories [8]. The classification performed by automated (unsupervised) or semiautomated (supervised) approaches are widely used in many LULC studies [1, 6, 13–16].

In this study, supervised method was used to classify LULC in the Khlong Kui watershed. Supervised classification employs samples of pixels that are already known informational categories to classify unknown pixels on an image. The class names were assigned into 12

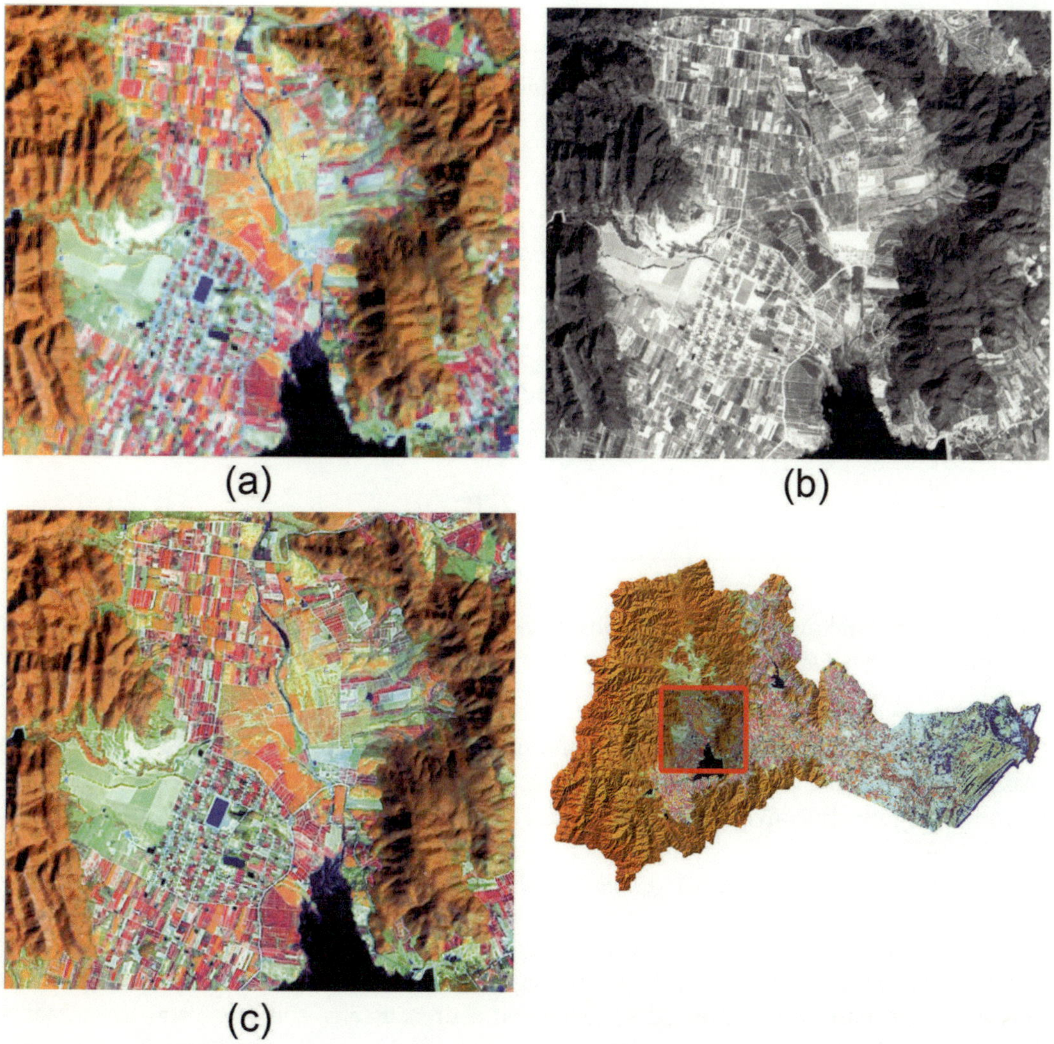

Figure 3. (a) Multi-spectral, (b) panchromatic channel and (c) pan-sharpening combination band 4-5-3 in R-G-B (Landsat imagery courtesy of NASA Goddard Space Flight Center and U.S. Geological Survey).

actual informational categories that are based on the 1976 USGS Land-Use and Land-Cover Classification [17] as (1) Urban villages (U11); (2) Cropland (A21); (3) Orchards (A22); (4) Rice field (A23); (5) Rangeland (R31); (6) Deciduous forest (F41); (7) Evergreen forest (F42); (8) Coastal forest (F43); (9) Water and reservoirs (W51); (10) Wetland (W61); (11) Barren land (B71) and (12) Beach (B72).

2.3.5. Ground truth and field checking for land use classification

Ground Truth and Field Checking for LULC classification was conducted during 10–20 April 2014 by identifying 100 locations as samples including main LULC as forest types, agricultural crops, rangeland and village area. The NEXUS 7 (Acer Tablet) with Android 4.4 combined online Google map for navigating to the sample location by using 3G internet connection and GPS-GLONASS L1 receiver brand ASTECH model Promark 100. These samples were then applied for image classification accuracy assessment by generating classification confusion matrices and accuracy report.

2.3.6. *Accuracy assessment*

Accuracy assessment is an essential requirement of image classification, and it can be resulted by the confusion matrix. Confusion matrices quantitatively compare the relationship between the classified images and the reference data which contains field survey, high resolution digital map and/or thematic maps. After the confusion matrix is generated, overall accuracy, producer's and user's accuracies, omission and commission errors, and Kappa statistics [1, 15–18] can be written as shown in Eq. 1.

$$K = \frac{N\sum_{i=1}^{r} x_{ii} - \sum_{i=1}^{r}(x_{i+} \times x_{+i})}{N^2 - \sum_{i=1}^{r}(x_{i+} \times x_{+i})} \tag{1}$$

where N is the total number of sites in the matrix, r is the number of rows in the matrix, x_{ii} is the number in row i and column i, and x_{i+} is the total for column i, x_{+i} is the total for row i, [1].

LULC change analysis was conducted in three temporal periods: 1991–2005 and 2005–2014. Cross-tabulation table and cross-classification image were used for the change analysis. The cross-tabulation table presents the unchanging and changing frequencies of each LULC type by comparing pixels from the earlier classified image to the later one.

2.3.7. *Identification of the potential risk areas of soil erosion*

According to the literature, soil erosion of a land surface is caused by various factors. These factors include topography (e.g., slope orientation, steep and length), soil cover (e.g., trees, grasses, water, bare soil and paved surface), soil character (e.g., soil mass, soil components and soil materials), and climate (e.g., rainfall amount and intensity, temperature and wind) [7, 12, 19]. In this study, we chose four different factors based on data availability to identify the potential risk areas of soil erosion in the Khlong Kui watershed. These factors are (1) slope, (2) LULC, (3) soil parent material and (4) rainfall. To construct the model, we executed two processes: (1) variable ranking and layer creation and (2) model development.

2.3.8. *Variable ranking and layer creation*

The factors were categorized into six thresholds based on a review of the literature. The threshold categories were ranked from 1 as lower risk of soil erosion to 6 as higher risk of soil erosion as shown in **Table 1**.

2.3.9. *Model development*

The model was constructed using Multi-Criteria Modeling (MCM). MCM is an influential efficient technique for managing different types of ecological modeling for decision-making and

	Slope	LULC	Parent material	Rainfall	Weight calculation
Slope	1	7/5	7/3	7	0.3496
LULC	5/7	1	5/3	5	0.2496
Soil material	3/7	3/5	1	3	0.1496
Rainfall	1/7	1/5	1/3	1	0.0496

Table 2. A pairwise comparison matrix of the relative importance of erosion factors.

environmental planning [20, 21]. Weights for the erosion factors were derived from pairwise comparison by ranking the importance of each factor and comparing them with another as shown in **Table 2**.

Slope was ranked (as 7) as the most important factor because steep slope areas usually have high potential for soil erosion. However, different types of vegetation cover can prevent erosion; hence, LULC was ranked (as 5) as the second most important factor. Soil material and rainfall were ranked (as 3 and 1) as the third and the fourth important factors. The Fuzzy Logic method (IDRISI Software Ver.17) calculates the weights which were obtained by the relative importance matrix as 0.3496 for slope; 0.2496 for LULC; 0.1496 for soil material and 0.0496 for rainfall as displayed in **Table 2**.

The weights were then used to create two equations using the attribute calculator tool from software QGIS Ver. 2.6 as given in Eq. 2:

$$\text{Risk scores of soil erosion} = [\text{Slope}] \times 0.3496 + [\text{LULC}] \times 0.2496 + [\text{Soil\_Material}] \times 0.1496 + [\text{Rainfall}] \times 0.0496 \quad (2)$$

The final risk scores of each model were standardized in percentage of potential risk by Eq. 3:

$$\%\textit{potential risk of soil erosion} = \frac{X - Min}{Max - Min} \times 100 \quad (3)$$

where X is the final risk score, *Min* is the least score and *Max* is the highest score [12]. Finally, the percentages of potential risk of erosion were divided into five classes: very low (<20), low (20–40), moderate (40–60), high (60–80) and very high (>80) potential risk of soil erosion.

3. Results and discussions

3.1. LULC classification

The water and reservoir, deciduous forest, evergreen forest, rice field, urban and village land categories presented good classification performance during the study period. Based on the ground truth and field check data of 100 samples for LULC types, the classification assessment with the confusion matrices were generated for evaluating the overall, producer and user accuracy of each LULC types. As we can see in the confusion matrix as shown in **Table 3**,

Classified data	Reference data												Classified overall	Producer accuracy (%)
	U11	A21	A22	A23	R31	F 41	F 42	F 43	W 51	W61	B 71	B 72		
U11	3	0	0	0	0	0	0	0	0	0	0	0	3	100.00
A21	0	11	0	1	0	0	0	0	0	0	0	0	12	91.67
A22	0	3	10	1	0	0	0	0	0	0	0	0	14	71.43
A23	0	1	2	8	0	0	0	0	0	0	0	0	11	72.73
R31	1	0	0	0	5	0	0	0	0	0	2	0	8	62.50
F41	0	0	0	0	0	28	1	0	0	0	0	0	29	96.55
F42	0	0	0	0	0	0	19	1	0	0	0	0	20	95.00
F43	0	0	0	0	0	0	0	0	0	0	0	0	0	No data
W51	0	0	0	0	0	0	0	0	2	0	0	0	2	100.00
W61	0	0	0	0	0	0	0	0	0	0	0	0	0	No data
B71	1	0	0	0	0	0	0	0	0	0	0	0	1	0.00
B72	0	0	0	0	0	0	0	0	0	0	0	0	0	No data
True overall	5	15	12	10	5	28	20	1	2	0	2	0	100	
User accuracy (%)	60.00	73.33	83.33	80.00	100.00	100.00	95.00	0.00	100.00	No data	0.00	No data		
Overall accuracy: 86%														
Overall Kappa statistic:	0.831													

Table 3. LULC classification 2014 Feb 02: confusion matrix.

water and reservoir archived 100% for both producer and user accuracy because the signature of water is sufficient difference from vegetation and other land cover types. Deciduous forest and evergreen forest are also classified with high accuracy as 96 and 95% for producer accuracy and 100 and 95% for user accuracy, respectively. The cropland (91.67% producer accuracy, 73.33% user accuracy) and orchard (71.43% producer accuracy and 83.333 user accuracy) categories had moderate classification performance. Rangeland (62% producer accuracy) had lower accuracy as it was mixed with barren land and urban village type. The wetland category had poor classification performance except in the 2005 classified image where it had high accuracy performance. The uncertainty of classification among forests, agricultural lands and wetlands occurred due to similar spectral reflectance of green vegetation. This confusion usually occurs when using moderate spatial resolution images such as Landsat satellite images to classify areas that have heterogeneous LULC [22].

For the overall classification accuracy of the 1991, 2005 and 2014 images, a satisfactory accuracy of more than 80% was achieved with 100 reference samples. LULC classification resulted in overall accuracy at 80% for 1991, 83% for 2005 and 86% for 2014 and Kappa Statistic at 8.83, 0.79 and 0.76 for 2014, 2005 and 1991, respectively as seen in the confusion matrix as shown

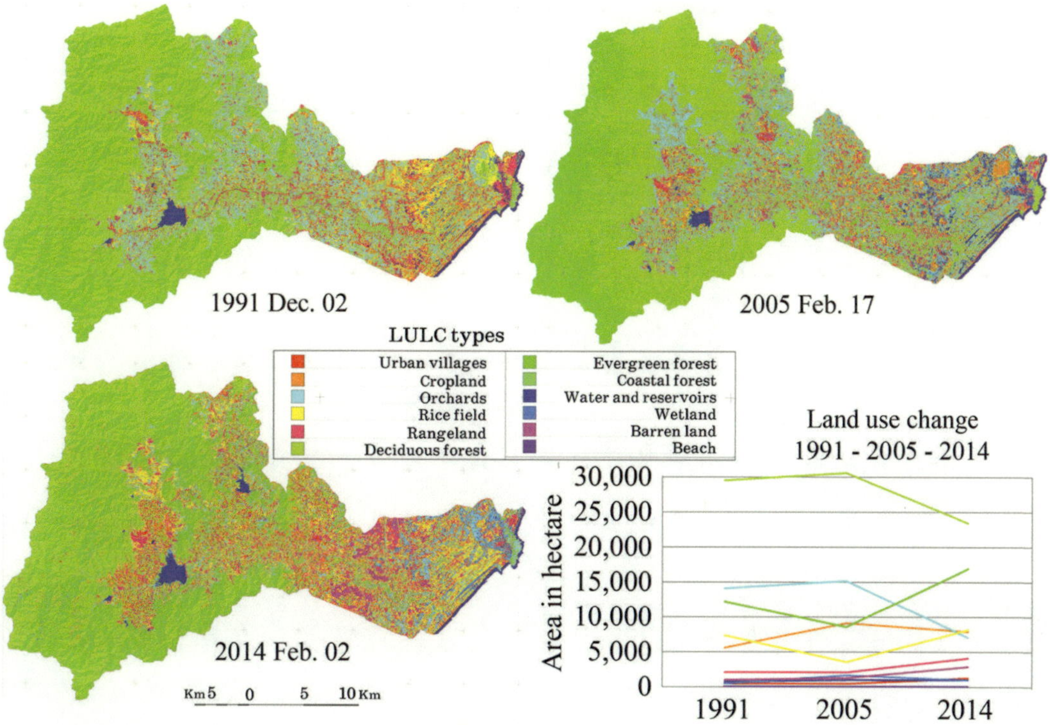

Figure 4. Land use land cover classification in 1991, 2005 and 2014.

Code	LULC types	Area in percentage			
		1991 Dec. 02	2005 Feb. 17	2014 Feb. 02	Change 1991–2014
U11	Urban villages	0.65	0.64	1.66	1.003
A21	Cropland	7.63	12.39	10.64	3.015
A22	Orchards	19.09	20.53	9.69	−9.404
A23	Rice field	9.93	4.92	11.05	1.120
R31	Rangeland	2.79	2.83	5.60	2.811
F41	Deciduous forest	40.12	41.42	31.84	−8.281
F42	Evergreen forest	16.64	11.66	22.96	6.327
F43	Coastal forest	0.04	0.02	0.02	−0.015
W51	Water and reservoirs	0.96	1.43	1.32	0.364
W61	Wetland	0.70	2.34	1.27	0.576
B71	Barren land	1.44	1.81	3.92	2.475
B72	Beach	0.01	0.01	0.02	0.009
	Total	**100.00**	**100.00**	**100.00**	

Table 4. Summary of land use land cover changes from 1991 Dec. to 2014 Feb. in Khlong Kui watershed.

	Code	U11	A21	A22	A23	R31	F41	F42	F43	W51	W61	B71	B72	Total
	LULC in 1991 area (ha)													
LULC in 2005	U11	46.42	62.93	117.41	66.08	45.88	26.42	1.55	0.54	11.07	6.53	36.05	0.61	421.47
	A21	90.43	1753.38	3480.39	1789.65	502.52	1036.73	91.31	0.97	76.10	62.26	233.24	0.41	9117.36
	A22	84.06	2041.34	6151.68	2644.72	439.83	3761.01	132.05	0.38	50.87	123.95	205.70	0.05	15,635.64
	A23	65.84	601.58	1058.18	607.91	316.80	224.69	21.29	4.57	46.94	31.50	183.26	0.38	3162.92
	R31	37.13	344.12	694.40	293.92	155.12	172.13	5.20	0.07	6.68	6.53	90.63	0.02	1805.92
	F41	12.74	369.77	1865.99	621.56	61.70	20,262.30	7930.80	0.92	17.03	37.06	32.04		31,211.90
	F42	1.13	12.13	150.71	28.08	4.93	4217.90	3780.32	0.09	2.72	5.92	2.68		8206.59
	F43				0.11	2.63		0.14	6.50	1.22	0.00	0.20	0.23	11.03
	W51	29.45	68.24	56.61	208.35	76.93	12.83	18.79	5.72	606.93	116.42	33.62	2.12	1235.99
	W61	32.51	243.47	271.46	604.71	149.06	78.77	63.68	2.09	66.98	86.65	61.04		1660.43
	B71	76.01	183.26	415.80	176.69	144.25	87.91	2.57	1.82	16.47	9.90	106.90	0.70	1222.27
	B72	0.86				0.32		0.07	2.30	0.29	0.00	0.09	4.57	8.48
	Total	476.55	5680.22	14,262.62	7041.78	1899.95	29,880.67	12,047.74	25.97	903.30	486.70	985.43	9.07	**73,700.00**
	LULC in 2005 area (ha)													
LULC in 2014	U11	46.60	214.18	205.58	129.33	102.89	37.78	2.48	0.25	14.81	23.99	73.31	0.18	851.36
	A21	52.83	1943.06	3319.56	609.71	321.55	812.57	43.43	0.79	33.77	178.63	213.32	0.56	7529.77
	A22	48.04	1049.63	2934.59	406.98	150.26	1549.49	129.02	2.18	276.23	545.47	127.55	0.56	7220.00
	A23	82.42	2805.39	3847.86	718.74	472.82	616.55	27.50	0.32	43.38	192.35	275.02		9082.33
	R31	65.77	1240.56	1128.38	437.56	341.01	198.11	12.76		22.77	95.24	164.95		3707.11
	F41	16.07	852.37	2934.05	246.04	149.54	15,104.70	4503.02		0.11	14.02	96.41		23,916.32
	F42	3.92	90.36	380.43	29.16	12.42	12,625.00	3473.91	0.32	11.41	126.92	14.69	0.05	16,768.58
	F43	0.34			4.19	0.02	0.05	0.00	5.04	3.22	0.41	2.72	0.61	16.58
	W51	29.50	138.17	77.24	92.36	32.99	23.87	6.37	0.29	668.69	53.46	46.62	3.15	1172.71
	W61	14.51	101.14	130.91	141.32	15.32	54.74	1.40	0.00	110.23	286.99	43.63		900.18
	B71	62.08	678.47	674.69	346.39	207.92	174.51	5.83	0.27	67.97	139.23	160.74		2518.09
	B72	0.65	3.08	0.18	3.13	0.14			1.49	0.63	1.06	3.33	3.31	16.99
	Total	422.71	9116.40	15,633.47	3164.90	1806.86	31,197.36	8205.69	10.94	1253.22	1657.76	1222.29	8.42	**73,700.00**

Table 5. Land use land cover change analysis from 1991 to 2014.

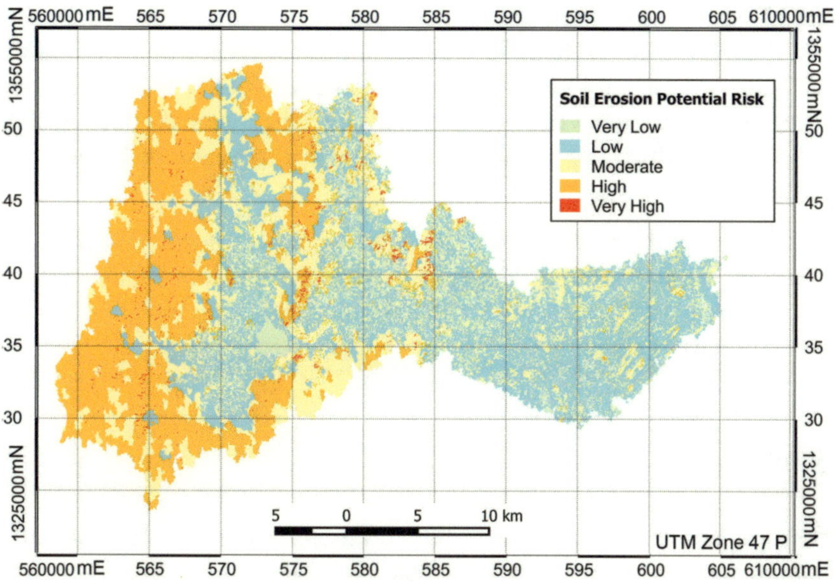

Figure 5. Khlong Kui watershed: Potential soil erosion risk map.

in **Table 3**. These overall accuracies are decreased for old dated data (1991 and 2005) due to many changes in forest and agricultural land use in comparison with those identified during ground truth period. The LULC types are presented in **Figure 4**, and the statistics of area is calculated in **Table 4**.

3.2. LULC changes and analysis (1991–2014)

The gains and losses are shown in **Figure 4**, and the cross tabulation of the changes between 1991 and 2014 (**Table 5**) is reliable with the previous two periods (1991–2005 and 2005–2014). Although there were gains in evergreen forests (6.32%) from croplands, orchards, barren land and deciduous forests, the great loss of deciduous forests (−8.28%) occurred due to conversion to evergreen forests, rangeland, barren land and croplands. Moreover, **Table 5** shows that the increase of evergreen forests, rice fields and croplands (3.01%) was mainly from deciduous forests. Most of the orchard losses (−9.40%) were converted to evergreen forests, barren land and croplands.

The major loss of coastal forest (−0.01%) was due to conversion to deciduous forests. Although the minor changes among LULC types could have followed by a result of agricultural activities such as shifting cultivation, crop rotation and infrastructure development and some of these changes could be added to the error of classification caused by similar spectral reflectance or mixed pixels from the various characteristic of LULC in the region.

3.3. Identification of the potential risk areas of soil erosion

Fuzzy Logic presented a major weighting factor in development of the model because mountains are a major landscape of the watershed. The results from Fuzzy Logic seem to be more conventional based on the topography of the watershed with very high risk (0.63%), high risk (32.00%), moderate risk (32.40%), low risk (31.21%) and very low risk (3.76%) as shown in **Figure 5**. The clusters of very high risk were consistent with the northern, central, eastern regions of the watershed as also presented in **Figure 5**. They were mainly located in mountainsides or hillsides, which are usually steep slope and boundaries between forests and highland crops.

In general, most areas of the Khlong Kui watershed had high potential risk of soil erosion due to the combination of mountainous topography and agricultural activities. High rainfall in high mountain area generated more areas of higher risk while low rainfall in low and flat area generated areas of lower risk.

4. Conclusions

This study used remote sensing and GIS techniques to assess land use and land cover (LULC) and its dynamics of change with identify the potential risk areas of soil erosion in the Khlong Kui watershed in 1991, 2005 and 2014. The Khlong Kui watershed was selected as the study area because this watershed has been experiencing deforestation and soil degradation due to the development of agricultural lands and urban areas. Moreover, the topography of the watershed, which includes mountains, hills and slopping lands, makes the Khlong Kui watershed an interesting region to examine potential risk areas of soil erosion. The key findings of the research are as follows:

4.1. Image classification and analysis

The major LULC of the Khlong Kui watershed are forests and agricultural lands. The study monitored an increase in orchards, croplands, evergreen forests, rice field and urban areas while a decrease in deciduous forests and wetlands in the watershed in 1991, 2005 and 2014. The overall accuracy assessment of the image classification was satisfactory in all three different years of satellite data acquisition.

4.2. LULC changes and dynamics

Deciduous forest, evergreen forest and orchards types were major drivers of land use and land cover changes. An increase of range land, croplands and evergreen forests were mainly derived from deciduous forests. The development of range land, barren land and crop land was related to an increase in infrastructure of the Khlong Kui watershed. There is a high probability of change from deciduous forests, wetlands and orchards to rice fields and croplands in 2014.

4.3. Potential risk areas of soil erosion

High-risk areas of soil erosion were primarily located in the northern and eastern regions of the watershed which are also with mountain ranges and hilly areas. High rainfall in high mountain area generated more areas of very high risk at 0.63% of the watershed. The change from forests to agricultural lands in the northwestern and northeastern regions of the watershed led to higher risk areas of soil erosion in the last 9 years.

4.4. Recommendation for further research

Due to limitation of research, financial budget and time, land use change and soil erosion model have lacked sample questionnaire for validation process. It is recommended for further research works that develop an additional surveying method to improve the soil erosion model to archive more accurate and creditable result.

Acknowledgements

The ASTER L1B data product was obtained through the online Data Pool at the National Aeronautics and Space Administration (NASA) Land Processes Distributed Active Archive Center (LP DAAC), USGS/Earth Resources Observation and Science (EROS) Center, Sioux Falls, South Dakota (https://lpdaac.usgs.gov/data_access).

The data used in this study were acquired as part of the Tropical Rainfall Measuring Mission (TRMM). The algorithms were developed by the TRMM Science Team. The data were processed by the TRMM Science Data and Information System (TSDIS) and the TRMM office; they are archived and distributed by the Goddard Distributed Active Archive Center. TRMM is an international project jointly sponsored by the Japan National Space Development Agency (NASDA) and the US-NASA Office of Earth Sciences.

This research was supported by the Department of Civil Engineering, Faculty of Engineering, Rajamangala University of Technology Rattanakosin, WangKlaiKangWon Campus, Prachuab Khirikhan, Donor. We would like to thank Rajamangala University of Technology Rattanakosin, WangKlaiKangWon Campus for its financial support and providing facilities to make this research possible. We would also like to thank the National Aeronautics and Space Administration (NASA), Land Processes Distributed Active Archive Center (LP DAAC), USGS/Earth Resources Observation and Science (EROS) Center, Sioux Falls, South Dakota (https://lpdaac.usgs. gov/data_access) for providing the ASTER L1B data product. We would also express our gratitude to the Goddard Distributed Active Archive Center that delivers TRMM for this research. TRMM is an international project jointly sponsored by the Japan National Space Development Agency (NASDA) and the US-NASA Office of Earth Sciences.

Author details

Patchareeya Chaikaew

Address all correspondence to: patchareeya.cha@rmutr.ac.th

Department of Civil Engineering, Faculty of Engineering, Rajamangala University of Technology Rattanakosin, Prachuap Khiri Khan, Thailand

References

[1] Jensen JR. Introductory Digital Image Processing: A Remote Sensing Perspective. New Jersey: Pearson Education, Inc; 2005. pp. 107-312

[2] Turner BL, Lambin EF, Reenberg A. The emergence of land change science for global environmental change and sustainability. Proceedings of the National Academy of Sciences (PNAS). 2007;**104**(52):20666-20671

[3] Delang CO. Deforestation in northern Thailand: The result of Hmong farming practices of Thai development strategies? Society and Natural Resources. 2002;**15**:483-501

[4] OAE. Office of Agricultural Economics: Land Use Statistics. 2014. Available from: http://www.oae.go.th/download/use_soilNew/soiNew/landused2554.html [Accessed: 2014-03-14]

[5] Longley PA, Goodchild MF, Maguire DJ, Rhind DW. Geographic Information Systems and Science. England: John Wiley & Sons, Ltd.; 2005. pp. 364-382

[6] Fromard F, Vega C, Proisy C. Half a century of dynamic coastal affecting mangrove shoreline of French Guiana: A case study based on remote sensing data analyses and field surveys. Marine Geology. 2004;**208**:265-280

[7] Sidle RC, Ziegler AD, Negishi JN, Nik AR, Siew R, Turkelboom F. Erosion processes in steep terrain—Truths, myths, and uncertainties related to forest management in Southeast Asia. Forest Ecology and Management. 2006;**224**:199-225

[8] BC J, Randolph HW. Introduction to Remote Sensing. New York: The Guilford Press; 2011. pp. 335-375

[9] TMD 2014. Thai Meteorological Department. Available from: http://www.tmd.go.th/en/province_stat.php?StationNumber=48500 [Accessed: 2014-03-02]

[10] Wu H, Adler RF, Tian Y, Huffman GJ, Li H, Wang J. Real-time global flood estimation using satellite-based precipitation and a coupled land surface and routing model. Water Resources Research. 2014;**50**:2693-2717

[11] FAO. Guidelines for Soil Description. Rome: FAO; 2006. pp. 126-284

[12] Masoudi M, Patwardhan AM, Gore SD. A new methodology for producing of risk maps of soil salinity, case study Payab Basin, Iran. Journal of Applied Sciences and Environmental Management. 2006;**10**(3):9-13

[13] Muttitanon W, Tripathi NK. Land use/land cover changes in the coastal zone of Ban Don Bay, Thailand using Landsat 5 TM data. International Journal of Remote Sensing. 2005;**26**(11):2311-2323

[14] Joao BC, Jose CP, Yosio ES. Land-cover mapping in the Brazilian Amazon using SPOT-4 vegetation data and machine learning classification methods. Photogrammetric Engineering & Remote Sensing. 2006;**72**(8):897-910

[15] Lillesand T, Kiefer RW, Chipman JW. Remote Sensing and Image Interpretation. United States: John Wiley & Sons, Inc; 2007. pp. 302-476

[16] Lu D, Weng Q. A survey of image classification methods and techniques for improving classification performance. International Journal of Remote Sensing. 2007;**28**(5):823-870

[17] Anderson JR, Hardy EE, Roach JT, Witmer RE. A Land Use and Land Cover Classification for Use with Remote Sensor Data. USGS Professional Paper 964. Washington, DC: US Government Printing Office; 1976

[18] Sirikulchayanon P, Sun W, Oyana TJ. Assessing the impacts of the 2004 tsunami on mangroves using GIS and remote sensing techniques. International Journal of Remote Sensing. 2008;**29**(12):3553-3576

[19] Sang-Arun J, Mihara M, Horaguchi Y, Yamaji E. Soil erosion and participatory remediation strategy for bench terraces in northern Thailand. Catena. 2006;**65**:258-264

[20] Clark JG, Jessica G. Tracking lyme disease: A hybrid approach to assess potential locations of risk in Jackson County, Illinois [Master thesis]. Carbondale: Southern Illinois University Carbondale; 2006. pp. 153-196

[21] Baja S, Chapman DM, Dragovich D. Spatial based compromise programming for multiple criteria decision making in land use planning. Environmental Modeling and Assessment. 2007;**12**:171-184

[22] Ibrahim O, Barabara K, Asan U, Gross CP, Hemphill S. Separation of citrus plantations from forest cover using Landsat imagery. Allgemeine Forst Und Jagdzeitung. 2007;**178** (11/12):208-212

Chapter 10

The Amazonia Third Way Initiative: The Role of Technology to Unveil the Potential of a Novel Tropical Biodiversity-Based Economy

Ismael Nobre and Carlos A. Nobre

Additional information is available at the end of the chapter

http://dx.doi.org/10.5772/intechopen.80413

Abstract

Abstract For the last two decades, the Amazon development debate has been torn between attempts to reconcile two rather opposing views of land use: on one hand, a vision of setting aside large tracts of the Amazon forests for conservation purposes (referred hereafter to as The First Way) and, on the other hand, seeking a 'sustainable' resource-intensive development, mostly through agriculture/livestock, energy and mining (referred hereafter to as The Second Way). The decrease of Brazilian Amazon deforestation from 2005 to 2014 (about 75% decline) opens a window of opportunity to conceive a novel sustainable development paradigm: The Amazonia Third Way initiative (A3W). It can represent a new opportunity emerging to protect the Amazon ecosystems and the indigenous and traditional peoples who are their custodians and at the same time develop a vibrant, socially inclusive biodiversity-driven 'green economy' in the Amazon by harnessing Nature's value through the physical, digital and biological technologies of the 4th Industrial Revolution (4IR). 4IR technologies are increasingly harnessing these assets across many industries from pharmaceutical to energy, food, cosmetics, materials and mobility, and making profits. A3W addresses ways to channel to the Amazon the benefits of the 4IR for the creation of bio-industries and local development as it protects the forests.

Keywords: Amazon, Fourth Industrial Revolution, Amazonia Third Way, Amazonia 4.0, Amazon sustainable development, land use

1. Introduction

It is more urgent than ever to find alternative ways to develop the Amazon. This realization comes with the science-based analysis that the Amazon may have come much closer to a tipping point than previously thought. Recent analysis [1] lends support to the idea that the whole Amazon system might flip to second stable climate-vegetation equilibrium, with degraded savannas covering most of the central, southern and eastern portions of the basin.

The drivers of such change are deforestation, climate change and increased forest fires. Given the simultaneous and synergistic impact of these drivers of change, total deforestation must not exceed 20–25% to avoid transgressing a potentially irreversible tipping point.

Global climate considerations also matter: CO_2 emissions from forest burning may well be the biggest unresolved global climate challenge. Without reductions in rainforest burning, including in the Amazon, international goals called for in ratified international Conventions for climate, biodiversity and water protection cannot be reached.

The heightened critical risk to the Amazon forests calls for intensifying the search for disruptive socioeconomic alternatives and transformations. For many decades, contradicting strategies to develop the Amazon have been at work: conservation (we call it the *'First Way'*) versus resource-intensive development (which we call the *'Second Way'*). Considerable efforts were made by successive governments and by NGOs to reconcile those two ways through agricultural 'sustainable intensification',—albeit with meager results. The question therefore remains how to unveil the potential of a forest-biodiversity economy in the Amazon.

We argue that a radically different *'Third Way'* for sustainable development of the Amazon is within reach. We propose to utilize modern technologies of the 4th Industrial Revolution to harness the biological and biomimetic assets of the Amazon's biodiversity. And we postulate that this *Third Way* can support a standing forest-flowing river bio-economy while being socially inclusive [2].

2. Methodological framework

The methodological approach of this study starts with a perfunctory examination of land use patterns in the Amazon. We examine two distinct models of land use pathways that in general terms may direct and define the maintenance or not of the Amazon forest. The first model is characterized by expansion of protected areas in the Amazon. It has been labeled 'The First Way'. In the other model, it is prevalent intensive natural resources exploitation. It has been labeled 'The Second Way'. In Section 3 of this chapter we briefly assess the overall results of these models in land use (for a comprehensive review, see [2]). We present updated literature data in support for current trends in land use changes, such as planned infrastructure, policies and evidence of ongoing land use processes and change.

We pose two research questions to guide the next phase of the study: Overall, current and planned patterns of land use are environmentally sustainable in the long run? If not, what would be an alternative way? The answers are developed from the basic concepts proposed by [2] for the so-called Amazonia Third Way (A3W), which is based upon a novel economic model. This rests on an innovative, knowledge-based standing forest-flowing rivers bio-economy, valuing the Amazon's renewable natural resources, biological and biomimetic assets, environmental services and biodiverse molecules and materials. A conceptual model of the A3W is proposed with the main drivers for its planning and implementation. Two of these drivers, namely Technological Drivers and Capacity Development, were considered key to the construction of A3W and are further developed in this work. The technologies of the 4th Industrial Revolution were coupled with core A3W guidelines, leading to the conceptual definition of the Amazonia 4.0. **Figure 1** shows a diagram of the methodological approach used in this work.

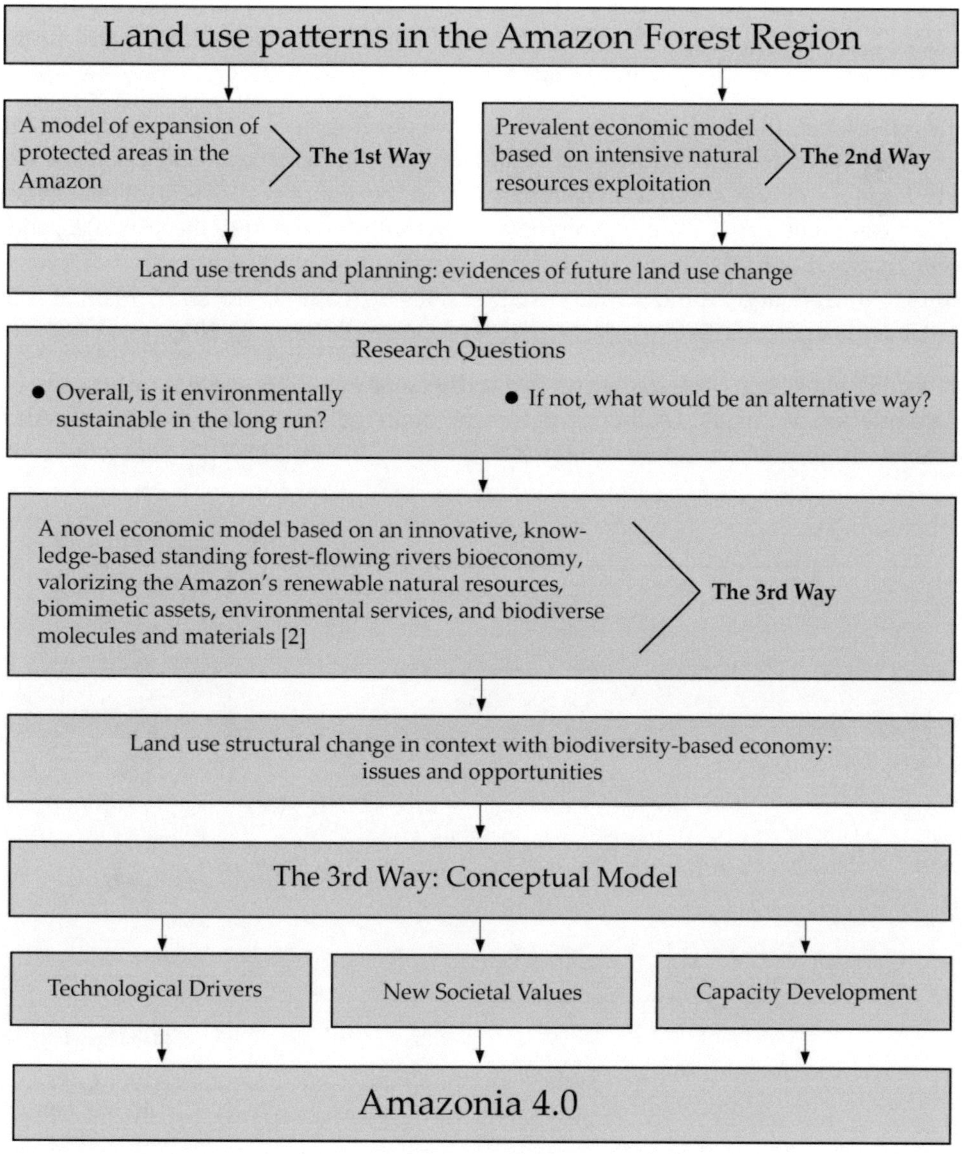

Figure 1. Methodological diagram for the conceptual development of the Amazonia Third Way.

3. Land use trends and planning: evidences of future land use change pathways

The Amazon forest biome has a total of 45.4% of its territory formed by protected areas and indigenous territories [3] as depicted in **Figure 2**. This large area where the forest is predominantly protected or managed in a sustainable way [4, 5] is the ballast that makes the First Way a possible model of land use for the Amazon. An effective example of the implementation of conservation policies by Amazonian governments is given by Brazil. In the 1990–2013 period, protected areas of the Amazon have grown from 11 to 125 million hectares and indigenous land have grown from 33 to 125 million hectares [6]. Indigenous territories and protected areas occupy 47.85% of the Brazilian Amazon [7].

On the other hand, the model of resource-intensive development (Second Way) rests mostly on economic activities that lead to the elimination of the forest and had cycles of intense growth for many decades. RAISG's 'Deforestation in the Amazon (1970–2013)' (see **Figure 3**) study indicates that up to 9.7% of the region have been deforested until the year 2000, and that between that year and 2013 that rose to 13.3%, which represents 37% increase in 13 years [9]. Given that, by and large, Amazon deforestation rates increased in the last 5 years, it is likely that total deforestation is close to reaching 16% of the whole basin by 2018.

Other studies show that protected areas and indigenous territories are not necessarily blocking deforestation completely. Although deforestation in indigenous territories in the Amazon remains relatively small, rates have grown 32% between 2016 and 2017 [7]. That points out that

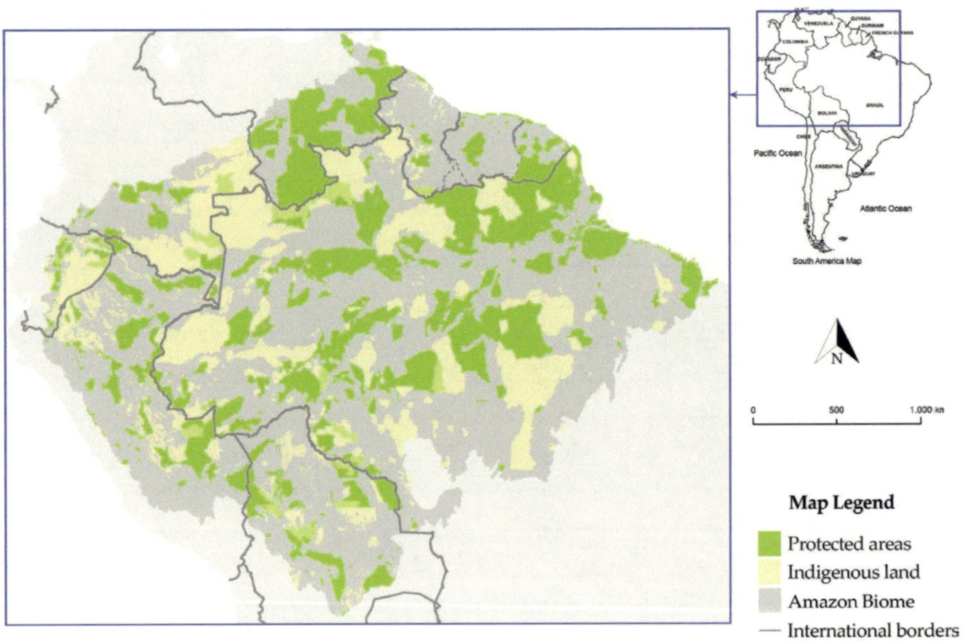

Figure 2. Protect areas in the Amazon basin. Source: Conservation International [8].

http://dx.doi.org/10.5772/intechopen.80413

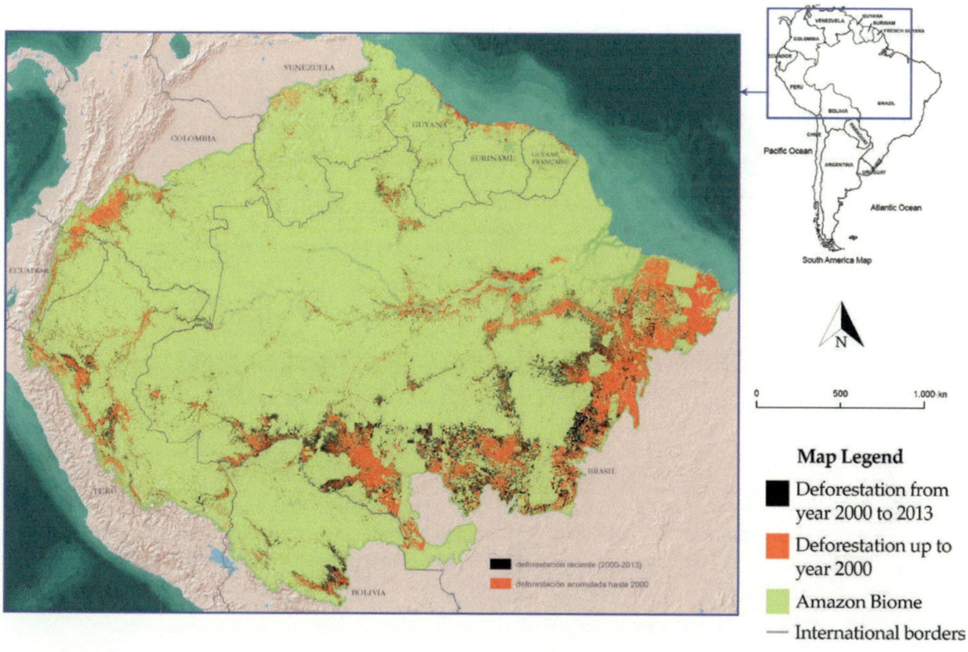

Figure 3. Mapping of deforestation of the Amazon forest biome for two distinct periods: the total accumulated up to 2000 (red color) and the increment from 2000 to 2013 (black color). Source: RAISG [9].

the barrier formed by indigenous land and other protected areas may vanish under the pressure of environmental crime and expansion of the commodities frontier, if adequate protection policies are not enforced. The increase of deforestation in some indigenous territories occurs at a time when the total rate of destruction of the Amazon rainforest fell by 16%, from 7892 km^2 in August 2015–July 2016 to 6624 km^2 in August 2016–July 2017. Notwithstanding the observed decrease, the level is still extremely high in absolute terms [7]. For the same period, the *Sistema de Alerta de Desmatamento* (SAD) from Instituto do Homem e Meio Ambiente da Amazônia (Imazon) detected an increase of 22% in the rate of deforestation in protected areas [10].

Besides the current evidences indicating that protected areas may not be a good proxy for permanent forest conservation because the prevalent model of intensive use of natural resources is a permanent dynamic force toward disrupting it, there are evidences that the future can be even more challenging for the First Way to ensure forest conservation. Official Amazonian countries' planned infrastructure developments indicate a huge increase in the construction of dams, roads, railroads and ports [11] throughout the Amazon basin. These types of infrastructure pose severe threats to the forestland through their construction and will almost certainly induce new developments of high deforestation profile.

Land use change in the Amazon: sustainability or deforestation?

In the Brazilian Amazon, which comprises 65% of the whole biome, deforestation figures from 2005 to 2017 show that a period of consistent decrease from 2004 to 2012 may be now reversed (**Figure 4**).

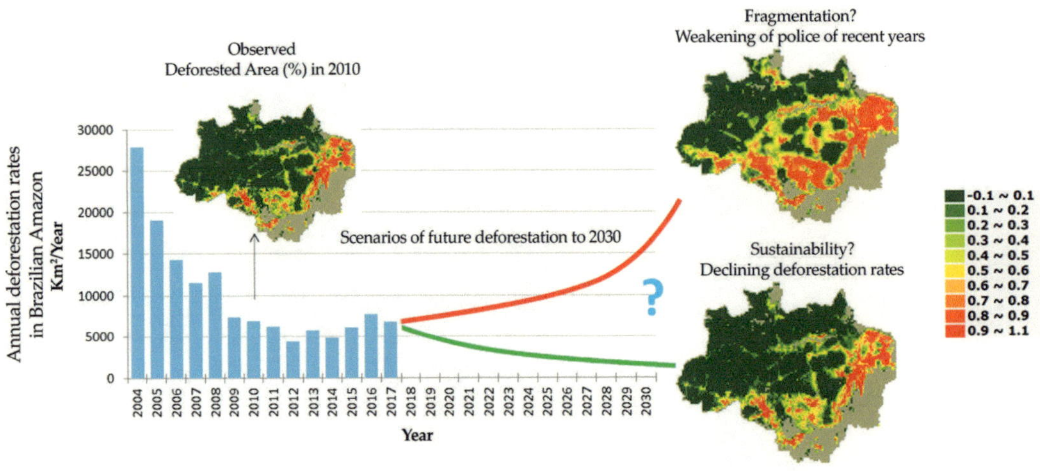

Figure 4. Annual deforestation rates in Brazilian Amazon (km^2) from 2004 to 2017 and map of fraction of land cover change for 2010 (left panel) based on PRODES data [14] and projections of two possible scenarios for the Amazon in the future up to 2030 [13]: one of large deforestation (called 'Fragmentation') and one of declining deforestation (called 'Sustainability').

Future land use change in the Amazon has been modeled [12, 13] for two rather opposed scenarios which lead to very different land cover changes (**Figure 4**). In one of them (the so-called 'Fragmentation' scenario), there is a continuous weakening of strict deforestation control policies successfully implemented from 2005 to 2012 in Brazilian Amazon and expansion of resource-intensive activities leading to agricultural and livestock expansion, resulting in over 50% of the Brazilian Amazon deforested by 2050. That is a scenario quite consistent with a progression in time of the Second Way. The other scenario in **Figure 4** (the so-called 'Sustainability' scenario) calls for continuation and strengthening of the environmental policies to bring deforestation rates close to zero in the near future. It is the land cover change scenario compatible with the Third Way.

The economic rationale to protect the tropical forests (The First Way or the 'Sustainability' scenario of **Figure 4**) rests to some degree upon the assumed low costs of maintaining intact forests as carbon storage and carbon sinks as a non-costly way to mitigate climate change in comparison to more expensive alternatives such as switching energy systems to renewable energy. Calculations for Brazil [15] estimate savings up to USD 100 billion/year to 2030 for Brazil to fulfill its NDC commitments to the Paris Accord if deforestation of the Amazon and Cerrado biomes can become smaller than 4000 km^2/year and the bulk of its commitment to reduce national emissions 43% relative to 2005 emissions by 2030 come from land use policy and not from rapidly switching the energy matrix to renewable energy. However, it is clearly short-sighted to view only the carbon pathways as justification to preserve tropical forests. In fact, the Third Way Initiative raises various limitations of such approach (see [2]) and proposes that, in addition to ecosystems services, the economic potential of tropical forests rests on their biological and biomimetic assets to a larger extent.

4. Identification of issues and opportunities for sustainable socioeconomic development

In this chapter, we analyze the issues and circumstances that have impeded to date socioeconomic development based on Amazon biodiversity assets to occur in large scale. We point out the major failures in dimensions such as concepts (imagination challenges), knowledge (research and information challenges) and implementation (governance and policy challenges & entrepreneurial capacity failures), and the lack of imagination of the potential of an innovative green economy based on nature that goes beyond the Amazon regional institutions. In the opportunity side, we present a summary of a major review in the scientific and technical literature, which identified more than 200 species of Amazonian plants with known potential to provide raw for an initial low-end bio-economy in the Amazon. Many biodiversity products of the Amazonian flora follow have established value chains. We did qualitative analysis on a sample of it to identify its main characteristics, problems, virtues and bottlenecks. This analysis included selected cases of innovative entrepreneurship leveraging relatively low-end technologies and evaluation of 25 enterprises that markets non-timber products of Amazonian biodiversity. The sample encompasses a range of segments, types, sizes and bio-assets processed.

4.1. Conceptual failures for sustainable tropical development

The challenges to achieving sustainable development in the Amazon can be broadly categorized in three categories, similarly to a conceptual framework laid out for planetary health [16]:

- conceptual failures (imagination challenges), such as the vision of the Amazon as only a source of commodities for the world and the lack of imagination to create alternative, less socially and environmentally damaging development pathways based on the Amazon's renewable natural resources (e.g., its rich biodiversity), with value added via technological innovations for an inclusive 'bio-industrial' model of development, generating higher income jobs and sustainable development.
- knowledge failures (research and information challenges), such as reduced amount of funding to research institutions in the Amazon, focus of research and monitoring systems on land use transformations, insufficient R&D investments by the private sector, and lack of innovative research, for instance, to unveil the hidden economic and societal value of biological assets, that is, a 'tropical model of development'.
- implementation failures (governance and policy challenges & entrepreneurial capacity failures), such as the failure of Amazonian countries' government to recognize the risks of current and past development policies and the inefficient implementation of a diversified economy by public and private actors and even the failure to share more equitably the benefits of the current resource-intensive economy, reducing social and income inequities.

The lack of imagination of the potential of an innovative green economy based on nature is not restricted to the Amazon regional institutions. Economic viability studies for the Amazon of serious institutions such as the World Bank almost completely ignore such potential. For example, recent studies [17] continue to see the value of forest products in an exclusively extractive way and assume very low returns. For example, less than $10 per year per hectare for non-timber products and just over $20 for sustainable selective logging. They ignore the concrete case of market success of agroforestry systems such as çaí, with proven annual returns of between $200 and $1000 per hectare [18], adding more than $1 billion annually to the regional economy [19].

The intense resource-based agribusiness, mining and hydropower in the Amazon generate wealth and little of that is reinvested to propel health and education improvements within the Amazon beyond what is called for in the licensing process. That is in part due to the regressive taxation system and in part due to historical inefficiencies in investments in public services. For instance, the highest average per capita income region in Pará—annual per capita income of close to R$50,000—is the iron ore-rich Carajás area, with overall income higher than national average. However, social indicators such as health and education services are no different than other regions of the State of Pará and much lower than national averages. In summary, very little of the wealth remains in the region and improves the wellbeing of the population.

The discourse on sustainability has been allowed to proceed as a sign of the times and to be aligned with global trends starting with the 1992 Earth Summit in Rio and to transmit an international aura of adherence, but in fact the concrete development policies for the Amazon never in fact deviated from the one devised by the military government out of geopolitical concerns: livestock and agricultural occupation to ensure sovereignty and exploitation of minerals, hydropower and fossil fuels as drivers for economic development.

The intense and swift expansion of the Brazilian agriculture frontier in the Amazon resulted not only in the growth of the country's GDP since the 1960s, but also in the rates of tree felling and greenhouse gas emissions—a consequence of conversion of forest landscapes into pasture for cattle raising and agricultural fields for grain production. Some numbers illustrate this human-orchestrated metamorphosis. Since 1997, more than 20 billion trees have been cut in the world's largest rainforest. In 2016, more than half of the 8000 km^2 of Amazon deforestation was transformed into new pastures. Currently, beef and dairy farming and production account for 45% of gross Brazilian GHG emissions.

The main public policies responsible for the sharp reduction in deforestation from 2005 to 2014 seem to have already reached their limit, so much so that deforestation has been growing in 2015 and 2016, even in a period of historic economic recession, demonstrating once again the decoupling of deforestation with economic growth, neither when GDP grows nor when GDP shrinks. The underlying reasons for continued land cover change are more complex than simply responding to global markets.

Unfortunately, we may not have a long window of time to change course with respect sustainable pathways for the Amazon. Tipping points not to be transgressed for forest-climate stability are in the horizon. The synergistic effects of land cover and climate changes, and with increased forest fires due to a combination of forest degradation, use of fire in agriculture and droughts, make the risks even greater. Earth system modeling [2] shows that the synergistic combinations of those drivers could lead to a relatively rapid transition to new forest-climate equilibrium with loss 50–60% of the forest over eastern, southern and central Amazon, replaced by degraded savannas and dry forests. The sense of urgency to avert a systemic risk to the Amazon forests must be kept in mind in the search for solutions.

4.2. Potential of a biodiversity-based bio-economy

The knowledge of nature, accumulated over 3.5 billion years of evolutionary processes, that finds in the Amazonian biodiversity one of its greatest showrooms, is a potentially very large bio-economic asset. The number of molecular substances with specific and usable functions is practically incalculable, since each existing species is itself a biochemical design laboratory. And most species are yet unknown and every 3 days, on average, one new species is discovered [20].

Even though a single substance with a desired function discovered by the study of living things in the Amazon could be biologically synthesized and produced industrially by laboratories to reduce costs or to provide quantities demanded for world consumption, the intrinsic knowledge that generated its form and function was stored in the forest and ready to be copied.

A review carried out in the scientific and technical literature as part of this work identified more than 200 species of Amazonian plants with known potential to provide raw for an initial low-end bio-economy in the Amazon. A reduced listing of the 20 very promising species that have been widely used, integrate local productive chains or show strong potential use in food, cosmetics, perfumery, medicinal, advanced materials and biotechnology have their distribution modeled. The listing includes rosewood (*Aniba rosaeodora*), Brazil nut (*Bertholletia excelsa*), cumaru/tonka (*Dipteryx odorata*), açaí (*Euterpe oleracea*) and rubber tree (*Hevea brasiliensis*) among other. A sample distribution for rosewood in the territory is shown in **Figure 5**.

Few of the biological assets of Amazonian biodiversity are known, others are being researched for their nutritional, structural, biochemical and market properties, to become products of future use.

A good example of this transition in the area of food is the açaí fruit of the *Euterpe oleracea* palm, widely and historically consumed only by local populations until the 1990s. From then on, it gained the world for its nutritional and functional qualities and its flavor, even with the operational difficulties of being a fresh, minimally processed fruit transported frozen from the vicinity of the forest to consumer markets elsewhere in Brazil and abroad (e.g., to the US and Japan) [18]. Its botanical genus (*Euterpe*) bears the name of one of the nine muses of Greek mythology, daughter of Zeus, who represents pleasure and happiness, as many consumers of açaí pulp may well attest.

Like açaí, many of the Amazonian biodiversity foods are traditionally consumed by the local population, with marked flavors and excellent nutritional properties, as well as functional foods and nutraceuticals in many cases. Camu-camu (*Myrciaria dubia* (HBK) McVaugh), for example, has 4 times more vitamin C than acerola [22]; murici (*Byrsonima crassifolia* (L.) Rich.), has excellent antioxidant properties [23], as well as açaí, that reached global markets. In addition to antioxidant activity and being a source of five types of carotenes, taperebá (*Spondias mombin* L.) is a rich source of vitamin A, at the rate of 100 g of fruit corresponding to more

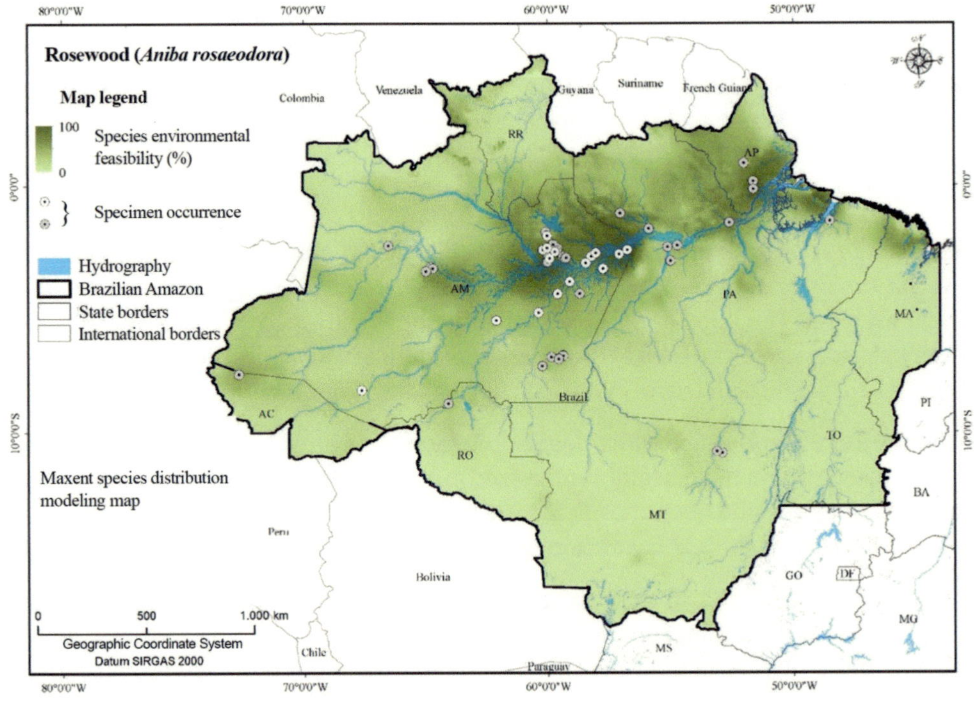

Figure 5. Geographic distribution for rosewood (*Aniba rosaeodora*) in the Brazilian Amazon [21].

than 37% of the daily needs of the vitamin [24]. Besides the well-known Brazil nut (*Bertholletia excelsa*), which is already a nut consumed worldwide for a long time, there are many other fruits and seeds of the Amazon with potential to gain new markets, such as cumaru-ferro (*Dipteryx odorata*); cupuaçu (*Theobroma grandiflorum*); uxi (*Endopleura uchi* (Huber) Cuatrecasas); graviola (*Annona muricata* L.); patauá (*Oenocarpus bataua* Mart.); guaraná (*Paullinia cupana*); priprioca (*Cyperus articulatus* L.); and bacuri (*Platonia insignis*), among many others.

The raw materials of Amazonian biodiversity are used in the industry of essences and oils to make cosmetic and perfumery products. As an example, the cumaru-ferro (*Dipteryx odorata)* fermented seeds produce an essential and industrial oil, while coumarin (*coumarinic anhydride*), which is an aromatic essence used as a narcotic and stimulant [25]. This oil is also used as a fixative in the perfumery industry [26]. Another example is andiroba (*Carapa guianensis*) available in the market in the form of essential oil, with anti-inflammatory, moisturizing, healing properties [27], being also sold for especially sensitive skin care cosmetics [28].

Açaí has also been studied and is used far beyond food: in the cosmetics sector its oil has properties for skin nutrition, revitalization and hydration, it contains omega 6, it is an antioxidant agent rich in polyphenols indicated for the formulation of anti-aging products [29, 30]. The anthocyanin present in large quantities in the açaí pulp was used in an application as a natural marker for teeth bacterial plaque [31] with large potential markets. In another development, nanoparticles of açaí oil are used to treat cancerous lesions [32]. Proving that applications of biodiversity raw materials tend to be innumerable, especially when combined with modern technological tools and cutting-edge research, a natural plastic was developed from açaí, with polyurethane produced from the seeds [33]. Discarding the abundant açaí berry seeds is a potential environmental problem in the pulp for food production cycle. The development of a plastic from the seeds also shows the possibilities of using by-products of a production chain in other associated chains for an even more efficient bio-economy with minimized externalities.

Other examples of uses of bio-composites are ucuuba (*Virola surinamensis*) from which a patented [34] butter is produced, which is capable of providing a matte effect in the skin. From the leaves and branches of the pau-rosa (*Aniba rosaeodora* Duckei), the linalool compound is extracted [35] which is one of the traditional components of the classic Channel No. 5 perfume. Currently, the following products of the Amazonian biodiversity for diversified products are on the market for cosmetics applications: Babaçu (*Orbignya oleifera*) oil, Buriti (*Mauritia flexuosa*) oil, Brazil nut (*Bertholletia excelsa*) oil, Copaíba (*Copaífera officinalis*) oil, Passionflower (*Passiflora edulis*) oil, Urucum (*Bixa orellana*) oil, Patauá (*Oenocarpus bataua*) oil, Pequi (*Caryocar brasiliense*) oil, Bacuri (*Platonia insignis*) oil, Cupuaçu (*Theobroma grandiflorum*) oil, Murumuru (*Astrocaryum murumuru*) oil and Ucuúba (*Virola surinamensis*) butter.

Research in the medical field confirms the value of many indigenous traditional medicines and goes beyond, with its own and advanced research methods [36]. As an example, we can mention the chichuá (*Maytenus guianen*sis Klotzsch ex Reissek) that presents anti-leishmaniosis [37] and anti-microbial [38] compounds; guaraná (*Paullinia cupana*) with its properties for the treatment of Alzheimer's disease [39], priprioca (*Cyperus articulatus* L.) with anticonvulsant properties [40], babaçu (*Orbignya phalerata*) with a cicatrizing compound [41], sacaca (*Croton cajucara* Benth.) with hypoglycemic properties [42] and as ulcer healing [43], pracaxi (*Pentaclethra macroloba* Willd.) with anti-hemorrhagic activity [44] and natural larvicide [45], in addition to estoraque (*Ocimum micranthum* Willd.) with its antifungal [46] and antioxidant [47] properties.

Quercetin is a flavonoid that has the ability to suppress free radicals and thereby help preserve the brain and heart, keep the immune system active, protect the body against cancer, and act to prevent diseases, especially neurodegenerative diseases such as Alzheimer's disease [48]. Quercetin, present in many foods but in low concentrations, is obtained from the natural purification process of the fava d'anta (*Dimorphandra mollis* Benth) [49]. And the uncera (cat's claw) (*Uncaria tomentosa* and *Uncaria guianensis*) and is largely used in the pharmaceutical industry [50]. Pilocarpine, an alkaloid with extensive use in ophthalmology [51], is extracted from jaborandi (*Pilocarpus microphyllus* Stapf ex Holm). These are many other examples of species already studied that integrate or can integrate local production chains in the production of drugs and phytotherapeutics.

But the biological assets also have application in industry, with emphasis on endophytic fungi (*Coniochaeta lignaria*, for example) with the capacity to degrade lignin in the cell walls of plant cells, with great potential for the bioenergy industry [52].

Another study with phytosterols isolated from endophytic fungus (*Colletotrichum gloeosporioides*), an Amazon fungus, offers potential sources of novel natural products for exploitation in medicine, agriculture and the pharmaceutical industry [53]. Microorganisms are an attractive source of new therapeutic compounds, they serve the ultimate readily renewable, and inexhaustible source of novel structures bearing pharmaceutical potential [54].

State-of-the-art research can unveil new and surprising uses even for forest assets that have been exploited for a long time. For instance, that is the case for natural rubber (*Hevea brasiliensis*). When combined with nanoclay composites using biomechanical technology, it results in an advanced material to be utilized as artificial skin (Biocure)—a patented active material that induces the formation of new blood vessels (angiogenesis) and new tissues (neoformation) on the surface on which it is applied [55]. Latex and clay compounds have also been developed to manufacture high-tech tire (run cooler, thus increasing tire durability and fuel economy), anti-rust coatings, tennis balls, gloves and masks [55].

4.3. Summary of Amazon value chains

The biodiversity products of the Amazonian flora follow well-defined paths between the origins of the raw material, to its processed form for final consumption or to be reprocessed into components for very high specialty products. The value-added paths of biodiversity products involve multiple steps and social and business actors, varying according to the nature of the raw material, the products to be processed and the location of the harvesting and processing regions. As a general rule, the production of the raw material, which may be fruit, seed, sap, or other part or component of the plants occurs in the rural environment. They may come from primitive areas of natural forest or managed agroforestry systems (SAF), such as natural forests with extensive extractive species and intercropped planted forests.

The rural area is home to communities where the first basic stages of preparation of the material collected or harvested for subsequent supply occur, such as cleaning, threshing, drying and other low-tech processes. Logistic processes such as the transport of the material from the collection and production sites to the pre-processing sites, storage and shipment to the processing centers also occur in the rural domain. In every aspect of this beginning of the value chain, there are opportunities for individual, family, cooperative or business based on local entrepreneurship.

After pre-processing, the materials are taken by boat or truck to companies or cooperative facilities in the Amazon or in another region in Brazil or in other Amazonian countries (e.g., Bolivia) where most or the entire product's actual processing takes place, in facilities with varying degrees of automation. From there it is ready for consumption, locally or in markets elsewhere in Brazil or abroad. Based on a comprehensive study we conducted with value chains of five plant species, we developed a conceptual diagram that represents the main places, environments and activities carried out throughout the whole transformation cycles, from inputs origin to final consumption, as shown in **Figure 6**.

Not all paths shown in **Figure 6** have fair remuneration in the value adding they represent. A 2005 study for cumaru (*Dipteryx odorata*) value chain in the State of Pará, Brazil, illustrates the problem [56]. The markup was 75.0% for the intermediary, 166.7% for wholesale companies in towns nearer production areas, and 233.3% for the wholesale companies from Belém, the State capital. The total markup from the beginning to the end of the market chain was approximately 500%. The price of the nut ranged from R$3.00 per kg for the collectors to R$18.00 per kg for the wholesale companies. It was observed that the exporting companies, which generated unequal gains within the chain, imposed the major additions to the product price. There were approximately 2700 families involved in cumaru nuts collection, exported mainly to Japan, France, Germany and China.

Another evidence of such imbalance in the value sharing was revealed by our study of five value chains. While Brazil nut (*Bertholletia excelsa*) seeds, mostly from manual forest extraction, come from dozens of places along the Amazon basin, the value aggregation of such yields takes place only on just a few locations furnished with processing plants, as shown in **Figure 7**.

Recent corporate social responsibility efforts focused on purchasing biodiversity products from communities or cooperatives have generated more balanced and fair-trade relations, as

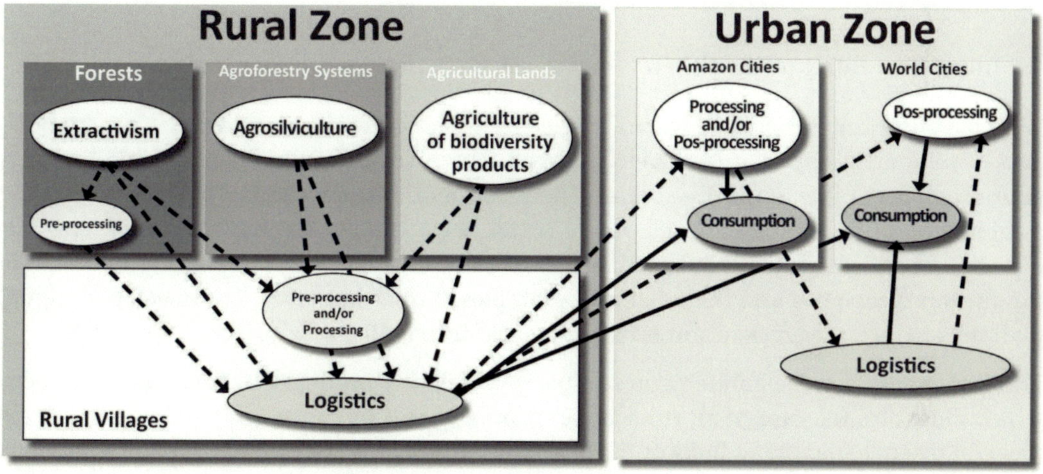

Figure 6. Conceptual diagram of the location of the basic stages of value chains of Amazonian biodiversity products. Solid lines mean the last stage of a product in the value chain [21].

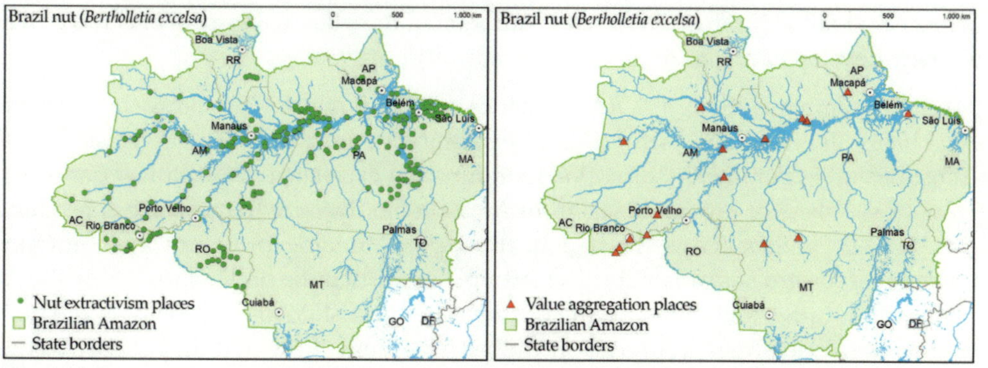

Figure 7. Differences between the many places where Brazil nut (*Bertholletia excelsa*) is collect in the forest (left map) and the few places where there is value aggregation of it (right map), in the Brazilian Amazon, found in a sample survey [21].

with the operations of a range of forest products purchased by the Natura company and açaí purchased by the Sambazon company. However, the typical market distortions in the values paid to the extractivist-producer by intermediaries still has to be resolved.

Agroforestry systems (SAF—Sistema Agroflorestal) are agricultural crops intercropped with tree species, used to restore forests and recover degraded areas. The SAF technology overcomes terrain limitations, minimizes degradation risks inherent in agricultural activity and optimizes the achieved productivity. There is a reduction both in soil fertility losses and pest attacks. The use of trees is fundamental for the recovery of ecological functions, since it allows the reestablishment of much of the relationships between plants and animals. The tree components are inserted as a strategy to combat erosion and the contribution of organic matter, restoring soil fertility. Two successful tropical agroforestry projects illustrative of this system in the Amazon are the CAMTA cooperative [57] in Tomé Açu, in the state of Pará and the RECA cooperative [58] in Abunã, in the state of Rondônia.

4.4. Innovative entrepreneurship leveraging relatively low-end technologies

With the advancement of consumer markets, technologies and business models, new business development opportunities have emerged from the products of Amazon biodiversity. Four examples of this innovative entrepreneurship model were selected to demonstrate the combination of technology and corporate social responsibility for the generation and fair distribution of benefits to all links and actors of value chains. Two of the examples illustrate production companies and the other two examples show companies that developed digital platforms to increase efficiency in transactions and traceability of biodiversity products.

The first example is the Tahuamanu company, a Bolivian producer of Brazil nut products, which illustrates the case of an Amazonian company that innovated by applying relatively low-end technologies, to all links of the Brazil nut productive chain, reflected in tremendous increases in productivity and benefits also to collectors at the base of the value chain. The 2016 severe El Niño-related drought in many parts of the Amazon may have wreaked havoc to the Brazil nut production that supplies the company. It is reported a 70% drop in harvest in 2017, responsible for laying off over 300 employees from his Cobija processing plants [59]. This unprecedented fall in production raise the question of the potential impact of climate change on the new development paradigm for the Amazon.

The second example is the NATURA cosmetics company and its bio-industrial operations. It is probably the most successful case of exploration Amazon biodiversity assets within the most desirable parameters of socio-environmental excellence. Natura has developed a network of suppliers of raw materials from Amazon biodiversity that organizes production of almost 3000 families across the region. It supports training programs and community empowerment toward sustainability. The example of the ucuuba butter shows how the combination of innovative R&D and training communities in sustainable exploitation can deliver good results. Ucuuba trees were used as timber for broom sticks and that was accelerating risks of tree extinction. Butter was developed out of the ucuuba seeds and that new product found its way in cosmetics of high added-value. Floodplain communities of the Marajó Island were trained to collect and pre-process the seeds for sale to Natura and to other companies which also process ucuuba butter. The net profit of those operations for those families is three times larger per year as compared to the only once income for felling the tree. Natura is also promoting the bio-industrialization in the Amazon itself. It opened the Ecopark, an industrial complex in Benevides, near Belém, state of Pará.

The third example is the FLORAUP digital platform that shows how information technology can be used to foster direct connection between local producers, from their remote locations in the forest, with potential buyers of their Amazon biodiversity products. After 1 year on air, the platform has only 57 registries, perhaps due to the relatively low digital connectivity of remote communities across the Amazon.

Finally, the fourth example is ORIGENS BRASIL, a production chain tracking digital platform. The platform allows anybody to know instantly the origin of the product that contains assets of Amazonian biodiversity since its raw material harvesting, its history and actors involved in the production. This is done simply by pointing a smartphone to the product packaging, which is equipped with a QR Code that accesses a remote live database. If one assumes that responsible consumers are an accelerating trend, such traceability platforms are in dire need for the Amazon.

4.5. Traditional bio-industries

Natural products developed on a sustainable basis have a long history in the Amazon since the rubber boom years. An increasing demand for these products for traditional and innovative uses in the food, cosmetics, perfumery and pharmaceutical industries has promoted new business opportunities in the Brazilian Amazon. As part of this trend, advances in biotechnology research have demonstrated a key role in expanding this potential, thus boosting the value chains that have as one of the main attributes the bio-industries focused on the processing of forest raw materials into biodiversity products.

This research evaluated 25 enterprises that markets non-timber products of Amazonian biodiversity. The sample encompasses a range of segments, types, sizes and bio-assets processed. From international corporations with more than 100 years in the market of extracting the finest Amazonian essences, to innovative indigenous entrepreneurship of collecting and selling forest's native species seeds in large amounts to support much needed reforestation efforts elsewhere.

These industries deliver a vast array of products: It ranges from an exfoliating agent of açaí seed (Beraca company) to a powder form of the same fruit for energy drinks (Yerbalatina Phytoactives and 100% Amazônia companies). The Amazon-based bio-industry is also well-defined and consolidated in the supplying chains of oils and essences. As early as 1921, the essential oil extracted from the pau-rosa (*Aniba rosaeodora*) wood, a native tree from the Amazon, which is rich in the aromatic compound linalool, was the main ingredient of the famous French perfume Chanel n° 5 [35]. From them on, the supplying of the finest and unique ingredients from the Amazon biodiversity thrived, adopting, mostly, adequate standards for social and environmental sustainability, which was not always the case with Pau-Rosa. Today, extracts of cumaru are present in the most famous and popular fragrances (Givaudan company) and the ingredients market for the cosmetics industry is supplied with essential oils of priprioca (Laszlo Aromaterapia & Aromatologia companies), pracaxi (Amazon Forest Trading company); copaiba (IFF—International Flavors & Fragrances company) and andiroba (Amazonoil company), among many other.

Another sector that has shown significant growth is the food, functional food and nutraceutical industries (e.g., Sambazon, Tahuamanu companies). Companies in this sector tap in the healthy food market and, by applying relatively low-end technologies, have put Amazon bioactives available worldwide at anyone's table. As a rule of thumb, most sectors have benefited from the adoption of newer and accessible technologies in their processing facilities. From Brazil nuts micro-factories for peeling seeds (COOPERACRE cooperative) to agrosilviculture producer's cooperatives focused on traditional bio-industries (CAMTA, RECA cooperatives).

In our study, we analyzed many products offered by the Amazon traditional bio-industries based on two defining axis: the amount of technology involved in the making of their products and the degree to which they are closer or further to their original state as furnished by Nature. It was a qualitative analysis and it shows status classes for these products. The diagram in **Figure 8** shows the result of this qualitative analysis.

As it might be expected, values such as environmental sustainability, social development and fair-trade are a matter of concern for virtually all operations, to a greater or lesser extent, from small chestnut cooperatives to the giants of the essences and cosmetics sector. Nevertheless, there are reports of large traditional bio-industry operations that required botanical resources at large scales that have driven transformation in the supplying of natural asset, once coming from extractivism or agroforestry systems, into an asset generated from monocultures in the agroindustry's usual patterns. It also disrupted traditional handmade extractive processes [60]. Accommodating increasing demands for bio-products with limitations inherent to Nature's carrying capacity and traditional and local people culture, needs and potentials

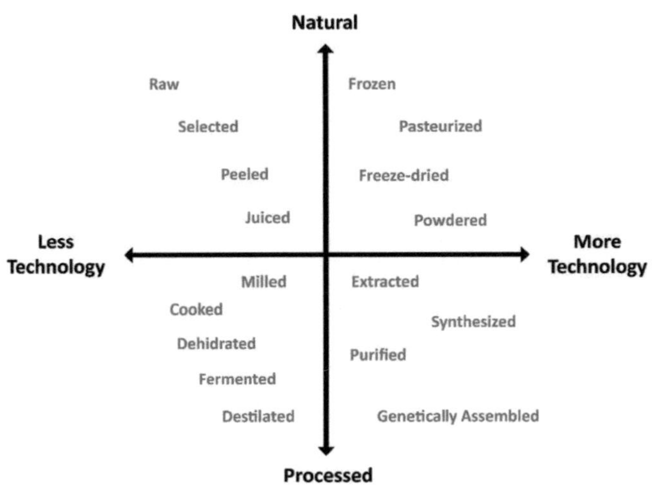

Figure 8. Diagram depicting status classes for Amazon bio-industry products based on the amount of technology involved in their making and the degree to which they are closer or further to their original raw material state as furnished by nature [21].

for insertion into new economic development paradigm is an imperative challenge for a real sustainable Amazon development strategy.

The industrial sector transforming biodiversity assets into available consumables act in the interface between biodiversity, biotechnology and bio-industry, which involves a complex system of partnerships between companies, universities, research institutes, official financial agencies, organized communities and cooperatives inside and outside the Amazon region.

5. Amazonia Third Way as a disruptive alternative

The *Amazonia Third Way* initiative is conceived as a disruptive social and technological transformation toward a sustainable Amazonian development path. It calls for *'an Amazon-specific Fourth Industrial Revolution innovation (4IR) "ecosystem". This system must be able to rapidly prototype and scale innovations that apply a combination of advanced digital, biological, and material technologies to the Amazon's renewable natural resources, biomimetic assets, environmental services, and biodiverse molecules and materials'* [2].

In support of socioeconomic development, systemic innovations will also apply to enhancing biodiversity-based value chains. Ideally, these would shape a unique 'Amazon-brand' able to conquer global markets [61–63].

The Amazonia Third Way Initiative promotes in-depth research on alternative pathways for sustainably developing the Amazon territory, in harmony with the twenty-first century's Zeitgeist. Forests in the Amazon are the result of evolution over millions of years. Nature has developed a wide variety of biological assets, which include metabolic pathways, and genes of life on land, in aquatic ecosystems, and in their natural products—both, chemical and material—in conjunction with biomimetic assets, that is, the functions and processes used by nature.

4IR technologies increasingly harness these assets across many industries from pharmaceutics to energy, food, cosmetics, materials and mobility. Indeed, they are making profits, but to date these profits have not been channeled back to conserve the Amazon and to support the custodians of nature—indigenous and traditional communities—and also urban population in the region.

Within a proper legal and ethical framework, the Amazonia Third Way Initiative offers unprecedented opportunities to local populations to develop a vibrant, socially inclusive 'standing-forest, flowing-river' green economy. By harnessing nature's value through physical, digital and biological technologies of the 4th Industrial Revolution, we can simultaneously protect the Amazon ecosystems and their traditional custodians.

The region is still largely disconnected from the main centers of technological innovation dealing with 4IR technologies and the advanced bio-economy. The Amazonia Third Way Initiative is conceived as a multi-level path toward a new inclusive bio-economy, combining a highly innovative, entrepreneurial and technological economy with the re-valuation of non-timber forest products and industries with low-end technologies.

5.1. Determinants of sustainable development pathways for the Amazon

The conceptual framework for the *Third Way* follows the overall structure of **Figure 9** for the determinants of sustainable pathways for the Amazon.

At the broader level, **first** we need to understand the nature of the socioeconomic and political drivers accounting for the rapid transformation of the Amazon in the last 50 years and the consequences of the resource-intensive development policies in action in contrast with the view of forest preservation and setting aside large tracts for conservation.

As mentioned before, the Third Way Initiative is not one more attempt to reconcile resource-intensive development with conservation. Instead, it will seek to implement the twenty-first century paradigm of knowledge societies to Amazon realities through research and development, entrepreneurship, twenty-first century skills and education, and fit for purpose sustainable development policies toward a standing forests-flowing rivers inclusive bio-economy.

Second, we deal with solution spaces, recognizing that an important effort has been done to identify and diagnose the risks to the Amazon of the current development actions and policies, including their fragilities. We are in urgent need to find feasible solutions of a different nature: driven by communities and by an entrepreneurial revolution powered by the Fourth Industrial Revolution and not only by powerful legacies, assisted by altogether more sustainable policies based on knowledge, be it scientific/technological or traditional.

Third, we discuss in more detail the role of some key enablers and catalysts to jumpstart sustainable pathways for the Amazon in two categories, those to enable a biodiversity-based

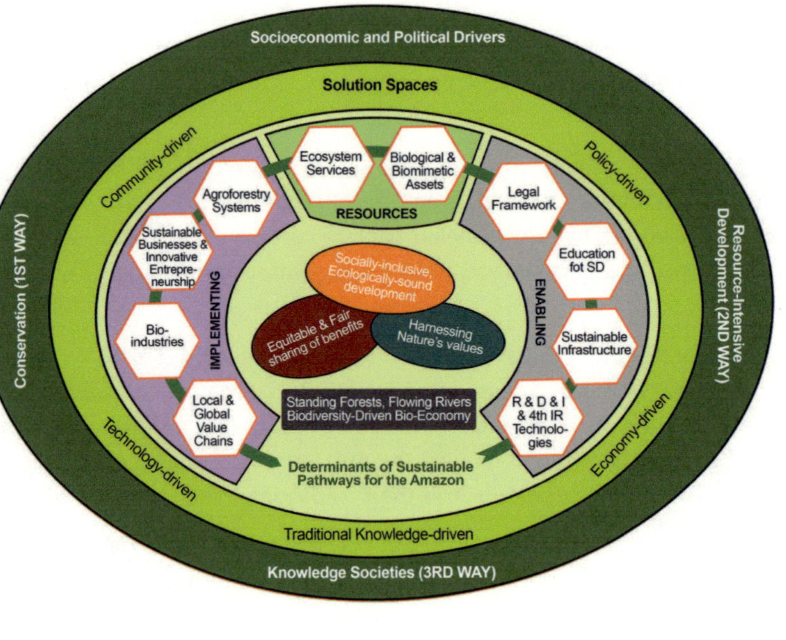

Figure 9. Determinants of sustainable pathways for the Amazon. The Amazonia third way initiative seeks *'to add value to the heart of the forest' by promoting a* novel sustainable development paradigm based upon harnessing biological and biomimetic assets of Amazon biodiversity.

development, namely research, development and innovation; harnessing the Fourth Industrial Revolution technologies to unlock the economic value of nature; and conducive regulatory framework; and those necessary to implement such novel paradigm, agroforestry systems; innovative entrepreneurship; bio-industries; product-based and knowledge-based value chains.

5.2. Fourth Industrial Revolution and innovation ecosystems in the Amazon

Within the Amazonia Third Way initiative, an approach has been developed to operationalize the principles and practices that will allow a proposed paradigm shift for Amazon sustainable development. It defines seven interconnected realms: (1) the existing natural knowledge; (2) the ability for learning from nature; (3) the capacity to applying biodiversity-based knowledge to human needs; (4) the capacity to producing biodiversity-based goods and solutions; (5) the insertion of biodiversity-originated products on a local-to-global bio-economy; (6) the fair sharing of socioeconomic benefits and life quality improvement for all; and (7) the rising of an Amazon Biome intrinsic valuing. With the advancements of 4th Industrial Revolution (4IR) technologies and its wide accessibility, we identified ways it can interact and make feasible a game-changing realization of such realms. We call '**Amazonia 4.0**' the prospects of realization of these seven elements by means of technological accessibility and resources, and market transformation made available by the 4IR.

The existing Natural knowledge is an initial condition of the system; it does not depend on any human technology. It is a source of information we inherited from evolutionary processes, occurring associated with 3.7 billion years old life on Earth. The A3W initiative targets to keep it going its course, valorizing it in many ways.

Learning from Nature is inherent to humans ever since we became a species (*Homo sapiens*) as a part of the Natural system. Ancient and traditional knowledge come greatly from observing and interacting with the natural elements. As we evolve, we became more apt to understand Nature's intrinsic knowledge with the building of science and its instruments. With 4IR technologies, which include biotechnology, advanced computing, genomics, nanosciences, materials science and advanced sensor platforms, we can learn from Nature in a depth and such fast pace never imagined before.

Applying knowledge from Nature to human needs is the next natural consequence. This is the realm of invention and innovation. 4IR technologies can boost invention and prototyping of new products and solutions. More than just facilitating invention, it creates demand for new solutions, advanced materials and innovative products.

Once a new biodiversity-based product or solution is developed, producing it in varying scales is the next outcome. It may utilize biodiversity inputs directly on its making or can only be sourced from biodiversity knowledge. To carryout industrial operation in the Amazon has been always a challenging, if not impossible, operation. With the changes brought by 4IR technologies and market demands, industrial equipment became smarter, lighter and customizable. It became possible to have plenty of electrical solar-powered energy in the forest, with equipment connected with satellite internet and local crews trained with virtual and augmented reality, for example. With 4IR technologies, including advanced sensors and AI, it is possible to control more precisely the use of natural resources to prevent possible negative impacts.

Insertion of biodiversity-originated products on a local-to-global bio-economy is a key for driving wide interest in conserving the bio-assets. Different than the traditional model of supplying commodities for further processing and generating value away from its origins, 4IR technologies and new manufacturing paradigm eases and redefines the possibilities to produce in close association with the local people on local environments, yet reaching global markets. Complicated logistic typical of a vast forest territory can be easily offset using self-flying cargo drones, for example.

Fair sharing of socioeconomic benefits and life quality improvement for all involved, including forest stakeholders and final consumers can be levered by 4IR technologies and social changes brought by the technological revolution. With distributed ledger technologies like blockchain and holochain, we propose the creation of the Amazon BioBank. It is a framework for attributing value to many instances of Amazon socio-biodiversity. Biological assets, biomimetic insights and discoveries, traditional knowledge, local people forest skills and other sources of resources will be registered in the Amazon BioBank digital platform through holochain distributed ledger technology [64]. The Amazon BioBank share common principles with the Earth Bank of Codes [65].

Aside from any specific technology, the ultimate, long-term result of these chain of events and realizations would be the rising of a socially shared Amazon Biome intrinsic value. The social valuing of Nature and its knowledge as an end in itself is an ideal state of relationship between humans and other elements of the natural system. By becoming acquainted and perceiving many times actual benefit from products and solution based on the Amazon biodiversity, made available by the chain of events depicted above, one can realize the value of the tropical forest. As a utilitarian value first, that over time may crystalize as core life, intrinsic value, forming the personal and social foundations to hold attitudes and behaviors that imply, support and demand conserving the Amazon Biome.

The 'innovation ecosystems' proposed in the Amazonia Third Way initiative are creative-productive arrangements based on the Amazon 4.0 principles that synergistically align several '*ignition powers*' for a novel Amazon bio-economy. Major research laboratories and universities are knowledge centers on biodiversity. Processes, molecules and genetic information with potential for diverse uses are discovered on daily basis. Start-ups are companies that specialize in rapidly transforming knowledge into business that tends to transform traditional consumer and service markets. Prospects for the industries with Internet of Things, or 4.0, announce new products to be created with computational tools, to be 'uploaded' and produced at any scale. Inventors and new businesses can idealize customized or niche-specific products, which are done automatically, even overnight. A dynamically well-developed and structured environment for locally rooted associations of (1) knowledge, (2) business and (3) production form the 'innovation ecosystems'. They are a way for transforming the biological wealth of the Amazon into economic wealth, locally anchored, with social benefits for communities and sustainable mechanisms for conservation of the forest.

5.3. Capacity development as a necessary condition for the Amazonia Third Way initiative

To begin to walk down the *Third Way* we need, above all, capacity development.

As results of the long-standing Program to Protect the Rainforests of Brazil (PPG-7) show, the lack of entrepreneurial skills has stood in the way of developing a non-timber bio-economy in the Amazon. Only with field-based knowledge and supporting academic curricula can tap into the Amazon's biological and biomimetic assets, and the mainstreaming of a standing forest-flowing river, biodiversity-based bio-economy be achieved. To do that, we propose the development of a capacity program 'Amazon Creative Labs' (ACL). The program is designed to promote technical, technological and entrepreneurial capacity development focused on non-timber products of the Amazon biodiversity, with training events carried out directly at local communities and towns throughout Amazon region.

We propose the launching of Amazon Creative Labs (ACLs)—laboratories for innovative experimentation set up throughout Amazonia. They will provide intensive training linked to local potentials to generate a virtuous insertion on bio-economy-related new opportunities. Typically, Creative Labs will be located in smaller communities, villages and towns, assembled on tents or on floating platforms packed with state-of-the-art equipment and technology for both, wide audience learning processes and core value chain local development.

Amazon Creative Labs will enable development of small-scale innovation ecosystems for co-design, co-development and co-creation of solutions and applications, serving as an effective interface with the knowledge and practices of the Amazon people.

The Amazon Creative Labs will operationalize sustainable 'Solution Spaces' (see **Figure 1**). It is of critical importance that the Labs be community oriented, joining technology and traditional knowledge, and designed to contribute toward a strong local and regional economy.

The Labs will promote capacity development activities focused on a number of products of Amazon biodiversity illustrative of an array of bio-economic and even bio-artistic applications, such as food, nutraceuticals, cosmetics, fragrances, pharmaceuticals, industrial oils, art crafts, bio-art, biomimicry, etc. Training activities can enable local communities to gather more information on the natural resources available to them, including the use of high-end technologies such as, genome sequencing.

The exposure to 4IR technologies will allow innovative concepts to emerge. With the assistance of technology experts on the one hand, and entrepreneurship specialists on the other, groups of participants from Amazonian communities, villages and towns will be invited to develop new applications and to prototype (at least digitally) such innovations. The Labs' creative environment will bring 4IR concepts like mass customization, democratized invention and smart & autonomous factories, powered by Industrial IoT, to a meaningful level with practical outcomes accessible at planned local and regional clusters of custom-sized processing and manufacturing plants.

Alongside communities—forest people, riverine communities and agroforestry farmers—young undergraduate or just graduated students interested in creating sustainable biodiversity-based businesses in the Amazon will be engaged. The expectation is that such 'on the ground' collaboration will give rise to new partnerships.

The Amazon Creative Labs design includes solar photovoltaic panels, convertors and batteries, for steady power supplying, and connection to broadband satellite internet. These features will allow digital, internet-connected equipment to work for prototyping potential applications of new products and processes. These infrastructures, operating in remote regions of the Amazon, are also proof of concept of how the newest available and accessible technologies can reach and benefit the whole spectrum of the social pyramid, from their everyday life to new work opportunities.

ACLs also include a focus on the realm of biomimetic, that is, the functions, processes and mechanisms of living organisms that, once learned, can provide insights and solutions for engineering new technologies and innovative products. They also leverage applications, including the high-end of genetic resources and genomics; prototype innovative processing of materials through the diverse links of value chains—raw materials, intermediate products, all the way to finished products.

To illustrate the potential of ACLs, we designed the three following conceptual examples of applications, based on currently available technologies and equipment. A final design should incorporate new technological solutions specifically tailored for solving implementation and scaling challenges and include consultation with local communities for accessing their specific needs, priorities and potentials.

A line of Amazon Creative Labs will deal with value chains feed by inputs from local biodiversity and an example of that is themed after nutraceutical Cupulate, a chocolate made from the seeds of Amazon fruit Cupuaçu, instead of cacao. From forest picking to creating a final product that combines basic Cupulate with other products of very high nutritional value, the lab also includes utilizing a 3D food printer for unique chocolate designs and precise dosage of the added natural micronutrients. A by-product of Cupulate-making is cupuaçu pulp, which is then freeze-dried in a value chain of its own. Heavy-lift electric-powered drones can help overcome logistics challenges the region poses, by easily and quickly taking loads of nutraceutical cupulate sculptures and bars to a nearby gateway.

Another example of ACLs focus is the Brazil Nuts value chain, known for the discrepancies between its higher cost for consumers and the low remuneration local people who harvest it from the forest receive. To change this, in one end, the ACLs will target extractivism issues, like processes precariousness that halts productivity and seeds' price, with accessible technological resources including GIS mapping, micro-controlled sensors arrays (for health safety on seed's harvesting and storing) and comprehensive traceability systems (origin and processes). At the same time, ACLs will carry out further locally based nut processing, using equipment that extracts oil and flour, by-products with greater trading value. With top technical education and processes precisely controlled with the aid of computers, sensors and biotechnological checks for sanitary standards, it becomes possible to output export-grade quality products strait from the forest vicinities. Those inputs also allow bringing to small villages the manufacture of even more processed products targeted to the natural cosmetics and nutraceuticals markets.

Another line of ACLs will tackle the potential of making Amazon local inhabitants aware of the genetic value of biodiversity and to take part in genome sequencing projects. The lab will take participants into a knowledge journey departing from the biodiversity that can be seen all the way to the microscopic and nanoscopic structures of it, and to the grasping of the molecular coding of life. To achieve this, the Lab will make use of optical and portable electron scanning microscopes and virtual and augmented reality gear, furnished with contents to experience and understand organic chemistry complex structures. At the end, participants will carry out actual DNA sequencing through ultra-portable genome sequencers, allowing for registering genomes of species and benefiting from the provisions of benefit sharing of the Nagoya Protocol of Access and Benefit Sharing (ABS).

6. Discussion and conclusions: envisioning the future for the Amazon

Systemic risks to the maintenance of the Amazon forest due to the synergistic combination of the main human drivers of change—namely regional climate change due to both deforestation and global warming, and augmented forest vulnerability due to fires—poses an urgent challenge to avoid an irreversible threshold being transgressed that would threaten to turn over 50% of the forest in degraded savannas in the second half of this century [2].

The natural resource-intensive mode of development (the Second Way) is the dominant mode of development and receives generous government subsidies for its continued advancement. Investments in conservation, forest restoration and a sustainable economy in the global tropics of about $20 billion annually receive less than 3% of total investments. The bulk of investments (around $770 billion annually) goes to the expansion of commodities frontier of cattle, grains, oil palm [66] and also to road, energy and mining infrastructure, which are also key drivers of deforestation [67]. One more detrimental effect of such path is the increasing rural violence in the Amazon. Brazil has the highest number of assassinated rural and environmental leaders since 2015, with more than 140 killings, mostly in the Amazon [68].

It is becoming crystal clear that trying to reconcile resource-intensive development with conservation is not leading to lasting and permanent solutions. Deforestation rates are still very high and do not show a tendency to go down near zero and rural violence is on the rise. Social inequalities in the Amazon remain high and are not improving at a fast pace at least to bring social indicators to the national averages of the Amazonian countries. Imposing strict conservation to protect large swathes of the forest has had clear successes over the last decades in the Amazon—about 50% of the Amazon forest is under some kind of protection. However, that in itself does not guarantee protection forever for tropical forests and eventually may affect the livelihoods of local population as is the case documented for Madagascar [69] who may bear a high cost for forest conservation.

The Amazon Third Way Initiative seeks to demonstrate the urgent need for a conceptual, educational and entrepreneurial revolution—a revolution based on knowledge, traditional and scientific. The current economy of meat, grain and timber in the Brazilian Amazon is less than $10 billion a year. The economy associated to biological assets of Amazon biodiversity in a few industries (food, cosmetics, oils, etc.) is already worth 30% of that and distributes income in fairer ways and benefits more of the local population. However, that is a tiny portion of the potential of a sustainable economy hidden in the biological and biomimetic assets of Amazon biodiversity that the Amazon Third Way initiative attempts to address and give visibility to. We will be estimating the real hidden economic value of these assets in a next phase of the initiative.

The Amazon forest is not a void of human presence. Diverse communities live all over the region. Even some communities of new settlers of the 1970s and 1980s have looked to find ways of generating income in agroforestry systems. There is rich traditional knowledge in many of indigenous and caboclo communities. Supporting the diversity of communities and economic pathways for a standing forest-flowing rivers economy is mandatory.

From a more general standpoint, sustainable development pathways based on natural resources exploitation should in principle put the local populations as priority. That is not the case for the Amazon currently (low HDI and other social indicators). Therefore, the Third Way Initiative also proposes that new sustainable paradigms have the development policy as a central tenet. The sustainable economy should first and utmost be means of wellbeing to the Amazonian people. That is not the case of the Second Way, where the Amazon is seen important for intensive resource exploitation for the Amazonian countries as a whole and taxation of the resource wealth should redistribute benefits as public services for all in the Amazon. However, a regressing taxation system does not realize that.

The Amazon has a number of good examples of biology laboratories and a number of entrepreneurship initiatives that beyond economic development target social responsibility and deployment of sustainable biodiversity value chains. They are true pioneers into the new era of sustainability. However, they are as yet a small minority. They may even accrue national and international visibility and are role models, but in critically insufficient numbers to create momentum economically and socially to give clout to the rupture needed to put Amazon on a different track.

The new model must rely on these existing good examples, on the diversities of forest communities across the Amazon, on state-of-the art knowledge generations laboratories and innovative entrepreneurship and build up from there.

In due course, one has to build up momentum for enhancing the policies that are necessary to uplift the Third Way; investment in zero-deforestation value chains; reducing the enormous subsidies for commodities that drive deforestation; but as importantly invest in knowledge generation through a network of advanced biology laboratories in the Amazon, in Amazonian Countries and internationally in association with private R&D labs and science-based start-ups and creation of innovation ecosystems throughout the regions. That is a pre-requisite to the development of local next generation bio-industries in towns and cities of the future.

By attracting venture capital and productive investments both for R&D and for industries, the political interest in the Third Way will rise in the eyes of governments to a tipping point in which government investments and subsidies will start to flow to this other type of economy, even on the absence of visionary governments that would see the potential of a new Amazon bio-economy and would design the pathways to reach it.

The implications of harnessing the Fourth Industrial Revolution to unlock the economic value of the Amazon's biological and biomimetic assets for governments, start-ups, corporations and R&D centers are profound. Partnerships among public and private R&D innovation labs to create a number of hubs of innovation throughout the region is necessary. This would accelerate new research and development leading to new products and innovations relevant for many industries locally and worldwide. Amazonian countries with immensely valuable natural assets would have an additional source of income to help protect these resources and support indigenous and traditional communities. These funds would create a new incentive on the part of communities and governments to protect rather than destroy natural habitats. The interest in understanding and sustainably using our biological and biomimetic assets could propel a new era of scientific exploration of life on the planet. Large new markets for sustainably sourced innovation could be created. Technology companies and start-ups seeking to demonstrate compliance with the Nagoya Protocol could be certified, through the transparency that distributed ledger technology offers.

In sum, development policy in the Amazon has historically taken two pathways. The first embraces nature conservation and protects large swathes of territory from any human activity. The second approach has focused on conversion or degradation of forests for the production of agricultural commodities like meat and soya or tropical timber at the forest frontier, and also mineral commodities and the build-out of massive hydropower generation capacity. These uses together have been historically responsible for the massive deforestation of the Amazon.

There is, however, a Third Way within reach in which we aggressively embrace high-tech innovation and look at the Amazon as a tremendous source of biological and biomimetic assets that can provide new, innovative products and services for current and new markets. System-level change in the Amazon as proposed cannot be executed single-handedly. On the contrary, we are proposing collaboration with leading public, private, academic and philanthropic actors for the journey ahead, engaging Indigenous and traditional communities across Amazonian countries, uniting the best capabilities of regulators, R&D centers, universities, technology start-ups and visionary companies all over the world.

The Amazonia Third Way can be the most effective Land Use Change Planning policy for the Amazon because it is fully based on a standing forest-flowing river bio-economy. If successful, this new development model can be applied to all tropical regions helping to preserve the Earth's great biological diversity. We have an important choice to make. The future of the Amazon and its impact on the planet lie so clearly in the balance. Time is not on our side, but we can still choose the Third Way.

Acknowledgements

This work has been supported by the National Institute of Science and Technology for Climate Change via CNPq Grant Number 573797/2008-0 and FAPESP Grant Number 2008/57719-9 and additional financial support by the Climate and Land Use Alliance (CLUA) and Moore Foundation. We express our thanks to Juan Carlos Castilla-Rubio and Luciana Castilla for their contributions to the development of the Amazon Third Way Initiative.

Author details

Ismael Nobre[1]* and Carlos A. Nobre[2,3]

*Address all correspondence to: nobreismael@gmail.com

1 Independent Consultant, São Paulo, Brazil

2 Institute of Advanced Studies/University of São Paulo, São Paulo, Brazil

3 WRI Brazil, São Paulo, Brazil

References

[1] Lovejoy TE, Nobre CA. Amazon tipping point (Editorial). Science Advances. 2018;**4**. http://advances.sciencemag.org/content/4/2/eaat2340/tab-pdf

[2] Nobre CA, Sampaio G, Borma L, Castilla-Rubio JC, Silva JS, Cardoso M. Fate of the Amazon forests and the Third Way. In: Proceedings of the National Academy of Sciences. Sep 2016;**113**(39):10759-10768. DOI: 10.1073/pnas.1605516113

[3] RAISG. Amazônia 2017—Áreas protegidas e territórios indígenas. Available from: www.raisg.socioambiental.org

[4] Adeney JM, Christensen NL, Pimm SL. Reserves protect against deforestation fires in the Amazon. PLoS One. 2009;**4**(4):e5014

[5] Crisostomo AC et al. Terras Indígenas na Amazônia Brasileira: Reservas de Carbono e Barreiras ao Desmatamento. Brasilia, Brazil: Inst Pesquisa Ambiental Amazônia; 2015

[6] ISA—Instituto Socioambiental. Terras Indígenas no Brasil. 2015. Available from: https://pib.socioambiental.org/pt/c/terras-indigenas/demarcacoes/localizacao-e-extensao-das-tis

[7] ISA—Instituto Socioambiental. Desmatamento cresce 32% nas Terras Indígenas da Amazônia brasileira, aponta ISA. 2017. Available from: https://www.socioambiental.org/pt-br/noticias-socioambientais/desmatamento-cresce-32-nas-terras-indigenas-da-amazonia-brasileira-aponta-isa

[8] Conservation International. Indigenous Land Protected Area. 2013. Peru. Available from: http://www.conservation.org/publications/Documents/CI_WCC2016_Amazonia-Mapping-Natural-Capital.pdf

[9] RAISG. Deforestación en la Amazonía (1970-2013). 2015. 48 p. Available from: www.raisg.socioambiental.org

[10] Fonseca A, Salomão R, Ribeiro J, Souza Jr C. Ameaça e pressão de desmatamento em Áreas Protegidas: SAD de agosto a outubro de 2017. p. 1. Imazon. Belém. 2017

[11] IIRSA. Informe de la Cartera de Proyectos del COSIPLAN 2017. Available from: http://www.iirsa.org/admin_iirsa_web/Uploads/Documents/CARTERA_DIGITAL_INGLES.pdf

[12] Aguiar APD, Vieira ICG, Assis TO, Dalla-Nora EL, Toledo PM, Santos-Junior RAO, et al. Land use change emission scenarios: Anticipating a forest transition process in the Brazilian Amazon. Global Change Biology. 2016;**22**:1821-1840. DOI: 10.1111/gcb.13134

[13] Aguiar APD, Ometto JP, Nobre CA, Lapola DM, Almeida C, Vieira IC, et al. Modeling the spatial and temporal heterogeneity of deforestation-driven carbon emissions: The INPE-EM framework applied to the Brazilian Amazon. Global Change Biology. 2012;**18**:3346-3366

[14] PRODES. Taxas anuais de desmatamento na Amazônia Legal Brasileira. 2017. Available from: http://www.obt.inpe.br/prodes/dashboard/prodes-rates.html

[15] Rochedo PRR. The threat of political bargaining to climate mitigation in Brazil. Nature Climate Change. 2018;**8**:695-698. Available from: https://www.nature.com/articles/s41558-018-0213-y

[16] Whitmere S et al. Safeguarding human health in the Anthropocene epoch: Report of the Rockefeller Foundation-Lancet Commission on planetary health. Lancet. 2015;**386**:1975-2018. DOI: 10.1016/S0140-6736(15)60901-1

[17] Soares-Filho BS et al. Economic Valuation of Changes in the Amazon Forest Area: Value Maps for Non-timber Forest Products (NTFPs). 1st ed. Belo Horizonte: Ed. IGC/UFMG; 2017. 82 p

[18] Brodizio EW. The Amazonian Caboclo and the Açaí Palm. Forest Farmers in the Global Market. Vol. 16. Advances in Economic Botany. New York, NY: The New York Botanical Garden Press; 2008. 403 p

[19] Costa FA. O açaí do Grão-Pará: arranjos produtivos e economia local, constituição e dinâmica (1995-2011). Núcleo de Altos Estudos Amazônicos. Originally presented as thesis requirement for promotion to titular professor, Universidade Federal do Pará. Belém, PA, Brazil. 2016

[20] Valsecchi J, Marmontel M, Franco CLB, Cavalcante DP, Cobra IVD, Lima IJ, et al. Atualização e composição da lista—Novas Espécies de Vertebrados e Plantas na Amazônia 2014-2015. Edição: Iniciativa Amazônia Viva da Rede WWF (Denise Oliveira e Sandra Charity), WWF-Brasil (Jorge Eduardo Dantas e Mariana Gutiérrez). Brasília, DF e Tefé. AM: WWF e Instituto de Desenvolvimento Sustentável Mamirauá. 2017

[21] Homma AKO. Extrativismo vegetal na Amazônia: História, ecologia, economia e domesticação. Embrapa. Brasília, DF. 2014. 468 p. Avalilable from: http://www.alice.cnptia.embrapa.br/alice/bitstream/doc/1016352/1/LivroExtrativismoHOMMAONLINE.pdf

[22] Yuyama K et al. Camu-camu: um fruto fantástico como fonte de vitamina C. Acta Amazonica. 2002;**32**(1):169-174

[23] Sousa MSB. Mecanismos de ação antioxidante de extratos de murici (*Byrsonima crassifolia* (L.) Kunth). Dissertação de Mestrado, Faculdade de Saúde Pública, Universidade de São Paulo, São Paulo. 2013. DOI: 10.11606/D.6.2013.tde-24052013-153230. Recuperado em 2017-06-16, de www.teses.usp.br

[24] Tiburski JH et al. Nutritional properties of yellow mombin (*Spondias mombin* L.) pulp. Food Research International. 2011;**44**(7):2326-2331

[25] Carvalho PER. Cumaru-Ferro. Comunicado Técnico 225. EMBRAPA. 2009. ISSN: 1517-5030

[26] Funasaki M et al. Amazon rainforest cosmetics: Chemical approach for quality control. Química Nova (São Paulo). 2016;**39**(2):194-209. DOI: 10.5935/0100-4042.20160008. Available from: <http://www.scielo.br/scielo.php?script=sci_arttext&pid=S0100-40422016000200194&lng=en&nrm=iso> [Accessed: Oct 29, 2017]

[27] Kimura VT et al. The effect of andiroba oil and chitosan concentration on the physical properties of chitosan emulsion film. Polímeros, São Carlos. 2016;**26**(2):168-175. DOI: 10.1590/0104-1428.2013. Available from: <http://www.scielo.br/scielo.php?script=sci_arttext&pid=S0104-14282016000200168&lng=en&nrm=iso> [Accessed: Oct 29, 2017, Epub June 14, 2016]

[28] Beraca. Product Catalogue [Website]. 2017. Available from: http://beraca.com/rainforest-oleo-de-andiroba.php

[29] Portinho JA, Zimmermann LM. Beneficial effects of açaí. International Journal of Nutrology. 2012;**5**(1):15-20

[30] Pérez V, Bokov A, Van Remmen H, Mele J, Ran Q, Ikeno Y, et al. Is the oxidative stress theory of aging dead? Biochimica et Biophysica Acta. 2009;**1790**(10):1005-1014

[31] Domingues AFN et al. Pigmentos antociânicos do açaí (Euterpe oleracea Mart.) como evidenciadores de biofilme dental. In: Pessoa JDC, Teixeira GH de A, editors. Tecnologias para inovação nas cadeias euterpe. Brasília: Embrapa; 2012

[32] Montes-Fuentes V. Terapia fotodinâmica mediada por nanoemulsão à base de óleo de açaí (*Euterpe oleracea* martius) para o tratamento de melanoma in vitro e in vivo. 2014;**xvii**:163 f. il. Tese (Doutorado em Biologia Animal)—Universidade de Brasília, Brasília, 2014

[33] Biofabris—Instituto de Biofabricação—Proj. no 2008/57860-3. Plástico de açaí. Revista Pesquisa FAPESP Edição n. 260. Outubro 2017

[34] Oliveira Dias AF et al. Cosmetic Composition Providing a Matte Effect, Process for Preparing Ucuhuba Butter and Use Thereof (US 20110256075 A1). 2011

[35] May PH, Barata LES. Rosewood exploitation in the Brazilian Amazon: Options for sustainable production. Economic Botany. 2004;**58**(2):257-265

[36] Viegas C Jr, Bolzani V da S, Barreiro EJ. Os Produtos Naturais e a Química Medicinal Moderna. Quimica Nova. 2006;**29**(2):326-337

[37] Meneguetti DU de O et al. Screening of the in vitro antileishmanial activities of compounds and secondary metabolites isolated from Maytenus guianensis Klotzsch ex Reissek (Celastraceae) chichuá Amazon. Revista da Sociedade Brasileira de Medicina Tropical. 2016;**49**(5):579-585

[38] Lima RA et al. Microbiological evaluation of isolated compounds from the bark of maytenus guianensis klotzsch ex reissek (celastraceae). Electronic Journal of Management, Education and Environmental Technology (REGET). 2016;**20**(1):592-603

[39] Trevisan MTS et al. Screening for acetylcholinesterase inhibitors from plants to treat Alzheimer's disease. Química Nova [online]. 2003;**26**(3):301-304. ISSN: 0100-4042. DOI: 10.1590/S0100-40422003000300002

[40] Bum EN et al. Anticonvulsant properties of the methanolic extract of *Cyperus articulatus* (Cyperaceae). Journal of Ethnopharmacology. 2001;**76**(2):145-150

[41] Amorim E et al. Efeito do uso tópico do extrato aquoso de Orbignya phalerata (babaçu) na cicatrização de feridas cutâneas: estudo controlado em ratos/Topic use of aqueous extract of Orbignya phalerata (babassu) in rats: Analysis of it's healing effect. Acta Cirúrgica Brasileira. 2006;**21**:67-76. (Suppl. 2). DOI: 10.1590/S0102-86502006000800011

[42] Farias RAF et al. Hypoglycemic effect of trans-dehydrocrotonin, a nor-clerodane diterpene from *Croton cajucara*. Planta Medica. 1997;**63**(6):558-560. DOI: 10.1055/s-2006-957766

[43] Hiruma-Lima CA et al. Effect of essential oil obtained from *Croton cajucara* Benth. on gastric ulcer healing and protective factors of the gastric mucosa. Phytomedicine. 2002;**9**(6): 523-529

[44] Silva JO et al. Antihemorrhagic, antinucleolytic and other antiophidian properties of the aqueous extract from *Pentaclethra macroloba*. Journal of Ethnopharmacology. 2005;**100**: 145-152

[45] Santiago GMP et al. Avaliação da atividade larvicida de saponinas triterpênicas isoladas de Pentaclethra macroloba (Willd.) Kuntze (Fabaceae) e Cordia piauhiensis Fresen (Boraginaceae) sobre *Aedes aegypti*. Brazilian Journal of Pharmacognosy. 2005;**15**(3):187-190

[46] Vieira PRN et al. Industrial Crops and Products. Chemical composition and antifungal activity of essential oils from Ocimum species. 2014;**55**:267-271

[47] Sacchetti G et al. Journal of Agricultural and Food Chemistry. Composition and Functional Properties of the Essential Oil of Amazonian Basil, Ocimum micranthum Willd., Labiatae in Comparison with Commercial Essential Oils. 2004;**52**(11):3486-3491. DOI: 10.1021/jf035145e

[48] Ansari MA, Abdul HM, Joshi G, Opii WO, Butterfield DA. Protective effect of quercetin in primary neurons against Abeta(1-42): Relevance to Alzheimer's disease. The Journal of Nutritional Biochemistry. 2009;**20**(4):269-275. DOI: 10.1016/j.jnutbio.2008.03.002 [Epub 2008 Jul 7, 2008]

[49] Roseghini R, Rocha DS, Clarêncio J, Costa SL, Costa MFD, Tardy M, et al. Flavonoid rutin alters the viability and function of mitogen-stimulated splenocytes and thymocytes

compared with non stimulated cells. Immunopharmacology and Immunotoxicology. 2007;**29**(2):271-285

[50] Heitzman ME, Catherine CN, Winiarz E, Vaisberg AJ, Hammond GB. Ethnobotany, phytochemistry and pharmacology of Uncaria (Rubiaceae). Phytochemistry. 2005;**66**(1):5-29

[51] Durrani AM, Davies NM, Thomas M, Kellaway IW. Pilocarpine bioavailability from a mucoadhesive liposomal ophthalmic drug delivery system. In International Journal of Pharmaceutics. 1992;**88**(1-3):409-415. ISSN: 0378-5173. DOI: 10.1016/0378-5173(92)90340-8

[52] Hargreaves PI. Bioprospecção de novas celulases de fungos provenientes da floresta amazônica e otimização de sua produção sobre celulignina de bagaço de cana/ Paulo Iiboshi Hargreaves. Dissertação (Mestrado em tecnologia de Processos Químicos e Bioquímicos)— Universidade Federal do Rio de Janeiro-UFRJ, Escola de Química, Programa de Pós-Graduação em Tecnologia de Processos Químicos e Bioquímicos. Rio de Janeiro. 2008

[53] Carvalho JM et al. Phytosterols isolated from endophytic fungus *Colletotrichum gloeosporioides* (Melanconiaceae). Acta Amazonica, Manaus. 2016;**46**(1):69-72. Available from: <http://www.scielo.br/scielo.php?script=sci_arttext&pid=S0044-59672016000100069&lng=en&nrm=iso> [Accessed: Oct 27, 2017]. DOI: 10.1590/1809-4392201500072

[54] Chandra S. Endophytic fungi: Novel sources of anticancer lead molecules. Applied Microbiology and Biotechnology. 2012;**95**:47-59. DOI: 10.1007/s00253-012-4128-7

[55] Rippel MM. Borracha natural e nanocompósitos com argila. Quimica Nova. 2009;**32**(3): 818-826

[56] Silva TM da et al. O mercado de amêndoas de Dipteryx odorada (cumaru) no Estado do Pará. Floresta [S.l.]. 2010. ISSN: 1982-4688

[57] CAMTA. Sistema Agroflorestal de Tomé-Açu—SAFTA [Website]. 2017. Available from: http://camta.lwsite.com.br/voce-conhece-o-safta

[58] RECA—Associação dos Pequenos Agrossilvicultores do Projeto Reca [Website]. 2017. Available from: http://www.projetoreca.com.br/site/

[59] ERBOL—Periódico Digital. Firma castañera paralizará operaciones y más de 300 se quedarán sin empleo. Available from: http://www.erbol.com.bo/noticia/economia/06042017/firma_castanera_paralizara_operaciones_y_mas_de_300_se_quedaran_sin_empleo

[60] Pinheiro CUB. Extrativismo, cultivo e privatização do jaborandi (Pilocarpus microphyllus Stapf ex Holm.; Rutaceae) no Maranhão, BrasilExtrativismo, cultivo e privatização do jaborandi (Pilocarpus microphyllus Stapf ex Holm.; Rutaceae) no Maranhão, Brasil. Acta Botanica Brasilica. 2002;**16**(2):141-150. DOI: 10.1590/S0102-33062002000200002

[61] Nobre CA. Brazil: Boost pro-forest economics. Nature. 2014;**510**:210

[62] Homma AKO. Em favor de uma revolução tecnológica na Amazônia. Tropical Plant Pathology. 2012;**37**:1-17

[63] Messias FB, Nascimento EP. Economia criativa para o desenvolvimento sustentável da Amazônia. In: Anais do SICASA e ANPPAS Amazônia. Anais Manaus (AM): UFAM/ANPPAS; 2016

[64] MetaCurrency. The CEPTR Project. 2018. Available from: http://ceptr.org/ [Acessed: May 18, 2018]

[65] World Economic Forum. Harnessing the Fourth Industrial Revolution for Life on Land. Towards an Inclusive Bio-Economy. Fourth Industrial Revolution for the Earth Series. 2018. Available from: www3.weforum.org/docs/WEF_Harnessing_4IR_Life_on_Land.pdf [Accessed: May 18, 2018]

[66] Haupt F et al. Progress on the New York Declaration on Forests. Finance for Forests. Goals 8 and 9 Assessment Report. Available from: forestdeclaration.org. New York. 2017

[67] Sonter L et al. Mining drives extensive deforestation in the Brazilian Amazon. Nature Communications. 2017;**8**:1013-1018. DOI: 10.1038/s41467-017-00557-w

[68] Global Witness. Defender of the Earth: Global Killings of Land and Environmental Defenders in 2016. Global Witness Org. 2017. Available from: www.globalwitness.org

[69] Poudyal M, Jones JPG, Sarobidy Rakotinarivo O, Hockley N, Gibbons JM, Mandimbiniana R, et al. Who bears the cost of forest conservation? Peer Journal—Life and Environment. 2018;**6**:e5106. DOI: 10.7717/peerj.5106

Section 4

Land-Use Strategies, Dynamics and Policies

Chapter 11

Planning, Power, and Politics (3P): Critical Review of the Hidden Role of Spatial Planning in Conflict Areas

Raed Najjar

Additional information is available at the end of the chapter

http://dx.doi.org/10.5772/intechopen.78779

Abstract

This chapter discusses theoretical reviews about urban space theory, the paradoxical roles of spatial planning, and introduces a revolutionary definition for sustainability, namely, the four-dimensional spatial sustainability (4DSS) model. Interestingly, the empirical section in this chapter underlines the links that emerge when addressing spatial critical transformations accorded by interconnected spatial relations when attached to conflict areas, mainly: planning, power, and politics, the (3P) concept. Theories pertaining to spatial planning and sustainable development have substantially evolved during the past century. However, both of these themes still remain underestimated and require further investigation when exploring conflict regions. Spatial planning in conflict zones requires forming fast-changing spatial policies accompanying the creation of irreversibly altered urban fabrics that generate in many cases drastic challenges for inhabitants, especially for the indigenous residents when considered a minority group. Therefore, clarifying the relationships between the 3P and 4DSS is a central issue in this chapter. Understanding these relationships reveals the range of political influence upon the role of planning and its objectives. In Jerusalem, the aforementioned interrelationships have generated a deeply divided city, where dramatic spatial and demographic changes have adversely affected the lives of Palestinians, threatening their presence and, by consequence, their identity.

Keywords: space, spatial planning, sustainability, four-dimensional spatial sustainability (4DSS), planning, power, and politics (3P), regressive planning, progressive planning, conflict area, Jerusalem

1. Introduction

The philosophy of understanding space in terms of the temporal dimension correlates with the actions or activities within a space transforming them into places and moments. In that sense, relationships give rise to order. Hence, the temporal relations give rise to a temporal order and the spatial relations to a spatial order. What's more, it is not easy to make a plausible preliminary list of basic spatial relationships in terms of its designated entities [1]. Few concepts are more crucial to understand our world today than power, especially when politics and planning emerge as key spatial determinants in conflict affected areas, thus the spatial identifications shall ultimately be intensified.

Arguments concerning the theory of space under conflict have reappeared in the agenda of many scholars and theorists in recent years. Conflict in its broad definition means a struggle or clash between opposing forces, armed aggression, widespread violence, and widespread human rights abuses. It represents a state of opposition between ideas and interests. Spatially, "conflict areas" are zones where "conflict" is prevalent. The area may be a region, a country, an area within a country, or an area that crosses one or more country boundaries. In the context of conflict areas, urban fabric is in a continuous process of change. Immigrants cluster together and mix with others; ethnic and racial groups are segregated in ghettos and slums, and the nonmarginalized are able to displace to more habitable places; poor families are forced to look for other quarters, because of rabid urban restructuring; and classy housings are developed on the most attractive spaces, in order to attract the rich. Thus, spaces are divided. Likewise, separated neighborhoods indicate the presence of different urban fabrics within a city. These all have different features; they may be modern or undeveloped, secured or risky, deprived or privileged, clean or dirty, old or new, or contrasted on countless other aspects. Remarkably, there is a certain contrast in these spaces: between luxurious and marginalized zones and between places where only those wealthy can afford to live and places where the deprived are forced to live because of the lack of alternatives elsewhere [2]. These cities are described as divided cities, dual cities, polarized cities, fragmented cities, and partitioned cities [3].

The relationship between division and planning is prominent, but questionable. In contested spaces, deliberate and discriminatory actions against the weaker population occur. As such, planning in many cases is inequitable, implicitly biased and reflects not what it promises to be. It is used as a control tool over the marginal group, rather than a tool for positive change. Hence, in conflict areas, urban planning has to be re-conceptualized to go beyond the narrow framework of physical land use planning. Therefore, the 3P concept [planning, power, and politics] can help to address this as it is a dynamic process that underpins the ambiguity of the spatial modality created in conflict areas and correlates that concept with a pertinent case study: Jerusalem. During the last decades, many conflict cities witnessed hypersegregation, ethnic separation, and persistent racial discrimination, as what has been seen in Belfast, Nicosia, Beirut, and Mostar. However, the case of Jerusalem is more passionate, as it presents the case of deeply divided city due to the intensity of the ethnic conflict it has faced for more than 70 years, and eventually, its perception as a frontier city [4]. And so, the urgent need to assess and reread the space in Jerusalem in the context of changing socio-political power arises. Consequently, it is important to scrutinize such influence upon East Jerusalem's (EJ) urban space after the Israeli occupation of 1967 and to explore this interaction between planning and politics, where the latter has strongly and directly influenced the former.

2. Space, a conflictual concept

Transformation of a given status is not, unquestionably, one of the core themes of philosophy. The abstract character of the philosophical work in the past and present is rooted in the social conditions of existence. Social struggles rage when various socio-political dimensions merge together. Transformation of space produces tension between power and different social strata forming what is known as the battle over performance space [5]. Presentations of space outline a complex set of variables and power and politics surface as important indicators. The role of space in the producing cultural and political power has been largely ignored in cultural theory and criticism. Focusing on power as a spatial presentation helps researchers to precisely theorize the manifold social reproduction processes [6].

Space is classified into two main categories: mental space (experiential) and physical space (existential). Experiential space, in a merely metamorphic sense, refers to a mental image or nonphysical representation for time or duration, area or extension. Whereas, existential space has three dimensions considered as a volume not an area. According to many, this is illusionary. Invoking space as a metaphor rather than a physical "quantifiable" subject is problematic, because invocations of space habitually adopt space as known, specified, and unproblematic [7]. Space is exponentially correlated to social relations and is the convenient medium of power that is socially constituted through material relationships that enable an explicit political interaction. The historic spatial relationship of powers defines history as pure reflection of spaces which evidently would be the history of power [8]. The basic explanation of power represents influence, forte, movement, and strength. In this regard, space can be expressed as an active, nonstatic or limited object; it is rather a result of relations that are themselves dynamic and continuously changing [9].

Instability addresses, implicitly, a conflict of powers regardless of their form, be it physical, natural, political, or social for example. Spatial configurations thus constitute unequal relations and, therefore, the emergence of differences and the quest for power. That "mess" of relations is useful for theorizing the "unbalanced powers" and "unequal relations" of a space in terms of social complexity (classes, races, segregation, etc.). Therefore, terms differentiating strong and weak powers, such as dominant and marginal relations, respectively, arise. Hence, the "differences" emerging out of spatial relationships are addressed in social and the cultural theory [10]. The spatiality of powers (re)-constitutes our social references and identities. Space and spatial relations should be considered as active components in the unequal and heterogeneous production and distribution of social references, politics, and powers, which altogether highlight place configurations.

For a better interpretation of the socio-spatial relationships between space and place, it is important to refuse considering the framework of social identities as the sole background against which all other investigations of social or cultural relations occur. This is key as social markers are constantly varying parameters, and they are also continually altered, disputed, and reproduced. Space comprises an active and constantly changing site of power; however, the theory of "politics of location" does not critically capture that phenomenon [6].

Recalling the meaning of space highlights the necessity to underline the changing characteristics of space in social and physical aspects. To exemplify this abstract concept, space representation helps clarify this subject. Capturing a physical presentation of space, such as the city, seamlessly outlines how space is a lively non-static mass, rather a dynamic organism. The philosophical interpretation of the city concept asserts the need to understand the relationships carried out within this "closed container" before analyzing its components, and to accept that "the city" is not merely a container. Another pertinent explanation introduces the city as a place where there is still a recognizable concentrated, teeming, dynamic expression of urbanism. It is a place that becomes very enjoyable for its inhabitants and lots of visitors every year [11]. In theory, the fundamental meaning of space or place is a relative norm, which is highly correlated to social and cultural concepts; it depends on the cognitive images of a place conceived by the manifold experiences and backgrounds of people.

The interest in place and space has significantly grown during the last century; it is reflected by the development of the so-called new regional geography [12]. In consequence, presentations of space and the development of place related themes became far sighted during the 1980s and repeated invocations about spatial perspectives within the geographical imagination. The dialectic mode of thinking facilitates understanding the paradoxical nature of space. The corner stone of dialectics attempts to philosophize what the world is without detaching its components for the purpose of analysis and presentation [13]. Dialectical reflections commonly address the question of change regarding various spatial questions: interrelationships, interconnections and interactions, processes, activities, flows, relations, and eventually contradiction. Accordingly, dialecticians often conceptualize "dynamism" as the basic framework to all matter and thus "stability" is an irrelevant status that necessitates explanation.

Philosopher Sir Isaac Newton elaborated that space is absolute, proper to itself, and independent of the objects it contains. According to the dialectical mode of argumentation, the complex composition of space—notably spatial relations, power, politics, productions, and phenomena—is conceived just as a single entity, that is to say with the quality of wholeness. Wholeness (totality) could be demonstrated as *"the way the whole is present through the internal relations of each of its parts"* [14]. Although it is not possible to comprehend multiple interrelated elements of a whole without understanding how the elements relate to each other within this whole, totality is signified in its wholeness as: *"a need to look on the world as an undivided whole"* [15]. However, other philosophical approaches oppose dialectical thinking and contrast obviously with the notions of wholeness, considering separate objects by splitting thoughts and problems into parts and in rearranging these in their rational order. This mode of argumentation represents the Cartesian method, which is a scientific philosophy that explores the reality via mechanical and mathematical representations, and also perceived to be merely as the "method of doubt" [16]. From this regard, space could be perceived as being autonomous or a passive empty container independent of physical characteristics [17]. Wholeness, as such, amounts to nothing more than the sum of the parts. Conversely, dialecticians reject this approach of detaching the diverse features of reality. Instead, the dialectical philosophy confirms the unity of knowledge and the total character of reality. Space, therefore, in this logic is a unity containing within itself different aspects.

Needless to say, the border line between the total and the part is undetectable. Not only is the concrete character of space and place in terms of their real ontological status, therefore, debatable, but also their distinction depending on the comprehensive integration for grasping the interconnected spatial relationships, social, power, politics, and processes among them. Indeed, critical philosophy highlights politics, among the heterogeneous and conflictual elements of space as an internal parameter and major player; thus the overall production process of space and place is genuinely a political event [18]. This conclusion is coherent with the concrete foundation of dialectics: the contradiction. Spatial contradictions of urban spaces born of political conflicts are played out between social benefits, economic powers, and political forces, which express themselves in place, an element of space. Yet, dialectically, these elements are divergent components of the same unity; however, the significance of these qualitative aspects of place and how they, in turn, shape space, cannot be downplayed.

3. Sustainability: a new revolutionary definition

Rethinking the traditional definition of sustainability, especially during the recurrent global and regional challenges—notably massive immigration, urban shrinkage, dissolving heritage, climate change, poverty, and injustice, etc.—has become more critical and progressively urgent. Spatial development could be addressed as a normative, but challenging, response to community and human needs. However, troubling debates have been gaining ground and therefore bringing attention to the multidimensional consequences presented during historical development of modern society. The Industrial Revolution witnessed rapid transformation in both the norms of knowledge and community urban growth patterns. Unfortunately, the fast mode of production and unregulated urban growth resulted, in many cases, in social degradation, poor living conditions, and environmental concerns [19]. Reconsidering the norms of urban growth and the models of development thus became an urgent issue. Consequently, the science of sustainability has shown up introducing significant challenges for planners and policy makers as well. Sustainable development is, therefore, a major concern with reference to the crucial need to protect the global environment while attaining a better life for people. The concept of sustainability continues to attract more attention; thus, sustainable development is presented in more than two hundred definitions, while featured on more than 8 million web pages and the number keeps rising [20].

Sustainable development is thoroughly tied to the environmental concerns that continue to introduce changes in knowledge and the sciences. The world has changed rapidly due to the conception of sustainability; however, most of the challenges that gave impulse to the introduction of the concept have not yet been solved. On contrary, the irreversible loss of natural resources, rapid depletion of certain energy resources, troubling climate change, and social injustices are observed. Originally, the interest about sustainability intensified in the 1980s. In 1987, the classical definition of sustainable development was drafted in "Brundtland Report - Our Common Future" as the paths of human progress that meet the needs of the present, without compromising the ability of future generations to meet their own needs. The Brundtland Report incorporated components of sustainability within the economic and political context of international development, as well as combining ethical norms of welfare, democracy, and environment [21]. Sustainability science has revealed advanced development over the last decade [22]. It explores long-term relationships and implications between large-scale socioeconomic and ecologic systems, the complex restructuring processes lead to degradation of these systems, and probable associated risks to human well-being. Hence, sustainability probes the natural and social systems, questioning their interactions, most notably achieving needs' balance of present and future generations while improving well fare and preserving the planet life support systems.

The interdisciplinary mode of thinking has recently succeeded in linking the traditionally separate intellectual fields of critical social theory and environmental science. However, responses to meet the increasing demands of a growing population in an interconnected but unequal world have undermined the earth's essential life-support systems. Sustainability is therefore a concept that provides new visions for the national and international development and formulates new solutions for the recurrent socioeconomic needs. Nevertheless, the world's present development path is not sustainable [20]. The ecological transformations accompanied by the development processes are considered chief global challenge along with the intense alterations underway in socioeconomic and cultural life. Key indications of such consequences are global climate warming, urban sprawls, degradation of biological diversity, deprivation and increase in poverty levels, and the excessive exploitation of resources with unmatched rates of pollution [23]. In this context, the current concept of sustainability is vulnerable to the same criticism of the vague idealism proffered against comprehensive planning [24].

The "classical" definition of sustainability, in that context, could be conceived as a guiding tool that connects the "present" of a spatially referenced activity into a "future" projected and thus desired, status. Still, classical sustainability does not cover *the past*, in particular, within its temporal analysis. Moreover, it could be debated that the aforementioned classical definitions for sustainability in both, be it theory-driven or action-driven directions, neither integrate nor incorporate the cultural and historical aspects of space and place within the framework of the development process. Thus, it is also arguable that adopting the "classical" norms of sustainability disregard the cultural identity and historical aspects and could, therefore, lead toward a critical cultural transformation, degradation, or even evanescence and disappearance.

To overcome the existing challenge and to recover that gap in the classical conception of sustainability, it is necessary to put forward a revolutionary model for spatial sustainability that not only conceives sustainability as a guidance tool aiming at steering the development wheel toward environmental protection, economic growth, and social equity but also at integrating the cultural-historical factors in a more holistic temporal analysis as well. In other words, this requires shifting the "classical sustainability" concept into a more comprehensive model, namely, the "four-dimensional" spatial sustainability (4DSS) presented in **Figure 1**. This adds a new comprehensive time-based dimension allowing the integration of the past of a referenced space by evaluating its cultural and historical identity and to assess the consequential cause-effect impacts of the projected, future development. Furthermore, considering the cultural-historic dimension within the development process enhances the integration toward a well-balanced time scale focusing on understanding a past-present trajectory before linking it into the future. Hence, the 4DSS considers the following four substantial dimensions for investigation and integration—social, environmental, economic, and cultural-historic—forming together the SEEC orbit of sustainability, centered by gravity zone of tension. In practice, considering the dynamic interactions while integrating these four-dimensional perspectives

http://dx.doi.org/10.5772/intechopen.78779

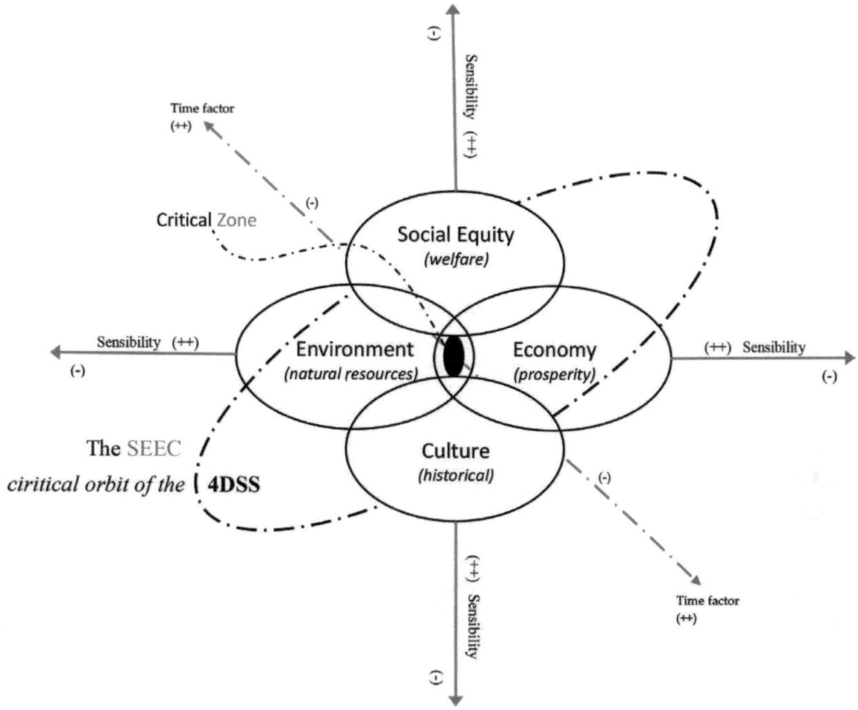

Figure 1. The four-dimensional spatial sustainability (4DSS) model presenting the critical orbit of sustainability (SEEC) where tension zone located at center [author].

to ensure balanced spatial sustainability creates a "critical conflictual zone" centered in the core of the orbit of sustainability, the SEEC. The proposed 4DSS model aspires to offer a fascinating, holistic way of evading these conflicts highlighting the time-factor and the gravity scale, but they cannot be resolved so simply.

4. Spatial planning

In order to understand the notion of spatial planning, it would be useful to begin by understanding a definition of this term. A brief review of the literature pertaining to "planning" in terms of physical development would immediately reveal that the word has a wide variety of meanings. Gradually, the terms "urban and regional planning" or "town and country planning" and "city planning" as they are called in the UK and North America, respectively, are ideally represented by the term spatial planning. Spatial planning is not a static notion that presents a single concept, procedure, or tool; it is rather a well-structured mix of all that must be comprehensively functioned if desirable outcomes are to be achieved [25]. From this regard, the perception of spatial planning indicates the necessity for integrating several spatial sectorial policies in order to create positive synergies. The emerging compound nouns of planning have attracted academics to explore the meanings and semantics of those flexible names. For instance, the terms "land use planning," "regional planning," "town planning," and "urban planning" are often used interchangeably, and in many cases will depend on the reference country, but do not always have the same meaning. In the United Kingdom, Australia, and New Zealand, the term "town planning" is common. Meanwhile, in the United States and Canada, the term "urban planning" is more familiar. However, in Europe, the preferred term is increasingly "spatial planning." Spatial planning is perceived as going beyond traditional land use planning to bring together and integrate policies for the development and use of land with other policies and programs which influence the nature of places and how they function [20].

It could be argued that the definition of spatial planning depends on the basic historical and institutional differences between the various settings where planning is practiced. Hence, there are variances for the perception of planning according to the spatial reference that originates the definition [26]. Italian intellectuals, for instance, perceived "planning" as an element of the city seen as a work of art. Alternatively, the British have regularly concentrated on the regulation in different scales of physical development. Meanwhile, American researchers have often referred to planning as a loose concept, dealing essentially with public and private policy efforts. Generally, there are two major, but contradictory, definitions for planning, depending upon the aims and tools of planning, or in other words, based upon the "role" of planning. The concepts pertaining to the power of planning to act either as a progressive or as a regressive agent of change and the probability of using planning as a "control tool" instead of "reforming tool" principally upon ethnic minorities are briefly presented in the following sections.

4.1. Progressive planning

Planning, in its conventional sense, simply refers to the process of setting goals, developing strategies, and outlining activities and schedules to accomplish desired objectives [27]. The term progressive planning refers to that sort of planning that acts as a means of positive change to achieve more urban justice, economic growth, equity, and stability. Achieving these goals has proven very demanding. Specifically, after the early start of the Industrial Revolution, the world went through rapid transformation processes, moving from simple agricultural communities into massively urbanized ones. This quick transformation resulted, in many cases, in the emergence of unhealthy living conditions, social dilemmas, and environmental hazards. Accordingly, planning was born as a way to heal the ills of urbanization and then evolved into an organized field of human activity; thus it was normative reaction to the exigent of ameliorating the deteriorated living conditions [28]. This fundamental explanation of planning inspired urban planners to introduce "ideal" concepts such as utopianism, liberty and equity, economic reform, and improvement of living conditions. These basic thoughts formed the foundation of planning theories. Therefore, progressive planning has been conceived as a problem-solving activity that relates knowledge to action in different ways, and thus is optimally characterized as reformative norm [29]. Spatial planning represents the interrelationship between the concepts of space and place. It explores how such concepts reflect the shift in geographical thought to a dynamic, discontiguous, relational conceptualization of spatiality [30]. It is a multidisciplinary, hermeneutic discipline, which integrates the integration of many other disciplines in order to explain spaces and eventually to optimize strategic mechanisms in developing spaces toward a more sustainable and equitable living conditions [25].

4.2. Regressive planning

Regressive planning is a concept used for the cases in which planning is oriented to function as a "control tool" in order to achieve oppressive objectives. More specifically, regressive planning is utilized to serve a specific social stratum and neglect or even restrain the other strata. This sort of planning is critically practiced at places where there is conflict, political instability, or racial disputes. It presents a considerable degree of uncertainty and vagueness, therefore affecting its legitimacy, ability to create consensus, and sustainability in real contexts. Consequently, this widespread uncertainty of planning concepts continues to raise doubts regarding a presumed disciplinary status and even professional conception for planning and its expected, and in many cases, unpredictable roles.

From the viewpoint of physical development in conflict areas, the interactive power relationships clearly exist between contingent styles of planning and their institutional and cultural contexts which illustrate, to some extent, the differentiation in planning tools and the variety of outcomes [31]. Examining the theoretical and empirical studies regarding planning practices in different contexts could help clarifying a solid core of common trends and problems constituting a series of challenges, dilemmas, and limitations that are valid in different institutional, government, economic and administrative frameworks [32].

Understanding the context in which planning is transformed into what can be understood as "imperfect planning," addresses substantial exploration of particular scopes that reveal how planning is used as a socio-graphical control tool. These scopes are four-fold [20]:

- *Territorial scope*: It is also known as the spatial context and it reflects upon space, geography, time and people. It presents the territorial policies and ordinances utilized as a powerful tool of control over minorities, particularly in deeply divided societies, where ethnic groups often reside in their own regions.
- *Power relations and decision-making scope:* This is also known as the methodological scope and it includes the statutory aspects that determine the formal relationship between the regime and the public. It is employed in order to marginalize specific groups, thus enhancing segregation and exclusion of ethnical or minority groups from the active and real participation in the process of decision-making.
- *Socioeconomic scope:* This focuses on serving the economic interests of the dominant party and thereby contributes to create weaker groups of people who become more dependent on the dominant party, who in turn manipulates the regime to increase its influence and power.
- *Cultural scope:* It deals with the influence and effect of planning on the multiple cultures and identities within a space. It is critically utilized through the planning strategies that are practiced by the dominant ethnic group who often aims to minimize and alienate the other ethnic cultures.

5. Conflict city of Jerusalem

Few cities evoke such a sharp and expressive response from so many people all over the world as does Jerusalem. Sacred to at least three major faiths—Judaism, Christianity, and Islam—Jerusalem has been a source of inspiration to adherents of these religions for thousands of years [33]. Jerusalem has therefore been a focal point for world powers during many different eras as shown in **Figure 2**.

Thus far, Jerusalem is considered a contested, frontier, polarized, and deeply divided city [4]. Until 1917, Jerusalem was an "Ottoman Province". After WWI and in particular after the Battle of Jerusalem in December 1917, the British military captured Jerusalem city and considered it to be the capital of their Mandate in Palestine. The League of Nations, through its 1922 ratification of the Balfour Declaration, designated the United Kingdom to administer the Mandate for Palestine and help establish a Jewish state in Palestine [34]. During the successive three decades of the British Mandate (1917–1948), many areas in Jerusalem looked into the construction of new garden suburbs mainly in the northern-western direction. Then, at the end of the 1948 Arab-Israeli War, Jerusalem was divided for the first time in its history. The first spatial division of Jerusalem was set out by the Armistice Agreement of 1949 between Israel and Jordan cut through the center of the city creating the western and eastern parts from 1949 until 1967. During that time, West Jerusalem (WJ) was controlled by Israel, while EJ was controlled by Trans-Jordan. In 1949, Israel declared WJ as its capital.

The next dramatic moment, the 1967 Six-Day War, had dramatic consequences for what followed. Israel had, unilaterally, annexed 70.5 square kilometers of the occupied Palestinian territory (oPt) including EJ, which presents 6.5 square kilometers of the total. Israel's domestic jurisdiction was extended to EJ through Amendment No. 11 of the 1967 Law and Administration Ordinance. The city's illegal unification and its controversial status as the eternal capital of Israel were declared through the Basic Law in 1980. However, the status of United Jerusalem as Israel's eternal capital has not been officially recognized by most of the international community, and nearly all countries maintain their embassies in Tel Aviv. However, in December 2017, the president of the US violated the UN resolutions and announced his controversial decision to recognize Jerusalem as Israel's capital a political discourse that intensifies the tension rather enhances the peace process. Consequently, the UN General Assembly has decisively backed a resolution effectively calling on the US to withdraw its recognition of Jerusalem as the capital of Israel and voted overwhelmingly to ask nations not to establish diplomatic missions in the historic city of Jerusalem. Consequently, the Assembly adopted the resolution "Status of Jerusalem," by which it declared "null and void" any actions intended to alter Jerusalem's character, status or demographic composition; and stated that any decisions and actions which purport to have altered the character, status or demographic composition of the Holy City of Jerusalem have no legal effect and must be rescinded in compliance with relevant resolutions of the Security Council [35]. These acts are contrary to international law. Israel, therefore, continues to violate international law, going against United Nations resolutions and agreements with

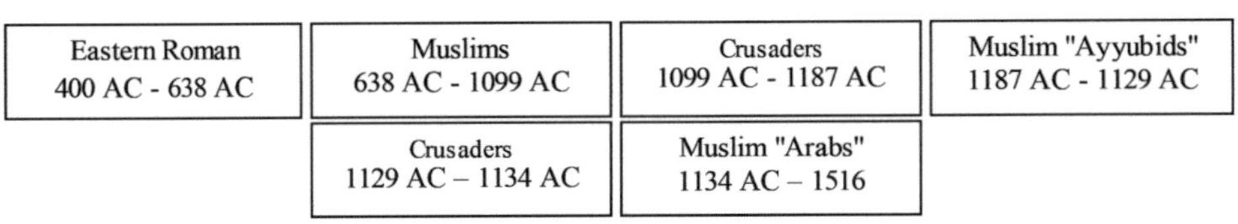

Figure 2. Historical powers that controlled Jerusalem before Ottomans [author].

Palestinians. Nonetheless, the Palestinians consider EJ as the capital of a future Palestinian state. Palestinians also refer to the UN Security Council's Resolution 252, which considers as illegal the confiscation of land and other actions that tend to alter the legal status of Jerusalem. The status of Jerusalem and of its holy places remains contended up to date.

The wall encircling the Old City of Jerusalem, spatially defined the city during a long period. The geographical location of Jerusalem gives the city high geo-political and logistical values for its proximity to other regional capitals such as Amman (85 km), Damascus (290 km), Beirut (388 km), Cairo (528 km), and Baghdad (865 km). This centrality of that position accords Jerusalem with a unique logistical characteristic and is one of its distinguishing geo-political features. The present city of Jerusalem has grown beyond the Old City. After 1948, the city expanded toward the north and west where the Israeli government established modern and massive Jewish neighborhoods, whereas since the year 1967, Israel has concentrated its settlement construction works in the eastern part of the city, imposing therefore a new Jewish demography inside the Arabic Palestinian neighborhoods. The Israeli-Palestinian conflict over Jerusalem has been dramatically intensified since Israel occupied EJ in 1967. Jerusalem has been described as a deeply divided city due to the intensity and persistence of the ethnic conflict it has faced for decades [36]. Furthermore, the future perspectives of Jerusalem's status are unpredictable due to the competition between the Palestinian and Jewish ethnonational identities [37]. Additionally, Jerusalem has also been characterized as a frontier city. Frontier cities are not only polarized along ethnic and ideological lines, but also are disputed foremost because of their location on fault lines between ethnic, religious or ideological entities [38]. Accordingly, given its spiritual, cultural, and historical values, Jerusalem outlines the core of the Palestinian-Israeli conflict.

6. The planning context in Jerusalem

Jerusalem is a unique case study in terms of its historical development, especially during the last century when administrative control of the city changed several times. Within five decades (1917–1967), Jerusalem was controlled by four distinctive regimes, namely, Ottoman, British, Jordanian, and Israeli. During these radical administrative transformations, Jerusalem experienced rapid and varying development modes, which together have produced different challenges for its spatial characteristics, most particularly, in terms of the fast changing composition of the city population and urban fabric. Hence, the overall experience in the field of physical planning in Jerusalem offers unique and special aspects of profound interest for any scholar in urban planning, spatial socio-political relations, history, and human geography.

Today Jerusalem reflects two divergent images. The first is the timeless of one of the most historic cities in the world, while the other is that of one of the most modern cities in the world. These two contradictory images of the city are accompanied with by the heterogeneousness of the population, arriving mainly during the last century. The successive administrations in Jerusalem have created an extensive maze of rules and regulations, making the planning system complex and in many ways inefficient. The historical powers that had characterized the official planning system in Jerusalem since the Ottoman period are listed chronologically in **Figure 3**.

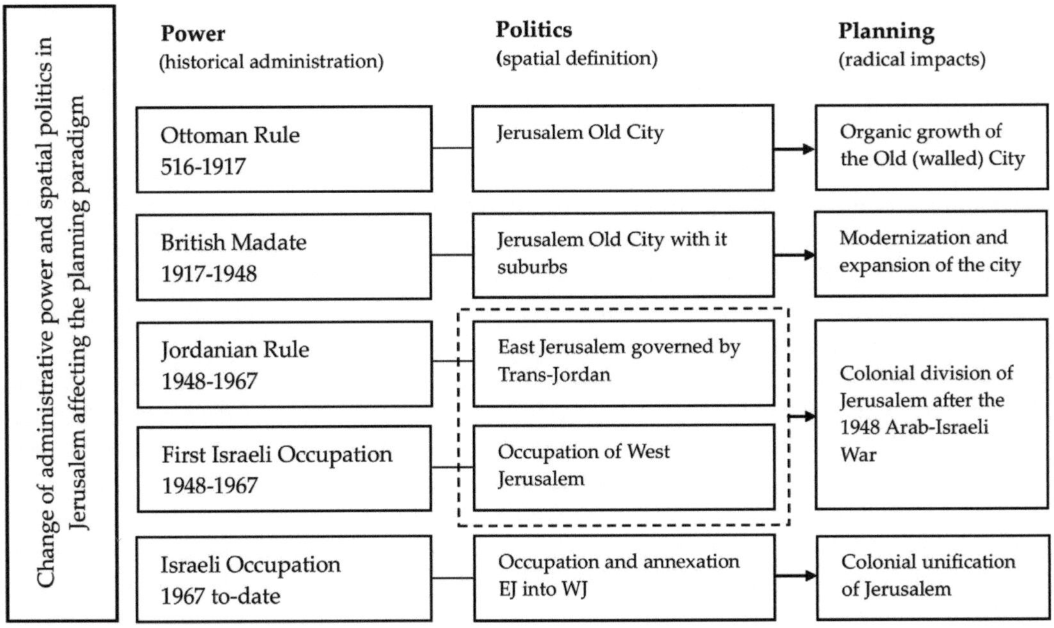

Figure 3. Historical administrative development in Jerusalem since Ottoman rule [author].

The previous figure reflects the quick and dynamic transformations of the administrative authorities in Jerusalem. During the rule of each authority, Jerusalem was "spatially" defined in a completely unique and different way. The smallest delimitation was certainly during the Ottoman rule, in which Jerusalem was mainly developing within the boundaries of the Old City, which is accurately defined by its inspiring encircling wall. The Old City's internal narrow roadway system that forms a maze of alleys and stairways hide a treasure of historical, cultural and spiritual heritage that reflects 5000 years of passionate history condensed in barely 1 square kilometer. The Old City greatly outlines the features of the fortified cities built during the reign of the Ottoman Sultan Suleiman, the Magnificent in the early sixteenth century. The political borders of the city then changed following the succession of the administrations, successive control power. The spatial frontier was first set outside of the walled area during the British Mandate and it continued to expand until the Israeli occupation, today encompassing three times more than its mandatory perimeter. Today the spatial appearance of the city reflects divided communities and segregated neighborhoods as shown in **Figure 4**.

Israel occupied WJ after the termination of the British Mandate in 1948. In 1967, after the Six-Day War, Israel unlawfully annexed EJ to its territory. Since then, Jerusalem has been subjected to extensive Israeli planning policies aiming at expropriating more of the Palestinian lands and expelling native Palestinians from EJ. Hence, spatial planning in Jerusalem consists of two contradictory approaches based upon the ethnic and cultural identity of the residents:

- The *"Progressive planning paradigm"* practiced in WJ and in the Jewish Settlements (JS) spread to EJ, which aims to improve the welfare of the Jewish people, who are now the dominant group, and their neighborhoods by creating more convenient, equitable, healthful, efficient, and attractive places for the existing and future Israeli Jewish generations.

http://dx.doi.org/10.5772/intechopen.78779

- The *"Regressive planning paradigm"* applied in the Palestinian Arab Neighborhoods in EJ limits the current and constrains the future development of the Palestinian residents who are now the minority group.

The dynamic interactions between planning, power, and politics (3P) have produced paradoxical spatial development patterns in Jerusalem as conceptualized in **Figure 5**. Accordingly, urban spaces in Jerusalem are produced in two contradictory modes of production. These unequal planning progressive/regressive modes reflect two-sided planning paradigms of the current Israeli policies. Although Israel declared the city of Jerusalem as a "unified" city in its political boundaries, it is still "separated" in terms of its spatial context and urban fabric. WJ presents an "active and dynamic" space for the Jewish residents, whereas the opposite is represented in EJ, namely, an "inactive and fragmented" space for the Palestinian residents. This contradiction in the city atmosphere is guided by the Israeli central government and thus maintained deliberately by the political power of the state.

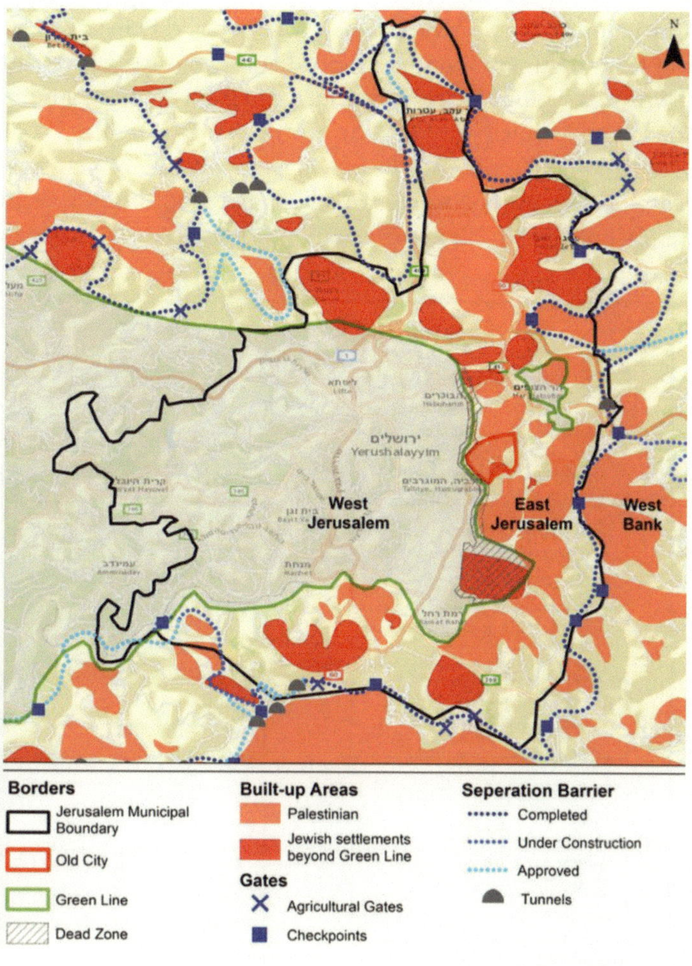

Figure 4. The appearance of the divided city of Jerusalem, 2014 [3].

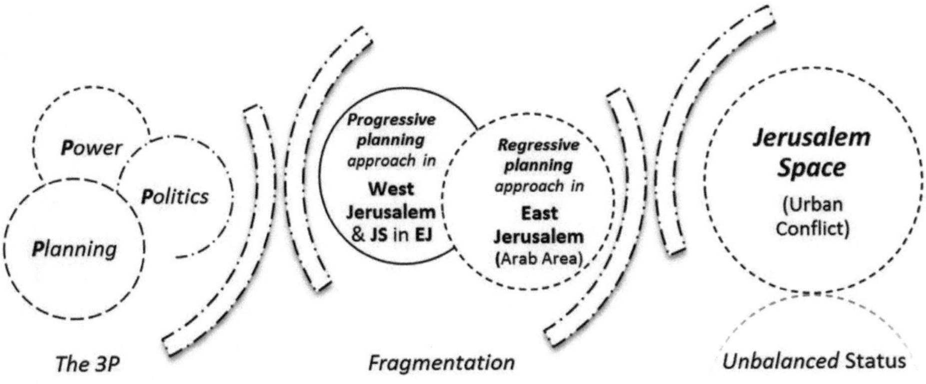

Figure 5. Spatial reflection of the 3P concept in Jerusalem [author].

The 3P concept demonstrates how the paradoxical Israeli spatial planning policies in Jerusalem formulate two contrasting societies within the same spatial governorate, namely, the Palestinian Arab community, and the Israeli Jewish society. However, the earlier deprived community suffers from fragmentation of its social and geographical contexts in contrast to the later society, which is well connected and integrated by spatial continuity and physical infrastructure. The Israeli Municipality of Jerusalem has intensified the complexity between the Palestinian urbanized neighborhoods in EJ, via a complex set of planning tools and regulations. Spatial regressive planning, besides mismanaged land use policies, is the chief challenge in that sense, as outlined in **Figure 6**, which presents the systematic unjust Palestinian

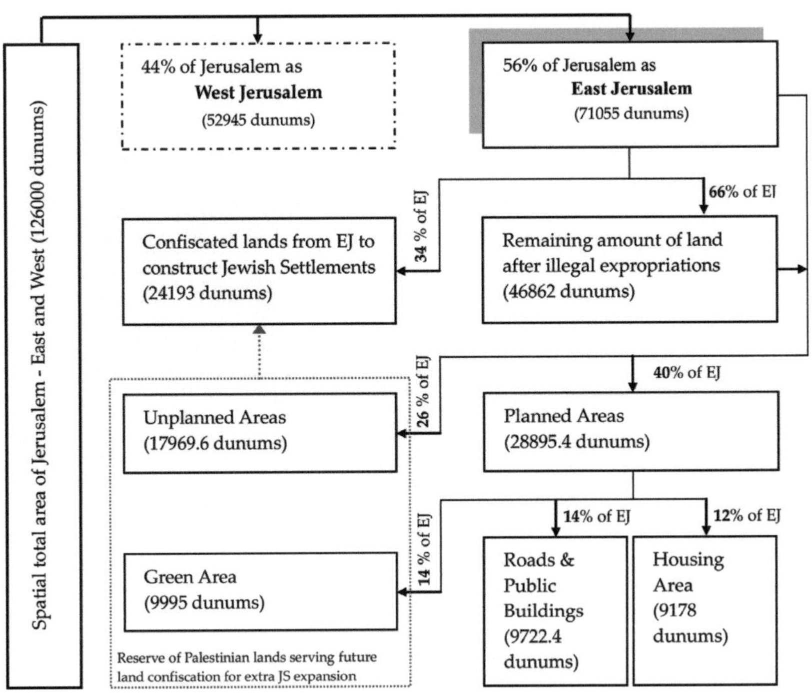

Figure 6. Analytical diagram of the Israeli regressive land use planning in EJ [author].

land cut-off in the Israeli planning system. It is noteworthy mentioning that only 40% of the total area of EJ has valid approved plans. However, due to regressive Israeli sophisticated planning regulations, Palestinians in EJ are neither able to develop most of the planned areas for their critical needs nor the other unplanned zones. Approximately 34% of EJ is confiscated for Jewish settlements, while 26% is still unplanned zones. Accordingly, Palestinians living in EJ face serious challenges in finding enough room for their future development and expansion. Thus, more than 74% of the total area of EJ is part of the "static sphere" where vacant Palestinians' lands are prevented from any kind of development by the regressive Israeli planning policies and transferred into the future for the purpose of establishing Jewish settlements, see **Table 1**.

The Israeli planning policies treated the Palestinian residents of the city as unwanted immigrants and worked systematically to drive them out of the area. Hence, the Israeli government created systematic mechanisms for expropriating the vacant Palestinian lands and limit their future development. One of these is the regressive land use planning policy by which huge areas are designated as green spaces in the Palestinian local town plans. Approximately 35% of the total planned area of EJ is zoned for this purpose. As such, construction is completely forbidden in open landscape areas, where the permitted usage only includes forestry, groves, agriculture, and the use of pre-existing roads.

Unlike open public land, open green spaces are not expropriated from their owners and remain private property unless the Israeli government decides to confiscate these green lands for the purpose of either expanding the boundaries of existing Jewish settlements. This is what happened in Shufat Arab Town, which is surrounded by lands designated as green lands from which lands were expropriated to expand Reches Shufat's Jewish-only settlement shown in **Figure 7**. They can also be used for constructing new Jewish settlements as what happened in the Har Homa Jewish Settlement shown in **Figure 8**.

The adopted Israeli planning policies in Jerusalem have aimed at constraining the future development of the Palestinian residents. Demographically, during more than 70 years of Israeli colonization in Palestine, the Jewish community has grown up amounting today nearly 800% of its original size in 1922 as per the British Mandate Census for that year Palestine's population was characterized as 88% Muslim and Christian Arabs and 12% Jewish. Immigration

Space	Planning and urban management			% according to EJ total area	% of planned area
East Jerusalem	Expropriated Palestinian lands			34	—
	Remained after expropriation	Unplanned areas		26	
		Planned areas	Housing	12	30
			Roads and public buildings	14	35
			Green areas	14	35
	Total			100	100

Table 1. Regressive land use planning in EJ [author].

Figure 7. (Above)—Reches Shufat Jewish-only settlement established over expropriated Palestinian green lands, (below)—Shufat Plan #3456A in Jerusalem General Outline 3000B shows the Palestinian built-up area of Shufat town surrounded by green areas and Jewish settlements in all directions [author].

accounts for most of the increase in the Jewish population at that time, while the increase in the non-Jewish population was due to birth rates [39]. By the end of the British Mandate, immigration influxes saw the Jewish population increase to more than six times more than it was before the Mandate period [40] as presented in **Figure 9**. Hence, the regressive biased Israeli planning policies targeted the Palestinian presence in critical life aspects as presented in **Tables 2** and **3**.

Figure 8. Palestinian green spaces in EJ are expropriated by Israel and used illegally to construct Jewish settlements. Above: Abu-Ghneim Green Mountain transformed into Har Homa Jewish Settlement [author].

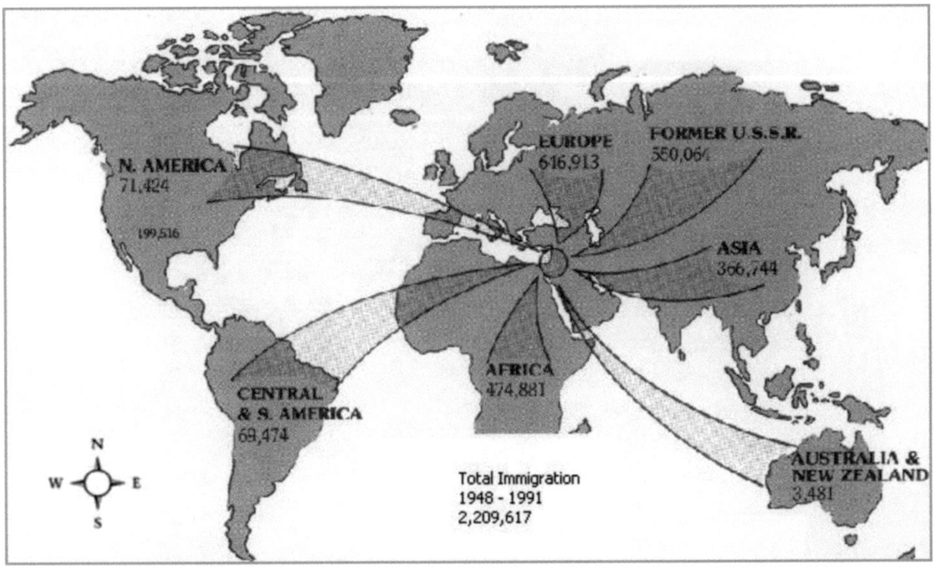

Figure 9. Massive Jewish immigration to Palestine between 1948 and 1991 [20].

The dark side of Israeli planning is evident in EJ where Palestinians live with substandard living conditions. The dual planning criteria clearly manifest the discriminatory treatment of the Palestinians. Israel has closed many Palestinian service-providing organizations in EJ aiming at eliminating the Palestinian identity. It guarantees for Israel the socio-economical and institutional subordination of the Palestinians' life aspects. Israel has used land use planning

Unequal Israeli housing planning policies in Jerusalem					
Construction densities in 1968 (units per dunum)		Average housing density (person per room)		Housing policies for Palestinian Arabs in EJ*	Housing policies for Israeli Jewish in EJ settlements
Jewish	Palestinian	Jewish	Palestinian	• One housing unit was added for each additional 9.7 Palestinian residents thus 10,473 units during 1967–1997 • 5354 housing units were added between 2000 and 2011 (21.9% of total)	• One housing unit was added for each additional 3 Jewish residents thus 70,692 units during 1967–1997 • 19,068 housing units were added between 2000 and 2011 (78.1% of total)
6.1	2.2	1.1	2.2		
Population living in densities ≥3 (person per room)		Population density (person per dunum)			
Jewish	Palestinian	Jewish	Palestinian		
2.4%	27.8%	21.7	14.6		

*Israel adopts aggressive house demolition policy against Palestinians. In EJ, Israel demolished 759 Palestinian housing units and left 4151 Palestinians homeless during 2001–2018 [20, 41].

Table 2. The discriminant Israeli planning policies in Jerusalem—East and West [author].

http://dx.doi.org/10.5772/intechopen.78779

Service	EJ (service for Palestinians)	WJ (service for Israelis)
Status of sewage network (km)	67	650
Number of buildings not linked to sewage network	2620	70
Status of roads (km)	87	680
Status of pavements (km)	73	700
Number of social care centers	3	20
Area/number of public parks	324 (dunums)/45	5216 (dunums)/1087
Average number of persons per public park	7362	477
Number of family health centers	5	32
Average number of children per center	68,882	1821

Table 3. Comparison of municipal services in EJ and WJ [18].

as a control tool to direct the Palestinian development in an "unsustainable" manner, since all the approved plans in EJ are designed to make the land, as much as possible, unavailable for Palestinian future growth. Thus, Israel has utilized political engineering through urban planning that fragments Palestinian neighborhoods in EJ. Israel continues fostering this political engineering, which intensifies ethnic separation between Palestinian and Israeli by adopting sophisticated physical segregation policies on the ground: flying checkpoint, permanent checkpoints and, eventually, the Separation Wall. This Wall was constructed illegally and in direct violation of the International Law. The wall ethnically divides two communities living in one city. It forms segregated clusters and discrete spaces. Further social disintegration, displacement and fragmentation of Palestinian families have taken place due to the construction of the Separation Wall, shown in **Figure 10**. The Separation Wall disconnects the Palestinian in the oPt from what used to be their economic hub, and in turn, disrupts the entire Palestinian economy by constricting the flow of income. All these policies have created a uniquely political architecture in Jerusalem that delineates aggressive military and security morphologies.

Figure 10. Divided Palestinian communities surrounded by the Israeli Separation Wall in EJ [author].

7. Conclusion

The growing impact of the 3P concept in Jerusalem has created shocking realities on the ground. It has generated severe adverse impacts on Palestinian life. Since the illegal annexation of EJ in 1967, the Israeli regressive planning policies in EJ have been targeting the Palestinian presence via imposing complex spatial planning policies aiming largely at marginalizing the Palestinian communities, forming deprived spaces for the Arabs, and minimizing Palestinian demographic and cultural identities. Israel has effectively frozen most of the Palestinian vacant lands in EJ preventing therefore Palestinians from any kind of development there. Instead, these vacant lands are systematically, and illegally, confiscated for the purpose of constructing Jewish settlements, thus changing the physical Palestinian landscape, as well as altering the cultural identity of space and the demographic character of the city. According to the current Israeli regressive spatial planning policies, 74% of the total area of EJ is zones where Palestinians are not allowed to utilize for their basic or urgent development; of which 34% is expropriated lands thus deducted from Palestinian EJ lands and annexed, illegally, for the advantage of the Jewish population; 26% is unplanned, and therefore undeveloped areas, and 14% is green areas systematically subjected into future Israeli expropriation for the purpose of illegal construction of additional Jewish settlements.

It is evident how the Israeli regressive planning has forced Palestinians in EJ to suffer in satisfying their essential daily needs. The illegal expansion of Jewish settlements and the continuous spreading of inspection checkpoints (**Figure 11**) have damaged the social and urban profiles and shrunk the space available for Palestinians to live and work, and therefore has deepened a general feeling of insecurity. On contrary to international law, Israel constructed the Separation Wall to isolate Jerusalem from the rest of the occupied Palestinian territories severing therefore the city from its socio-economic support base. Thus, Palestinian neighborhoods in EJ have collectively faced sudden rupture of their social, environmental, cultural, and economic life aspects, that is, all the 4DSS dimensions are adversely impacted. Hence, Palestinians in EJ face definitively unsustainable mode of development. The role of politics in shaping architectural space in the Palestinian areas in EJ is underestimated relative to the significant effects it has over life aspects. The examination of the analytical spatial context reveals the extent to which politics and power were evident in producing divided urban forms, in the conflict areas. Indeed, "politics" has played a significant role in defining the

Figure 11. Israeli check point at EJ northern entrance in Qalandia [author].

lifestyles of the Palestinians by forcing them to meet regular challenges. The regressive Israeli planning policies with all their inevitable consequences against the Palestinians continue and include: land expropriation, Palestinian neighborhood fragmentation, massive construction of Jewish settlements, restrictions concerning Palestinian building, destruction and confiscation of homes, lack of adequate public infrastructure, prejudicial land and zoning laws, changing residency rights and permits, and construction of the Separation Wall. These are concrete and sorrowful facts indeed. Palestinians suffer in consequence.

Author details

Raed Najjar

Address all correspondence to: raedfnajjar@gmail.com

Post-Doctoral Associate Researcher in Spatial Planning, Urbanism, and Sustainability, TU Dortmund, Dortmund, Germany

References

[1] Fraassen C. An Introduction to the Philosophy of Time and Space. New York: Columbia University Press; 1985

[2] Kempen R. Divided cities in the 21st century: Challenging the importance of globalization. Journal of Housing and the Built Environment. 2007;**22**:13-31

[3] Caner G, Bölen F. Urban planning approaches in divided cities. A|Z ITU Journal of Faculty of Architecture. 2016;**13**(1):139-156

[4] Hasson S, editor. Jerusalem in the Future: The Challenge of Transition. Jerusalem: The Floersheimer Institute for Policy Studies; 2007

[5] Thiong'o N. Enactments of power: The politics of performance space. The Drama Review. Massachusetts: MIT Press; 1997;**41**(3):11-30

[6] Shome R. Space Matters: The Power and Practice of Space, Communication Theory. International Communication Association; 2003;**13**(1):39-56

[7] Smith N, Katz C. Grounding metaphor. In: Keith M, Pile S, editors. Place and The Politics of Identity. New York: Routledge; 1993. pp. 67-83

[8] Foucault M. Power/Knowledge. New York: Batheon Books; 1980

[9] Massey D. Spaces of politics. In: Allen J, Sarre P, editors. Human Geography Today. Cambridge, UK: Polity Press; 1999. pp. 279-294

[10] Grossberg L. The space of power, the power of space. In: Chaimbers I, Curti L, editors. The Postcolonial Question. New York: Routledge; 1996. pp. 169-188

[11] Gruen V. The Hart of Our Cities. New York: Simon & Schuster; 1964

[12] Pudup M. Arguments within regional geography. Progress in Human Geography. 1988; **12**:369-390

[13] Ollman B. Dialectical Investigations. London: Routledge; 1993

[14] Lefebvre H. Dialectical Materialism. London: Jonathan Cape; 1968. 111 pp

[15] Bohm D. Wholeness and the Implicate Order. London: ARK Paperbacks; 1980. 11 p

[16] Grosholz E. Cartesian Method and the Problem of Reduction. Oxford University Press; 1991

[17] Smith N. Uneven Development. Oxford: Basil Blackwell; 1984

[18] Lefebvre H. Critique of Everyday Life. Vol. 1. London: Verso; 1991

[19] Vitousek P, Mooney H, Lubchenco J, Melillo J. Human domination of the Earth's ecosystems. Science, New Series. 2007;**277**(5325):494-499

[20] Najjar R. Spatial planning, urban land management, and political architecture in the conflict areas—Jerusalem case study [thesis]. Dortmund: TU Dortmund; 2013

[21] Naess P. Urban planning and sustainable development. European Planning Studies. 2001; **9**(4):503-524

[22] Spoelstra S. Sustainability research: Organizational challenge for intermediary research institutes. NJAS—Wageninge Journal of Life Sciences. 2013;**66**:75-78

[23] McNeill J. Something New Under the Sun: An Environmental History of the 20th Century World. New York: Norton; 2000

[24] Campbell S. Green cities, growing cities, just cities? Urban planning and the contradictions of sustainable development. Journal of the American Planning Association. 1996;**62**(3):296-312

[25] Albrechts L. From traditional land use planning to strategic spatial planning: The case of Flanders. In: Albrechts L, Alden J, da Rosa Pires A, editors. The Changing Institutional Landscape of Planning. Aldershot: Ashgate; 2001

[26] Yiftachel O, Alexander I, Hedgcock D, Little J, editors. The Power of Planning: Spaces of Control and Transformation. The Netherlands: Kluwer Academic; 2001. pp. 1-19

[27] Reiter W, Sanders F. Physical Planning: Policies, Methods and Techniques. Delft: Delft University of Technology; 1998

[28] Cherry G. Cities and Plans. London: Blackwell; 1988

[29] Hall P. Cities of Tomorrow: An Intellectual History of Urban Planning and Design in the Twentieth Century. London: Blackwell Publishing; 1988

[30] Healey P. The treatment of space and place in the new strategic spatial planning in Europe. International Journal of Urban and Regional Research. 2004;**28**(1):45-67

[31] Booth P, Breuillard M, Fraser C, Davis D, editors. Spatial Planning Systems of Britain and France. London: Taylor and Francis; 2007

[32] Rodwin L. Cities and City Planning. New York: Plenum Press; 1981

[33] Kanaan A. Jerusalem: Political and Religious Status of Jerusalem among Muslims. Amman: The Royal Committee for Jerusalem Affairs; 2005

[34] Mendelsson D. British Rule. Jeruslem: Department for Jewish Zionist Education, The Jewish Agency for Israel; 2007

[35] UN. Meetings Coverage and Press Releases. 37th Meeting (AM) Tenth Emergency Special Session; GA/11995. 21 December 2017

[36] Hasson S. When two rights collide: Some lessons from Jerusalem. In: Roger L, David S, editors. Geographies and Moralities: International Perspectives on Development, Justice and Place. Oxford: Blackwell; 2004. pp. 149-164

[37] Khalidi R. Palestinian Identity: The Construction of Modern National Consciousness. New York: Columbia University Press; 1997

[38] Kotek J. Divided cities: In the European cultural context. Progress in Planning. 1999;**52**: 227-237

[39] Justin M. The Population of Palestine: Population History and Statistics of the Late Ottoman Period and the Mandate. New York: Columbia University Press; 1990

[40] Morris B. Revisiting the Palestinian exodus of 1948. In: Rogan E, Shlaim A, editors. The War for Palestine: Rewriting the History of 1948. UK: Cambridge University Press; 2007

[41] B'Tselem. Demolition of Houses in East Jerusalem: 2004-2018. Jerusalem: The Israeli Information Center for Human Rights in the Occupied Territories; 2018

Chapter 12

A Critical Assessment of the Adaptive Capacity of Land Use Change in Chile: A Socio-Ecological Approach

Daniela Manuschevich

Additional information is available at the end of the chapter

http://dx.doi.org/10.5772/intechopen.80559

Abstract

Land use and land cover change (LULCC) can be defined as a socio-ecological system (SES): social, economic, and political processes in interaction with ecological processes result in a given land use trend. Instead of forest recovery, Chile has been identified as a case of a forest transition dominated by commercial plantations. This chapter aims to examine the process LULCC in Chile from a socio-ecological perspective. Drawing upon frameworks of SES by Scheffer et al., this chapter analyzes the adaptive capacity of LULCC in Chile. First, SES concepts are presented. The next section is a summary of the political and economic process that underpinned the plantation transition in Chile and its consequences on the landscape. In light of SES theory, the 518,174 hectares wildfire observed in 2017 is a consequence of the lack of adaptive capacity. Nevertheless, Chile's LULCC is unlikely to change due to abovementioned dynamics. Finally, this chapter discusses the implications for policy making and the global forest transition discussion. In summary, using the case of Chile, this chapter aims to contribute to SES theory and forest policy, seeking sustainable futures based on a systemic view.

Keywords: socio-ecological, forest transitions, global south, tree farms, policy

1. Introduction

Adaptive capacity can be defined as the ability of a system to adapt to changing internal demands and external circumstances [1]. Social and ecological systems are interlinked systems where outcomes result from the interaction between social and ecological dynamics. Dynamics among social and ecological systems can result in a socio-ecological system that has high adaptive capacity—this is a system able to respond to changing condition—or a system that, despite the intensity of the stimuli, does not respond. Whether the stimulus is climate change, a natural disaster, a new societal preference, a new invasive species, an adaptive SES would respond to the new demands and stresses.

Land use and land cover change (LULCC) can be defined as a socio-ecological system (SES). LULCC is the result of the interaction between natural and human systems [2]. Social, economic and political processes in interaction with ecological processes result in a given land use trend [3]. Forest transitions describe a systemic land use trend change, where a geographic region switches from deforestation toward forest gains [4]. Chile has been identified as a case where, instead of forest recovery, the transition has been dominated by tree farms [5]. Despite the social demand for native forest protection that was raised after the fall of Pinochet's dictatorship, the native forest law, meant to foster and protect native vegetation, had no significant effect [6]. Moreover, in 2017 nearly 518,174 ha were burnt in a massive wildfire that lasted for at least 15 days, affecting three administrative regions. Nearly half of the area burned was fast-growing plantation [7]. Despite this disaster the Chilean State has created a new fund for tree farms. This chapter presents an analysis of the adaptive capacity of the land use trend observed in Chile from a socio-ecological perspective. Through this assessment, it is expected to draw lessons for other countries that are following the Chilean example, while providing deeper insights regarding theoretical questions of land use transitions in the Global South [8].

1.1. Afforestation in Chile

Similar to many developing countries, Chile had serious deforestation and related erosion problems in the twentieth century [9]. However, from 1973 to 2012, Chile expanded the extent of forestry plantations from 330,000 to nearly 2 million hectares (**Figure 1**) [9]. Chile has not only increased its afforested area during this 30-year period but also developed one of the most vigorous forestry sectors worldwide by increasing forestry exports more than a thousandfold in nominal terms, from US $36.4 million in 1976 to US $5.271 million in 2016 [10]. These policies have resulted in successful increases in tree cover, but not in native forest cover [11, 12]. Afforestation in Chile resulted from planting fast-growing trees under intensive management. These plantations use non-native of single species stands, such as *Pinus* and *Eucalyptus*. Tree farms in Chile can reach up to 1250 seedlings per hectare and require the intensive application of fertilizers and herbicides, while clear cuts are conducted every 18–20 years [9].

Compared to non-native species, native species have rarely been planted in Chile. The national statistics show that the rate of native species afforestation has been orders of magnitude smaller than tree farm afforestation. Moreover, in Chile the proportion of land area covered with native forest has diminished relative to tree farm areas. Between 1997 and 2011, the total covered by native forests increased by 169,008 ha but diminished by roughly 4% in its representation among total national forestry resources [13]. Although native forests still

http://dx.doi.org/10.5772/intechopen.80559

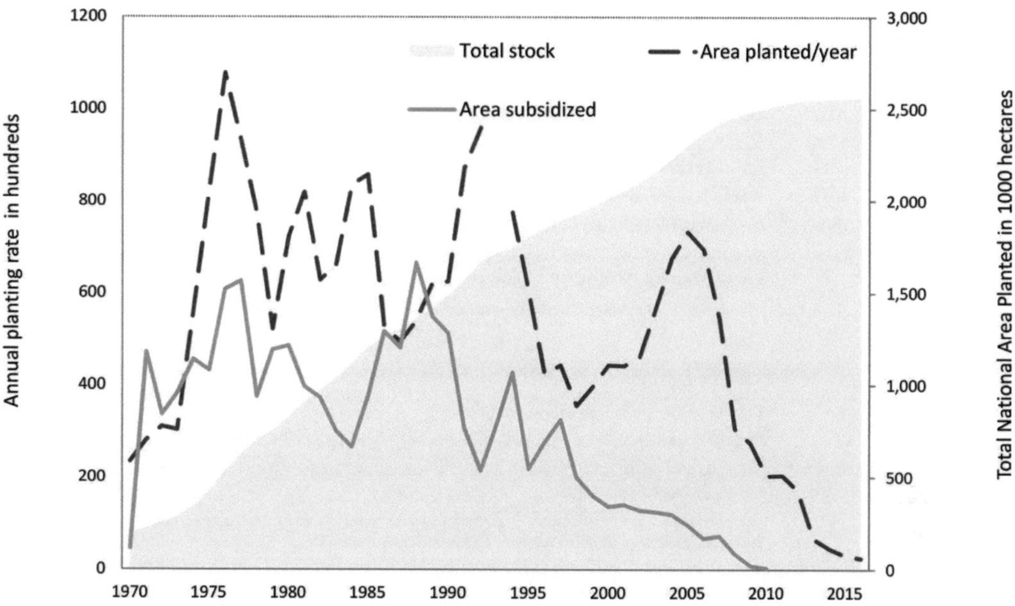

Figure 1. Tree farm national statistics for Chile. Solid area indicates the national stock for each year. Dashed line indicates the area planted with subsidy each year (DL 701), and the solid gray line indicates the area planted per year. (Sources: [7, 9, 74, 75]).

represent the majority of the national forestry resources, native forest are concentrated on the southernmost area of Chile where there is no overlap with biodiversity hotspots [14–16]. A closer look to the native forest subcategories shows that the most valuable native forest types are still diminishing. From an economic perspective, the most valuable woods (*Nothofagus* spp.) have not increased in area [13].

From an ecological perspective, valuable forest has also been lost. For example, mature forest that provides habitat for endemic and endangered forest specialist species, such as the families Rhinocryptidae and Berberidopsidaceae [17–19], has also diminished [13]. By differentiating between forest types and tree cover, comparisons indicate that Chile's afforestation policy has only increased the total area planted with non-native species, which has important consequences for ecosystem functions, such as water provision [20–22].

Even though afforestation came to solve serious erosion problem on the twentieth century (**Table 1**), today there are economic and political dynamics that limit SES adaptation to the new socio-economic conditions. The problem is not widespread erosion but water provision and regulation and resilience against climate change. Chile is already experiencing a mega-drought [23], and it is expected that the frequency of such events will increase [24]. The following section is an introduction, based on the work of Holling, Scheffer, and others, to some of the basic SES concepts.

Key milestones of the Chilean land use social-ecological system	
Deforestation and soil degradation	
1541	The capital of Chile, Santiago, is founded by Spaniard conqueror.
19th C	Native forests are burnt and used in construction, mining and railroads. Native forest is replaced by agriculture.
1880s	First trials of Pine and Eucalyptus plantations.
1931	Fist Forestry Law provides tax breaks for forested areas.
1940's	Several experts publish books concerning deforestation and erosion. Erosion is estimated in 4,000,000 ha mostly located in the coastal area [48,89]
1952	Foundation of the forestry trade association, called National Wood Corporation (CORMA)
State-leaded national program of reforestation: Stock of non-native trees increased in 330,000 ha	
1960	Creation of the public Forestry Institute (INFOR)
1964	Foundation of the public national research institute of natural resources (IREN)
1970	Foundation of the first public agency for public national afforestation (CONAF). CONAF had 18,000 employees in 1973.
1972	The Corporation for Development Production (government) operated the first two pulp processing facilities (Celulosa Arauco and Celulosa Constitución)
Restoration though shock therapy; by the end of 1989 the stock of non-native trees increased in 1,303,160 ha.	
1973-1990	INFOR, IREN, CONAF, and most of the processing facilities become private institutions, or self-financed. Bans on log exportation are removed as well as price bands of competing commodities (e.g.: wheat price bands). State-owned facilities and more than 60,000 ha of state-owned plantations are sold to privates (Arauco, Constitución, Inforsa).
1973	A coup d'état leaded by Pinochet takes the government.
1974	DL701: Main subsidy for tree plantation is decreed. The state pays 75-90% of the costs of any tree plantation or forest improvement, regardless of the tree species planted. Tax breaks are maintained in place, private property is secured by several policy reforms.
1975	Establishment of a state-backed system of credits for afforestation.
1980	The military-led government decrees a new political constitution
1981-1983	Economic crisis resulted reinforced concentration of facilities and lands into few companies financed by international investors
1988	National referendum ends the dictatorship
Rigidity trap: by the end of 2011 the stock of non-native tree increased in 1,063,241 ha, while the stock of native species increased in less than 67,847 ha.	
1990-present	In this period Chile signs several free trade agreements
1990	A week after the democratically elected government takes offices, the president decreed the prohibition to cut the Monkey Puzzle tree *(Araucaria araucana)* (Supreme decree 59)
1992	The first draft of the Native Forest law is sent to the House of Representatives
1993	The Native Forest bill passes to the Senate.
1994	Chile is provisionally accepted in the North American Free Trade Agreement (NAFTA)
1998	DL701 is renewed for small landowners (Act 19,561)
2001	Chile signs the Free Trade Agreement with the United States
2007	Policy stakeholders agree on a short version for the Native forest bill.
2008	The Native Forest Law was enacted (Act 20,283).
2011-2012	The DL701 is renewed for two years due to the earthquake (Act 20,488).
2015	Act 20,488 expires without renovation.
2017	January: Large wildfire affected south central Chile
2018	CORFO launched a public-private investment platform

Table 1. Key milestones in the Chilean land use socio-ecological system.

2. Socio-ecological framework

The socio-ecological system (SES) is an approach to understand the outcomes that emerge from the interaction between both social and ecological dynamics [25]. Historically, social and natural sciences have developed independently of each other, thus generating valuable but separated knowledge [25]. Therefore, policy is often based on a single discipline and fails to address the complexity of the socio-ecological systems. Neglecting complexity can easily result in unintended consequences, surprises, pathologies, or traps [26, 27]. In rigidity traps, institutions become highly connected, self-reinforcing, and inflexible so that "forces of power, politics, and profit are reinforced one another" [1, 26]. At the same time, rigidity traps are "accidents" waiting to happen. In rigidity traps interconnectedness is so high that any random event, such a as a fire or disease, can cause a system ripple [26]. By analyzing natural resource policy as a SES, it is possible to see those broader dynamics.

The adaptive change theory developed by Gunderson and Holling [26] was originated in ecological studies on population dynamics. The original ecological studies sought to explain why there could be multiple stable dynamics among two populations, as well as cycles of rapid change, including outbreaks and population decline [28]. The populations' studies found that changes in slow-changing variables, as well as stochastic events, can explain outbreaks and population decline and even new stable states [28–30]. Based on these studies, Holling and others elaborated on the idea of multiple alternate state in SES, focusing on the relationships between "slow" and "fast" variables [26]. Gunderson and Holling [26] have proposed that social variables are slow-changing variables that control change in SES. More specifically, Scheffer et al. [31] argue that large-scale cultural variables function as the slow variables in SES.

Scheffer et al. [31], hereafter SSEF (Scheffer's socio-ecological framework), proposed an explanation of how a SES might depart from an adaptive behavior by drawing on sociology, neoclassical economics, and systems ecology. Briefly summarized, SSEF starts by focusing on a hypothetical ecosystem where the relationship between the ecosystem integrity and the stress are noncontinuous and nonlinear.

The classical example is water turbidity and nutrient addition to a shallow lake. Low levels of nutrients result in low water turbidity. As more nutrients are added to the lake, turbidity increases up to a threshold where light levels are insufficient to support aquatic vegetation, and it disappears. After this point, turbidity continues to increase, but reducing nutrient concentrations does not result in more transparent water. SSEF works under the assumption that ecosystems do not have linear dynamics and restoration is much more than doing the opposite that damaged the system [31]. Because SSEF is a socio-ecological framework, it is important to consider how changes in the ecosystem affect total welfare of people. Total welfare is the third variable of the SES. In the SSEF, a hypothetical rational manager "knows" the optimal combination between ecosystem stress, ecosystem integrity, and maximum realizable welfare [31]. The realizable maximum welfare is different from the theoretical welfare because of the nonlinear relationship between stress and ecosystems integrity.

A more realistic account of SSEF includes the effects of politics. Different groups of people have different power, and, in some cases, political pressure from powerful groups influences the rational manager. The rational manager allows more stress over the ecosystems than what is optimal, resulting in a maladaptive behavior. In reality, this hypothetical rational manager, rather than a superhuman entity, is the outcome of the political struggle itself [31]. This means that, even in an ideal democracy, the regulating entity or rational manager will be elected in some political process.

SSEF elaborates further on this framework by incorporating ideas from discourse analysis and political ecology. Powerful parties not only influence the regulator but also can shape the meaning of the problem itself through signification, legitimization, and domination [31]. Literature on environmental discourse and policy has shown that power dynamics affects the way the environment and policy are constructed; this is how the environment and people are treated [32–34]. Power dynamics are evident at the international governance level [33], at the national level in countries with strong postcolonial legacies [35], in segmented societies [34], and in countries with recent dictatorships, such as Chile [36]. The main argument of this chapter is that the political-economic context, sustained by specific discourses (legitimization), works as the "slow variables," explaining the current lack of adaptive capacity of LULCC in Chile. The following section summarizes the political and economic changes that fostered timber plantations in Chile; this means 45 years of land use change through the lenses of the SSEF.

2.1. Soil degradation and the state-led national program of reforestation

By the beginning of the twentieth century, erosion and deforestation, partially inherited from the Spanish colonization, were important problem in Chile (**Table 1**). Since 1890, wealthy families started planting *Radiata pine* [37]. Studies done by one the first naturalists, named Albert, suggested that *Radiata pine* could be an appropriate species for reforestation due to its capacity of reaching harvesting size in just 15 years. In 1931 the first Forest Law, provided some tax breaks to forest owners. Nevertheless, the 1931 law did not have a large effect in terms of native forest conservation but did provide some incentive for non-native species. Between 1925 and 1936, 2.580 hectares were planted with non-native species [9]. By 1942, Elgueta [38] estimated that 4,000,000 ha were eroded and claimed that erosion was a national tragedy (**Figure 2**). As shown in **Figure 2**, the area is indicated as a continuous area of severe erosion. However, studies have shown that by 1955 and 1975, there were several large patches of native forest in the area, such as the land near Constitución or near Maule [5, 39]. It is unclear why this map was drawn in such a coarse resolution, but it is clear that the map shows erosive processes that covered a vast, continuous area, reinforcing the idea of an "environmental disaster."

The 1929 economic crisis inaugurated a period of imports substitution in several countries in Latin America, including Chile [40]. The import substitution model aimed to foster industries that would complement the agricultural and natural resource sectors [40]. The Chilean Development Corporation, CORFO, was created to foster the industrial sector. The CORFO, through loans and public-private investments, provided access to capital to extend tree plantation and develop the first forest industries [37]. Radiata pine was the chosen species due to its capacity of providing large returns in a short time. Between 1939 and 1942, nearly 5.492

http://dx.doi.org/10.5772/intechopen.80559

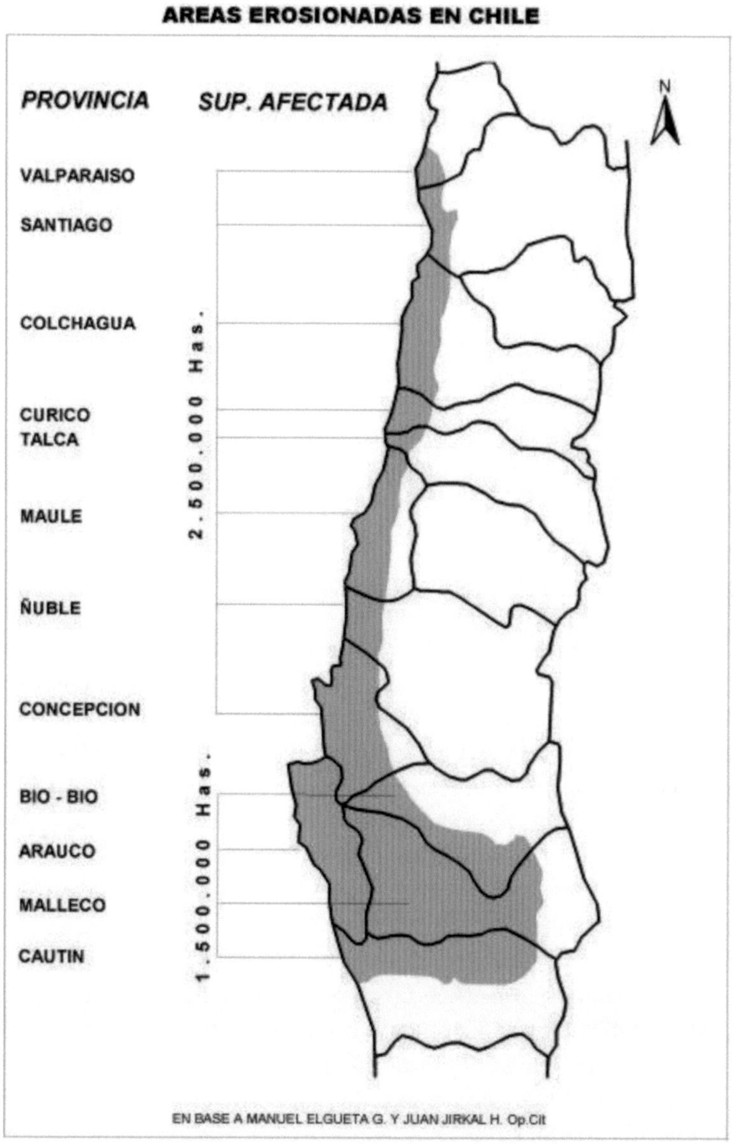

Figure 2. Map depicting the area under severe erosion in 1942 based on Elgueta and Jirkal. Notice the continuity of the eroded areas for over 12 provinces, accounting for 4,000,000 hectares (Reproduced with author's permissions).

hectares were planted. In 1950, the nascent pulp industry needed to secure the future supply of raw material [37]. The forestry industry requires long-term planning since it is dependent on tree's growth cycle. By 1952, several forestry entrepreneurs, including state-owned forestry companies, formed a single trade association called the National Wood Corporation (CORMA). The CORMA was created to advocate for the interest of the forestry companies [9]. In the next decade, a series of private and public-private forestry conglomerates were created, such as the Celulosa Arauco [37].

In the same period, the colonization of the frontier forest in southern and more isolated areas was a strategy to exert territorial presence and control. Colonization implied a new frontier for deforestation where this phenomenon wasn't already happening, implying further reduction of the area covered by native forest [9].

In the 1960s, the Frei Montalva administration led the agrarian reform. The CORMA managed to declare that the forested lands could not be legally expropriated under the reform. To protect their property, several landowners declared their land as forest land [9]. In parallel, the Frei Montalva administration created a national program of reforestation, the forest service (COREF, later called CONAF) and a national research institute of natural resources (IREN) [9]. Later, the democratically elected socialist president Salvador Allende took office in November 1970 and continued the afforestation program alongside the agrarian reform initiated in the previous administration. In September 1973, the president was overthrown by a military-led coup d'état. By 1973, at the national level, 330,000 hectares were planted with non-native plantations (**Table 1**). Although these LULCC figures are not negligible, the largest land use transformation was still ahead.

2.2. Renovation through shock therapy: 1973–1990

Between 1973 and 1990, Chile was ruled by a military dictatorship. The military took over in September 1973, shutting down the judicial and the legislative powers, destroying all opposition media, and executing or exiling political opponents. After the first years of the coup d'état, the objective of the military-led government was to restructure the economic system. The timber sector was seen as an important niche where Chile could develop a robust export market [37]. Following this logic, a series of policy changes were made. Land property rights were secured by a new constitution, and common land tenure was abolished. In 1974, the government passed the Decree Law 701 (DL 701) which states that the government would pay private owners between 75 and 90% of the costs of any tree plantation or forest improvement, regardless of the tree species planted, including both native and non-native species [41]. In terms of area, DL 701 came to subsidize large portions of the planted area at the national level. For example, in 1982 up to 88% of the planted area at the national level was subsidized through this law (**Figure 1**). The annual rate of afforestation passed from nearly 30,000 per year in 1972 to 85,772 ha in 1985. For the next 14 years, the average rate was 57,591 hectares per year (1976 to 1990, **Figure 1**). Until 1999, 94.2% of the subsidies were allocated to large and medium landowners (i.e., up to 200 ha or 800 ha in remote locations) [42].

CORMA's influence on the drafting of this decree was celebrated, as this trade association publication consigns:

> *It was in 1974 when the conditions were appropriate to enact a law that will actually encourage afforestation and really put the forest policies forward. Indeed, the Minister of Economy Mr. Fernando Leniz Cerda, who had served as president of the Chilean Wood Corporation (CORMA) for professional reasons, knew the forestry sector perfectly. [In addition] at that moment the Executive Director of the National Forestry Corporation was a forester with familial affinity [kinship] with the President of the Republic [Pinochet]. The DL 701 was developed and drafted exclusively by the Ministry of Economy,*

without external interference, taking advantage of some ideas and suggestions from the National Forest and the Chilean Wood Corporation, [and from] the Association of Foresters and the Universities.

("Temas de Fondo" Section p 127, CORMA vol 241, Nov-Dec emphasis added)

In other words, The CORMA had a monopoly in the design of DL 701 in 1974. Moreover, CORMA's connection with the military government was not only ideological but also involved kinship ties between President Pinochet and the CONAF executive director Julio Ponce Lerou, recently charged with corruption [43]. In the same venue, two more laws benefiting CORMA's interests were passed in this period: a decree allowing the exportation of wood products at any processing stage and Decree 600, which allowed foreign companies to buy Chilean external debt in exchange for state-owned companies [44]. During the 1981 economic crisis, the national currency was devalued relative to the US dollar as a means to reduce the external debt, which resulted in the bankruptcy of many forestry companies [9]. While small companies went bankrupt, large and medium forestry companies were taken over by the government and resold to holdings that could bring dollars to the country [9]. The removal of wheat price bands, as well as the removal of bans on logs export, also added to the profitability of tree farms [45]. The complete set of economic measures restructured the forestry sector and resulted in a massive rearrangement of land, companies, and facilities, which became concentrated in a few hands. As of 1997, 57% of forested land was owned by eight big companies [46].

Seven million hectares of native forest were lost in 30 years. Several studies indicate that at the national level in 1965, the nation had nearly 20 million hectares of native forest; meanwhile, in 1996 the national land cover assessment estimated a total of 13 million hectares [9]. A study including nine areas in south central Chile [39] estimated an average native forest loss of 31% for this period with an annual rate of deforestation of 2.9%. However, in the area where most of the processing facilities operate, nearly 72% of the area covered by native forest was then converted toward tree farms [39], which is the same area that harbors most of the plant diversity at the species level [16]. On average 20% of native forest loss was converted toward tree farms [39].

In summary, forestry plantations were seen as the best way to foster a competitive economic sector and recover the 4 million hectares eroded in south central Chile [47]. The tragedy of erosion, claimed in 1940, was skillfully used to extend tree farms, despite the erosion status of a particular area, as many of them were actually covered by native forest.

Looking through the lenses of adaptive theory, since 1973 the economic reforms conducted by the military-led government implied deeper systemic changes that generated an exceptionally stable land use trend. Yet the Chilean afforestation policy took place in the context of a dictatorship and resulted in a deep legacy of inequality and environmental risk sustained by very resilient political dynamics [48]. A single trade union (CORMA) was able to influence policy making without much counter balance during the dictatorship and afterward [36, 49]. This policy reform faced no political opposition, as was characteristic during the Pinochet dictatorship. Since economic reforms happened under a dictatorship, where political opposition was repressed, it would be expected that once the democracy had returned to Chile a more "balanced" policy would be in place in terms of native forest conservation.

2.3. Rigidity trap: 1990 and on

Rather than a change in the land use trend, due to the new social context and social demands, in this period the expansion of tree farms intensified, while the legal regulation for native forest exploitation came only after 16 years of congressional discussion. In this period the average area planted with tree farms per year was 43.663 hectares, reaching a maximum of 95.933 ha in 1992 (**Figure 1**). Meanwhile, the deforestation rate for the period 1990–2000 was 1.6% per year, then increasing to 2.4% per year between 2000 and 2010. Compared to the 1970–2000 period, the figures on hectares converted from native forest to tree farms on average were reduced by a third but still accounted for a 35% of the forest loss [39].

Firstly, tree farm expansion and native forest loss continued due to the open economy policies fostered by the left coalition *La Concertación*, the maintenance of the subsidy for tree farms and the delay and reduction of the regulatory power of the Native Forest Law. Free trade agreements were the key to access new markets for timber products. *La Concertación* political project was based on the continuation of the open market economy [50]. Therefore, between 1990 and 2010, 17 trade agreements were signed, such as the MERCOSUR (Common Southern Market in 1996), Canada (1997), Mexico (1999), Caribbean nations (2005), the European Union (2003), the USA (signed by Chile in 2001, ratified 2004), China (2006), India (2007), and Japan (2007), to name a few. This implied that Chile had access to a much larger market were both timber farm products and native forest chips could be sold[1]. Trade opening also fostered tree farms by making other land uses less competitive, such as wheat [45]. The economic reforms that started during the dictatorship were then intensified in the democratic governments.

Secondly, the subsidy for tree farms was maintained even though several studies indicated that tree farms were profitable enough without a subsidy [52, 53]. After the expiration of the subsidy in 1996, the political discourse turned to the "social" side of planting. In the previous period (1974–1996), the subsidy was concentrated in large landowners [54], so now a more social focus was needed. This discursive turn sought the promotion of tree farms among small and medium landowners: a frontier that could not be reached in the previous period. Thus, the DL 701, meant to last until 1995, was then replaced by another law [18, 55] to maintain the subsidy but reorienting it to small and medium landowners until 2008 [48]. Peasant's associations played a large role in the subsidy extension. The discourse uttered by peasant organizations, which have the closest contact with native forest, presented significant tensions of identities. On one hand, peasant's organizations understand themselves as a collective that has a historic relationship with land and forest, but those collective constructions were neglected in favor of a discourse of individual ownership. Given the political constitution of Chile, which only acknowledges individuals' rights, the farmers and peasants' organizations only get to claim political power when they articulate themselves as individual owners. In this

[1]For example, in 2016, Chile exported $2.8 billion dollars in wood pulp, and 66.7% went to Asia, where 45% went just to China. In contrast, 11 years before, in 1995 wood pulp exports were just a half, $1.4 billion dollars, and 45.6% went to Europe, while none of the countries concentrated more than 18% [52].

way, legacies from the dictatorship are legitimizing today's construction of forest policy [36]. The tree farm subsidy expired again in 2010, but the massive earthquake (8.8 Mw) in this same year was used to maintain the subsidy for another 2 years. The presidential message, asking the congress for an extension, was addressed as follows:

> *The extension of this initiative is even more important, considering the current situation in the country, marked dramatically by the earthquake and tsunami of last February 27 of the current year. This initiative will create jobs in forest areas strongly affected by the catastrophe that hit our country [54].*

The argument in 2010 implied that if the DL 701 was a job creation policy, then it could not be denied to the already vulnerable people. This new extension meant to last until 2014. Finally, the subsidy was not renovated in 2015 due to a large corruption case that involved the main forestry conglomerates, who are the natural buyer monopsony of forestry products [55]. Apparently, around 2010, tree farm expansion had reached a saturation point (**Figure 1**). Since 2006 the rate of plantations had diminished, while the stock of tree farms is constant. However, this statistical trend does not indicate a systemic direction change in terms of LULCC but the maintenance of the *status quo*. Since 2010 the total national stock has remained around 2.5 million hectares.

Thirdly, the Native Forest Law (NFL), created to respond to the social demand for native forest conservation, took 16 years to become a law and does not have an effect in promoting significant forest cover. In 1992, the Aylwin administration, the first democratic government after the dictatorship, decided to promote native forest conservation through the NFL, as a means to preserve the remaining forest (**Table 1**). The original version of the NFL included strict regulations for forest cuttings, subsidies for plantations, and a ban on native forest replacement by non-native species [54]. At the House of Representatives, the right-wing congressmen questioned the lawfulness of the bill, arguing that banning native forest replacement was a restriction over property rights, rights otherwise protected by Pinochet's constitution [56]. If a bill affected property rights, it would require 2/3 of the votes to pass through congress, which was very challenging, given Chile's bipartisan political system [57]. The law passed 16 years later, in 2008, called for minor regulations and promoted conservation only through a competitive fund for native species management (**Table 1**).

Despite the already mentioned effects of tree farm expansion on biodiversity, several ecosystem services, and the increasing demand for better forest conservation policies, there was no policy change. Rigid systems become "tragedies waiting to happen" [26]; these tragedies might be any random process, such as a drought or an insect outbreak. The rigidity means that the socio-ecological system becomes less able to deal with chance, reducing its own capacity for sustaining people.

The tragedy happened in January 2017. A rare combination of high temperatures and long drought fueled a large wildfire. The fire of January 15, 2017 was the first of a series of aggressive wildfires that lasted for at least 15 days. These fires affected 467,537 ha at the national level (**Figure 3**). Of the area burned, 248,204 ha were tree farms; this is nearly 12% of the national stock of timber farms. This wildfire affected 4,696 farmers, implying the loss of human lives, 850 houses, and the loss of the whole productive capacity of the farmers of the area: herd animals, fodder, crops, and beehives. This fire also affected 77,131 ha of native forest, 76,551 ha of pastures and shrubs, and 512 ha of wetlands [58]. Nearly 76% of the natural ecosystems affected are in a precarious conservation status (danger or critical danger) [7].

In June 2017, the CORFO presented a 13-million-dollar plan to reactivate the economy in the area affected by the wildfire. This plan included direct planting and restoration. The ecological restoration would receive 31% of the funds, while the rest of the money would be used to clean, take out the burned barks, sell them, and plant the same species that were burnt in the large wildfire. In this case, the planting would not be done through DL 701 but as direct contracting through the CONAF. The goal is to replant 40,000 hectares of tree farms in 3 years [58].

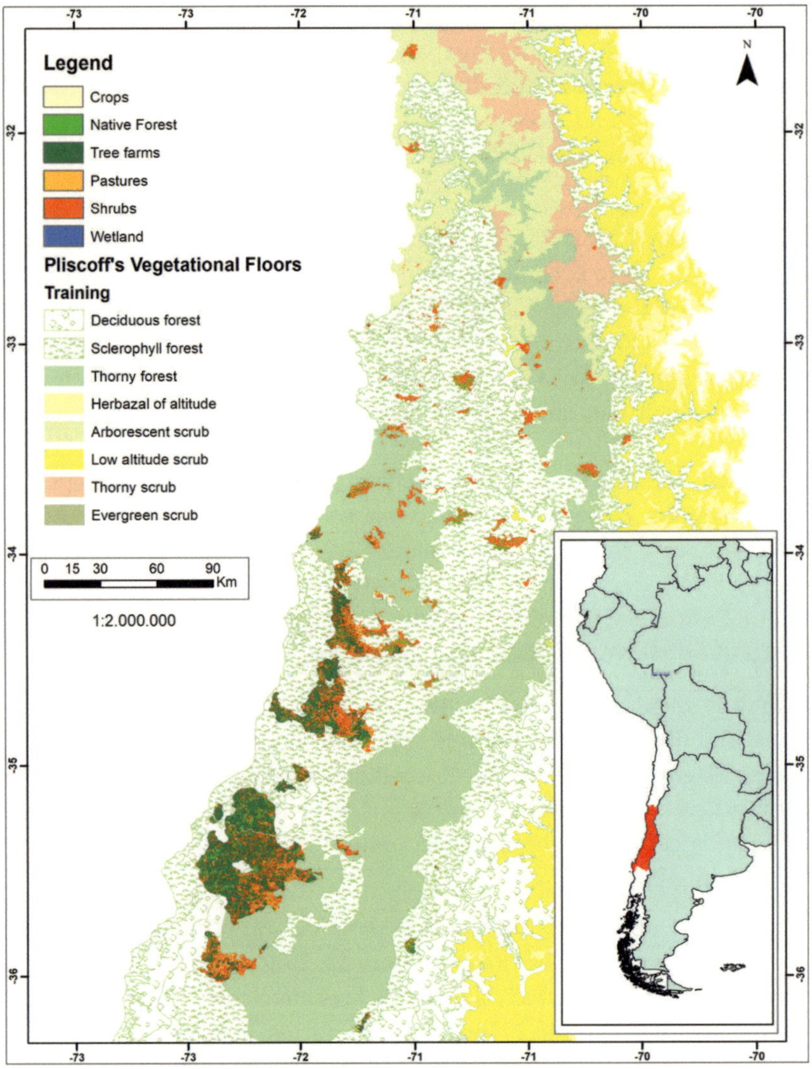

Figure 3. Area burnt in 2017 wildfire in Central Chile. The land use cover burn is indicted with solid bright colors, while vegetational formations (floors) are indicated with softer colors (Sources: [76, 77]).

Since the DL 701 will not be renewed in congress due to the corruption cases associated to the timber industry, the funds would need to come in another way. In September 2017 a new bylaw allowed the CORFO to create a new risk investment fund, called the risk investment funds for forestry and wood promotion. This fund was approved by CORFO's capital risk committee and the general comptroller of the republic [59]. In March 2018, the CORFO launched a platform for the forestry and wood sector [59]. This platform would be based in private and public investments, where the Chilean State, through CORFO, would add up to two-thirds of the fund for 30 years. This fund is meant to "achieve normal progress in the activities of the affected areas" [60]. As it happened in the 1960s, the CORFO articulated the capital needed to continue the tree farm planting, so the state of Chile backed the public- private ventures of tree farms expansion.

3. The adaptive capacity lenses

Using the idea of the rational manager from Scheffer et al. [31], this section will explore the questions: what is affecting the "behavior" of the rational manager? What is delaying adaptation in the Chilean SES despite the long-announced tragedy?

As stated at the beginning of this chapter, the political-economic context, which is sustained by specific discourses, works as the "slow variables" explaining the current lack of adaptive capacity of LULCC in Chile. The economic aspects of the timber plantations are clear. On one hand, the economic structure and the adaptation of specific tree species have configured a very profitable industry that would not have transformed the landscape without the strong support of the state. Now we see that even after the 2017 wildfire, state commitment for supporting these activities has been renewed by securing funds for the next 30 years, despite the change in the climate, vulnerability, and societal preferences. It remains to be seen whether this new fund would imply a second wave of forestry expansion, as it happened in 1975.

As proposed in the SSEF, the political articulation and the social construction of the environment is shaped by political dynamics and discourses. Given the political constitution of Chile, which only acknowledges individuals' rights, the farmers and peasants' organizations only get to claim political power when they articulate themselves as individual owners. The political constitution was utilized by opponents of the NFL to avoid any strong regulation for the conservation of the native forest. In this way, legacies from the dictatorship are legitimizing today's construction of forest policy. Historically, throughout the several policies that fostered timber farms, there was always a discourse that pivoted between the environmental tragedy, such as the erosion in the 1940s, the 2010 earthquake, or the need for fast recovery after the 2017 wildfire, as well as the social benefits of maintaining the DL 701 subsidy for small farmers. In summary, the political-economic variables legitimized, through a combination of strategies, the sustenance of a maladaptive behavior that has fostered landscape transformation for more than 40 years.

Few studies have simultaneously embraced both the social and ecological processes related to adaptive capacity and land use change. Allison and Hobbs found that ecological processes can be the slow variables leading to rigidity traps. In Australia, soil salt increase has led to decreased production, which now resulted in a rigidity trap [61]. In Alaska, regulation and economic cycles also played a central role. The Alaskan timber industry enjoyed 15 years of a policy monopoly until production decreased, the material interest of small contractors was affected, and federal environmental regulations came into place [62]. Both studies highlight the importance of spatial and temporal scales and processes. The influence of economic cycles on land use change goes beyond the scope of this work, nevertheless in the future should be addressed.

3.1. Alternative land use trends? Implications for forest transition theory

As mentioned before, tree farms in Chile are very profitable, and even a large subsidy for native plantations and no subsidy for tree farms would not have implied a large change in the land area demanded for native forest in the late 1990s [52, 63, 64]. Furthermore, in Chile, it is more difficult to plant native species than to plant tree farms. This is due to the technical difficulties of planting native species, which include summer drought, lack of affordable seedlings, non-native wild herbivores, lack of lay knowledge, and the fact that most of the research funding is focused in the development of research and management techniques for non-native species [65]. This suggests that tree farms are a very "resilient" land use, due to the economic, political, and technological context. Nevertheless, forest's natural regeneration can be a much more effective way of recovering the forest [66, 67]. Even though it is unlikely that a one-time payment for native forest management and conservation would initiate the reversal of a resilient land use trend—such as tree farms—forest can recover where tree farms are not planted [64]. Several studies show that if a land plot is not planted with three farms, the native forest recovers [64, 66, 67]. Thus the evidence suggests that tree farms are hindering actual forest recovery. These results can have interesting implications for broader questions regarding forest transition theory.

Forest transition theory seeks to explain why and when a country switches from forest loss to recovery. The theory initially developed by Mather [4] argues that countries recover their forests when they reach some level of economic development. Mather's explanation has guided much of the academic and policy-making discussion but has also obscured some relevant concepts, such as the difference between forest cover and tree cover. Although afforestation does provide some environmental benefits, tree farms do not necessarily provide the same benefits as native forest [68, 69]. Drawing from ecology, Putz and Romero [68] call for a critical assessment of forest recovery in the Global South. They call for more attention to the structure, function, and composition of forested ecosystems, rather than an oversimplified assessment of forest transition based solely on tree cover [68, 69]. Differentiating between forest cover and tree cover for assessing forest transitions becomes even more critical if trees are used mainly for industrial production, where inputs of energy and matter are inherently different from those of a natural system [20, 70–72].

Ecological theory combined with political ecology affords more insight on forest transition theory. Mansfield et al. [8] argue that forest transition theories "assume that the experience of the North reflects a general, desirable process that should be encouraged to 'diffuse' to

http://dx.doi.org/10.5772/intechopen.80559

developing economies" [8]. Instead, political ecology calls for a recognition of the place-specific factor that led to uneven patterns of development, production, and consumption [8]. Although Chile is often celebrated as a case of forest recovery [73], it is a case of tree cover increase, and this fact should not be ignored when designing policy recommendations for other countries, as it might shape the future of land use and its governance [73]. As mentioned above, the "success" of the afforestation policy was mainly the result of the larger changes rooted in specific events of the political and social history of Chile. Moreover, given the current situation, tree farming seems irreversible, highlighting the nonreversible nature of socio-ecological systems (Section 2). Forest recovery assessments must differentiate tree cover from forest cover, while also considering forest transitions as context dependent.

4. Conclusion, policy recommendations, and research significance

This chapter describes some of the mechanisms that have hindered adaptive capacity in Chile regarding forest policy and land use. Overall, Chile's case exemplifies how a land use trend can become resilient due to political, social, and economic processes. In the case of Chile, economic and political processes interact in a way that results in a rigidity trap across the social and ecological system. Current political and economic processes act as the "slow" variable and explain the current lack of adaptive capacity of LULCC in Chile.

To prevent rigidity traps, it is key to develop institutional mechanisms that break reinforcing dynamics between politics, power, and profits and foster change, diversity, adaptation, and learning [26]. More specifically for the case of Chile, it is difficult to claim that native forests would become a viable land use, without prior structural changes or some unimaginable surprise. Rather than providing panaceas for afforestation, Chile's afforestation case provides insights about the socio-ecological underpinnings of a short-term increase in tree cover and the implications of its success on adaptive capacity in the long term.

Through a transdisciplinary approach, this research has contributed to the literature on adaptive capacity, socio-ecological systems, and land use transition, providing useful insights for policy makers and for scientific and humanities scholars who want to address current environmental issues without hindering future capacity for adaptation.

Acknowledgements

This research is funded by the CONICYT FONDECYT Chile grant number 11150281. I would also like to thank Francisco Abarca for the map elaboration and Tania Manuschevich for her helpful review, as well as the anonymous chapter reviewers and editor.

Conflict of interest

I declare no conflict of interest.

Nomenclature

CONAF	National Forest Service (Corporación Nacional Forestal, private)
CORFO	Production Promotion Corporation (Corporación de Fomento de la Producción, public)
COREF	Afforestacion Corporation (Corporacion de Reforestación, public)
CORMA	National Wood Corporation (Corporación Chilena de la Madera, private)
DL 701	Decree Law 701
IREN	Natural Resource Research Institute (Instituto de Investigación de los Recursos Naturales, public)
MERCOSUR	Common Southern Market (Mercado Común del sur)
NFL	Native Forest Law (Ley de Bosque Nativo 20.283)
LULCC	land use and land cover change
SES	Socio-ecological systems
SSEF	Scheffer socio-ecological systems

Author details

Daniela Manuschevich

Address all correspondence to: dmanuschevich@academia.cl

Geography School, Universidad Academia de Humanismo Cristiano, Santiago, Chile

References

[1] Carpenter SR, Brock WA. Adaptive capacity and traps. Ecology and Society. 2008;**13**(2):40

[2] Turner BL, Lambin EF, Reenberg A. The emergence of land change science for global environmental change and sustainability. Proceedings of the National Academy of Sciences of the United States of America. 2007;**104**(52):20666-20671

[3] Lambin EF, Meyfroidt P. Land use transitions: Socio-ecological feedback versus socio-economic change. Land Use Policy. 2010;**27**:108-118

[4] Mather AS. The forest transition. Area. 1992;**24**(4):367-379

[5] Heilmayr R, Echeverría C, Fuentes R, Lambin EF. A plantation-dominated forest transition in Chile. Applied Geography. 2016;**75**:71-82

[6] Cruz P, Cid F, Rivas E, Neira E, Ladron de Guevara J. Evaluación de la Ley N°20.283 sobre recuperación del bosque nativo y fomento forestal. Subsecretaría de Agricultura. 2012:240

[7] CONAF. Análisis de la Afectación y Severidad de los Incendios Forestales ocurridos en enero y febrero de 2017 sobre los usos de suelo y los ecosistemas naturales presentes entre las. Sanitago de Chile: The Region de Coquimbo y Los Ríos de Chile; 2017. (Informe Técnico)

[8] Mansfield B, Munroe DK, McSweeney K. Does economic growth cause environmental recovery? Geographical explanations of forest regrowth. Geography Compass. 2010;**4**(5):416-427

[9] Camus P. Ambiente, Bosques y Gestion Forestal en Chile: 1541-2005. Santiago, Chile: DIBAM LOM Ediciones; 2006

[10] CORMA. Aporte del Sector Forestal [Internet]. 2018. Available from: http://www.corma.cl/corma.asp?id=5&ids=88 [Cited: 2013 May 15]

[11] Miranda A, Altamirano A, Cayuela L, Pincheira F, Lara A. Different times, same story: Native forest loss and landscape homogenization in three physiographical areas of south-central of Chile. Applied Geography. 2015;**60**:20-28

[12] Miranda A, Altamirano A, Cayuela L, Lara A, González M. Native forest loss in the Chilean biodiversity hotspot: Revealing the evidence. Regional Environmental Change. 2017;**17**(1):285-297

[13] CONAF. Sistema de Informacion Terriorial [Internet]. 2011. Available from: http://conaf2010.siigsa.cl/

[14] Armesto JJ, Rozzi R, Smith-Ramirez C, MTK A. Conservation targets in South American temperate forests. Science. 1998:1271-1272

[15] Tognelli MF, De Arellano PIR, Marquet PA. How well do the existing and proposed reserve networks represent vertebrate species in Chile? Diversity and Distributions. 2008;**14**:148-158

[16] Bannister JR, Vidal OJ, Teneb E, Sandoval V. Latitudinal patterns and regionalization of plant diversity along a 4270-km gradient in continental Chile. Austral Ecology [Internet]. 2011;**37**(4):500-509. DOI: 10.1111/j.1442-9993.2011.02312.x

[17] Díaz IA, Armesto JJ, Reid S, Sieving KE, Willson MF. Linking forest structure and composition: Avian diversity in successional forests of Chiloeì Island, Chile. Biological Conservation. 2005;**123**:91-101

[18] Alarcón D, Smith-Ramírez C, Hechenleitner P, Ramírez de Arellano P, Oliva M, Pinto M. Nuevas poblaciones de *Berberidopsis corallina* Hook.f. (Berberidopsidales: Berberidopsidaceae) en la región del Biobio, Chile: Ubicación y conservación de su hábitat. Gayana Botánica. 2007;**64**:217-231

[19] Paritsis J, Aizen MA. Effects of exotic conifer plantations on the biodiversity of understory plants, epigeal beetles and birds in *Nothofagus dombeyi* forests. Forest Ecology and Management. 2008;**255**(5-6):1575-1583

[20] Little C, Lara A, McPhee J, Urrutia R. Revealing the impact of forest exotic plantations on water yield in large scale watersheds in South-Central Chile. Journal of Hydrology. 2009;**374**:162-170

[21] Lara A, Little C, Urrutia R, McPhee J, Álvarez-Garretón C, Oyarzún C, et al. Assessment of ecosystem services as an opportunity for the conservation and management of native forests in Chile. Forest Ecology and Management. 2009;**258**:415-424

[22] Locher-Krause KE, Lautenbach S, Volk M. Spatio-temporal change of ecosystem services as a key to understand natural resource utilization in southern Chile. Regional Environmental Change. 2017;17:1-17

[23] Garreaud RD, Alvarez-Garreton C, Barichivich J, Boisier JP, Christie D, Galleguillos M, et al. The 2010-2015 megadrought in Central Chile: Impacts on regional hydroclimate and vegetation. Hydrology and Earth System Sciences. 2017;**21**(12):6307-6327

[24] Bozkurt D, Rojas M, Boisier JP, Valdivieso J. Climate change impacts on hydroclimatic regimes and extremes over Andean basins in Central Chile. Hydrology and Earth System Sciences Discussions. 2017. DOI: Org105194hess-2016-690. https://www.hydrol-earth-syst-sci-discuss.net/hess-2016-690/hess-2016-690.pdf

[25] Ostrom E. A general framework for analyzing sustainability of socio-ecological systems. Science. 2009;**325**:419-422

[26] Gunderson LH, Holling CS. Panarchy: Understanding Transformations in Human and Natural Systems. Island Press; 2002

[27] Holling CS, Meffe GK. Command and control and the pathology of natural resource management. Conservation Biology. 1996;**10**:328-337

[28] Holling CS. Resilience and stability of ecological systems. Annual Review of Ecology and Systematics. 1973;**4**:1-23

[29] May RM. Thresholds and breakpoints in ecosystems with a multiplicity of stable states. Nature. 1977;**269**:471-477

[30] Ludwig D, Jones DD, Holling CS. Qualitative analysis of insect outbreak systems: The spruce budworm and forest. The Journal of Animal Ecology. 1978;**47**:315-332

[31] Scheffer M, Westley F, Brock WA, Holmgren M. Dynamic interaction of societies and ecosystems-linking theories from ecology, economy, and sociology. In: Panarchy: Understanding Transformations in Human and Natural Systems. Island Press; 2002

[32] Arts B, Appelstrand M, Kleinschmit D, Pülzl H, Visseren-Hamakers I, Eba'a Atyi R, et al. Discourses, actors and instruments in international forest governance. In: Embracing Complexity: Meeting the Challenges of International Forest Governance a Global

Assessment Report. New York, NY, USA: International Union of Forest Research Organization; 2011. pp. 57-73. (IUFRO World Series Volume)

[33] Humphreys D. Discourse as ideology: Neoliberalism and the limits of international forest policy. Forest Policy and Economics. 2009;**11**(5-6):319-325

[34] Springate-Baginski O, Blaikie PM. Forests, People and Power: The Political Ecology of Reform in South Asia. London; Sterling, VA: Earthscan; 2007

[35] Dang TKP, Turnhout E, Arts B. Changing forestry discourses in Vietnam in the past 20 years. Forest Policy and Economics. 2012;**25**:31-41

[36] Manuschevich D. Neoliberalization of forestry discourses in Chile. Forest Policy and Economics. 2016;**69**:21-30

[37] Clapp RA. Creating competitive advantage: Forest policy as industrial policy in Chile. Economic Geography. 1995;**71**:273-296

[38] Elgueta M, Jirkal J. Erosión de los Suelos en Chile. Sud-America; 1942. 31 p

[39] Miranda A, Altamirano A, Cayuela L, Lara A, González M. Native forest loss in the Chilean biodiversity hotspot: Revealing the evidence. Regional Environmental Change [Internet]. vol. 17. 2016. Available from: http://link.springer.com/10.1007/s10113-016-1010-7 [Cited: July 21;17:2016]

[40] Silva E. The import-substitution model: Chile in comparative perspective. Latin American Perspectives. 2007;**34**(3):67-90

[41] Ministerio de Agricultura. Ley de fomento forestal. Decreto Ley. 1974:701

[42] Agraria CP. Síntesis Ejecutiva del Decreto de ley 701. Santiago; 2005

[43] SVS. SVS sanciona a personas, ejecutivos y corredora de bolsa en el marco de la investigación sobre. Sociedades Cascada [Internet]. 2014. Available from: http://www.svs.cl/portal/prensa/604/w3-article-17480.html [Cited: April 30, 2018]

[44] Ministerio de Economia. Estatuto de la Inversion Extranjera. DL600. 1974

[45] Niklitschek ME. Trade liberalization and land use changes: Explaining the expansion of afforested land in Chile. Forest Science. 2007;**53**:385-394

[46] Morales E. Changing Ownership and Management of State Forest Plantations: Chile. Cape Town, South Africa; 2003. Draft

[47] Hartwig F. La Tierra que Recuperamos. Santiago de Chile: Editorial Los Andes; 1994

[48] Reyes R, Nelson H. A tale of two forests: Why forests and Forest conflicts are both growing in Chile. International Forestry Review. 2014;**16**(4):379-388

[49] Arnold F. Native forest policy in Chile: Understanding sectoral process dynamics in a country with an emerging economy. International Forestry Review. 2003;**5**:317-328

[50] Garreton MA. Neoliberalismo corregido y progresismo limitado: los gobiernos de la Concertación en Chile 1990-2010. Sanriago de Chile: Arcis; 2012. (Colección Pensar América Latina)

[51] MIT. The Atlas of Economic Complexity by @HarvardCID [Internet]. 2018. Available from: http://atlas.cid.harvard.edu/explore/?country=42&partner=undefined&product=146&productClass=HS&startYear=undefined&target=Product&year=1995 [Cited: May 2, 2018]

[52] Cubbage F, Mac Donagh P, Sawinski Junior J, Rubilar R, Donoso P, Ferreira A, et al. Timber investment returns for selected plantations and native forests in South America and the Southern United States. New Forest. 2007;**33**:237-255

[53] Consultora-Novus. Evaluacion de resultados de la aplicacion del DL 701 de 1974. Santiago; 1998. p. 85

[54] Biblioteca del Congreso Nacional de Chile, editor. Historia de la Ley N° 20.488. Prorroga Vigencia del Decreto Ley N° 701, de 1974, y Aumenta Incentivos a la Forestación. Congreso Nacional; 2011

[55] UDEC. Análisis de la Cadena de Producción y Comercialización del Sector Forestal Chileno: Estructura, Agentes y Prácticas. Concepcion: Universidad de Concepcion, Departamento de Economia; 2009

[56] Political constitution of Chile. 2005

[57] Aninat C, Landregan J, Navia P, Vial J. Political Institutions, Policymaking Processes and Policy Outcomes in Chile. Inter-American Development Bank; 2006. Report No.: R-521

[58] Gobierno de Chile. Plan de Acción para la Recuperación del Patrimonio Natural y Productivo afectado por los Incendios de 2017. 2017

[59] Ministerio Economia, CORFO. Ejecuta Acuerdo N246 del Comité de Capital de Riesgo Adoptado en Sesion n8, de Fecha 25 de Agosto 2017, que Aprueba Programa de Financiamiento a Fondos de Inversion de Capital de Riesgo Para el Fomento Forestal y Maderero [Internet]. 2017. p. 17. Available from: https://www.leychile.cl/N?i=1114050&f=2018-01-25&p=

[60] CORFO. Corfo Lanza Nuevo Fondo de Inversión Para el Sector Forestal y Maderero [Internet]. 2018. Available from: https://www.corfo.cl/sites/cpp/sala_de_prensa/nacional/01-03-2018_nuevo_fondo_de_inversion [Cited: May 7, 2018]

[61] Allison HE, Hobbs RJ. Resilience, adaptive capacity, and the "lock-in trap" of the western australian agricultural region. Ecology and Society [Internet]. 2004;**9**(1). Available from: http://www.jstor.org/stable/26267657 [Cited: July 5, 2018]

[62] Beier C, Lovecraft A, Chappin S III. Factors influencing adaptive capacity in the reorganization of forest management in Alaska. Ecology and Society [Internet]. 2011;**16**. Available from: //WOS:000289081200018

[63] Sedjo R. Comparative Economics of Plantation Forestry a Global Assessment. Washington DC: Resources for the future; 1983. 161 p

[64] Manuschevich D, Beier CM. Simulating land use changes under alternative policy scenarios for conservation of native forests in South-Central Chile. Land Use Policy. 2016;**51**:350-362

[65] Donoso P, Otero L. Hacia una definición de país forestal: Donde se sitúa Chile? Bosque. 2005;**26**:5-18

[66] Birch JC, Newton AC, Aquino CA, Cantarello E, Echeverria C, Kitzberger T, et al. Cost-effectiveness of dryland forest restoration evaluated by spatial analysis of ecosystem services. Proceedings of the National Academy of Sciences of the United States of America. 2010;**107**:21925-21930

[67] Chazdon R, Uriarte M. Natural regeneration in the context of large-scale forest and landscape restoration in the tropics. Biotropica. 2016;**48**(6):709-715

[68] Putz FE, Romero C. Futures of Tropical Forests (Sensu Lato) [Internet]. PeerJ PrePrints. 2014. Available from: https://peerj.com/preprints/362/ [Cited: August 7, 2014]

[69] Chazdon RL, Brancalion PHS, Laestadius L, Bennett-Curry A, Buckingham K, Kumar C, et al. When is a forest a forest? Forest concepts and definitions in the era of forest and landscape restoration. Ambio. 2016;**45**(5):538-550

[70] Rivas Y, Oyarzún C, Godoy R, Valenzuela E. Mineralización del nitrógeno, carbono y actividad enzimática del suelo en un bosque de *Nothofagus obliqua* (Mirb) Oerst y una plantación de *Pinus radiata* D. Don. del centro-sur de Chile. Revista Chilena de Historia Natural. 2009;**82**(1):119-134

[71] Huber A, Iroumé A, Mohr C, Frêne C. Efecto de plantaciones de *Pinus radiata* y *Eucalyptus globulus* sobre el recurso agua en la Cordillera de la Costa de la región del Biobío, Chile. Bosque Valdivia. 2010;**31**(3):219-230

[72] Oyarzún CE, Frêne C, Lacrampe G, Huber A, Hervé P. Soil hydrological properties and sediment transport in two headwater catchments with different vegetative cover at the Coastal Mountain range in southern Chile. Bosque. 2011;**32**:10-19

[73] FAO. State of the World's Forests 2016 [Internet]. Food and Agriculture Organization of the United Nations. State of the World's Forests. 2017. p. 126. Available from: http://www.fao.org/publications/sofO/2016/en/ [Cited: May 15, 2018]

[74] ODEPA. Oficina de Estudios y Políticas Agrarias, Ministerio de Agricultura de Chile - Estadísticas - Por Macro Rubros [Internet]. 2013. Available from: http://www.odepa.gob.cl/sice/AvanceProductoParamsB.action;jsessionid=BEFB39A8CDF3E6A6B1DF8CC0799AE84E?rubro=otros-reportes&reporte=av_prod [Cited: March 5, 2013]

[75] Banco Central. Estadisticas económicas [Internet]. 2013. Available from: http://www.bcentral.cl/estadisticas-economicas/ [Cited: May 6, 2013]

[76] CONAF. Respuesta a Solicitud Transparencia AR003T0001007. 2017

[77] Luebert F, Pliscoff P. Sinopsis Bioclimatica y Vegetacional de Chile. Santiago; 2006

Chapter 13

Comparative Approaches in Managing Wetland Environments and Land Uses: Rainbow Lake in Michigan, Guangzhou City in China, and Chinese Sponge Concept Case Studies

Zhen Wu, Jon Bryan Burley, Chun-Hua Guo, Na Li,
Yiwen Xu and Zhi Yue

Additional information is available at the end of the chapter

http://dx.doi.org/10.5772/intechopen.79323

Abstract

Environmental scientists, natural resources agencies, planners, landscape architects, engineers, and concerned citizens are interested in the impacts that land uses within watersheds have upon lake water quality and water runoff volume. For the past 40 years, much has been discovered and many North American water bodies from small to large can be reliably modeled and studied, employing phosphorus as the identifier of water quality. We present an overview of the key features in this multi-disciplinary effort and illustrate how to apply the general method to Rainbow Lake, in Gratiot County, Michigan, the USA. In addition, we illustrate how these fundamental ideas are being employed at the Haizhu wetland park, a large wetland setting in Guangzhou, the People's Republic of China, and present Chinese planning and design efforts termed "sponge city" to address new ideas to reduce runoff and improve water quality.

Keywords: environmental design, landscape architecture, site hydrology, watershed studies, landscape research

1. Introduction

In the USA, watershed and site hydrology water quality modeling and runoff volume predictions made rapid advances in the 1960s and 1970s. By the 1980s, investigators could reliably predict both the volume and the quality of water runoff. However, this ability and technology has only slowly been adapted in other parts of the world. Nevertheless, such concerns have been of interest by humans for many millennia. Luo et al. describes concerns in managing the Yellow River in China, 5000 years ago, and more recent concerns in Japan [1]. The natural landscape was reconfigured with massive hill removals and topographic changes thought to reduce the impacts of flooding.

In Western Anatolia (Western Turkey), American environmental interest was invigorated by George Perkins Marsh (1805–1882). He wrote *Man and Nature: of Physical Geography as Modified by Human Action* [2]. He was stationed as an Ambassador to the Ottoman Empire. During his travels, he learned about the Meander River, a winding river in Western Turkey. Especially during the Greek and Roman eras, he learned about the deforestation of the surrounding hills and the ensuing erosion that expanded the Meander River's delta into the Mediterranean Sea (**Figure 1**) [3]. Along with his other observations in Europe, he developed his ideas about landscape stewardship. There was great concern that such adverse environmental impacts should not occur in North America. Individuals such as Gifford Pinchot developed a land management ethic and worked with Frederick Law Olmsted at Biltmore where Olmsted established the bass ponds to control soil erosion from the large construction site [3]. Today erosions control and detention ponds are commonplace.

French, English, German, Russian, and American engineers were engaged in developing methods and mathematical models to predict water flow in pipes, culverts, sewers, swales, creeks, and rivers culminating in such books as *Design Data Book for Civil Engineers* [4]. In

Figure 1. An image of the Meander River valley between the ancient Greek cities of Priene and Melitus. Both cities were Mediterranean ports during Greek Ionian times. Today the Mediterranean Sea in the figure is many miles to the right of the image (copyright 2006 Jon Bryan Burley, all rights reserved, used by Permission).

many respects by the end of World War II, the fundamentals of site hydrology and erosion control had been established for American landscapes. The effort was made possible because of the national rainfall data collected by the US Department of Agriculture across the country and various investigators had studied runoff percentages for numerous soils and cover types. Civil engineers now could predict estimated water flow for various types of storm events.

The understanding concerning the impacts to changes in watersheds was investigated. In the American West, the Mono Lake Basin illustrated the impacts of water removal to southern California [5]. Watershed hydrology was considered at the ecosystem level. The Colorado River was a case study in watershed ecosystems, as most of the water is harvested and very little flows out of the delta through Mexico [6]. Other investigators discovered large cataclysmic events, such as the massive great floods across western Washington during the end of the last ice age [7]. Ice dams blocked rivers, releasing water at an estimated 9.46 cubic miles per hour traveling at 58 miles per hour. The flow of water was over 10 times the flow of water in the world and 60 times great than the flow of water in the Amazon. It is estimated that this event may have occurred at least 40 times. Knowledge built concerning low- and high-water events and activities.

Refinements and modifications in planning and design continued. Albe Munson, a Professor at Michigan State University, both a landscape architect and a civil engineer, wrote the first noted American landscape construction book integrating site hydrology and landscape development [8]. Yet the book contained only simple estimates for yearly runoff. However, others later presented more usable equations to estimate pond sizing, pipe sizing, and swale sizing [9, 10]. Eventually extensive and quite comprehensive publications featured a wide array of knowledge in planning and design concerning water for landscape architects and engineers [11–13].

Soil scientists and hydrologists continued refinements in hydrological calculations. They considered models that were more reflective of field conditions such as saturated soil and frozen soil. Today, site hydrology calculations are more complicated than in the past with several methods and choices [14].

The integration of land use, human activities, and the forces of nature are being examined in a more integrated manner. The recent publication of a book titled *Third Coast Atlas: Prelude to a Plan*, illustrates this integration [15] The third coast is the coastline of the Great Lakes system and St. Lawrence seaway. It is longer than either the Atlantic Coast or Pacific Coast. Humans are attempting to integrate both large-scale issues and small-site detailed features in a more comprehensive manner.

One topic that has taken some time to develop was the prediction of water quality. By the late 1970s and early 1980s progress was made as investigators understood that the nutrients of phosphorus and nitrogen at excessive levels in the water reduced water quality. Wang (et al.) recently published an overview of this prediction approach where investigators are now attempting to develop treatments to intercede upon water quality [16]. The essential feature concerning improving water quality is to have the water come in contact with the substrate that removes the nutrient (phosphorus or nitrogen) from the liquid. Large volumes of standing water not in contact with the substrate will only be marginally treated. In addition, measures to limit the nutrients from entering free-flowing water in the hydrological system are also very effective and reduce the need for water treatment basins.

Two important seminal developments occurred in search to predict non-toxic water quality. Much of this work was conducted in Minnesota and Wisconsin, the USA, where there are thousands of lakes to study and manage. The first development was the recognition that nitrogen ions and suspended phosphorus in the water were key limiting nutrients influencing water quality. Investigators could approximate the amount of these nutrients in the water by simply placing a Secchi disk in the water and determining the depth at which the disk disappeared. The quicker it disappeared under the water, the more nutrient rich was the water and the poorer was the water quality [17]. The second development was the ability to assign phosphorus contributions to free-moving water by the land area and land cover type. By combining these contributions from the land to the hydrological calculations of water flowing through a watershed, the concentration of phosphorus in the water on a yearly basis could be estimated [18]. It was a long series of equations to estimate water quality, with each variable having a fair amount of variance. The reliability of such a series of equations to predict water quality was at first suspect because it was believed that the accuracy of each section of the equations could lead to a highly inaccurate prediction. However, as we introduce in the methodology and results sections, the set of equations produced a relatively accurate estimation of water quality for a water body.

Our intent in this investigation is to illustrate how this environmental prediction methodology is applied to management watershed in the states of Michigan (Rainbow Lake and Minnesota (Sauk River Watershed), the USA. In addition, the chapter discusses how the quest to manage water in rural and urban environments continues in parts of the world, employing two examples from the People's Republic of China (PRC). The chapter also includes statements and comments from an interview (May 2017) of Dr. Jon Bryan Burley, FASLA, a landscape architect who has been engaged in teaching and writing academic papers concerning landscape for over 40 years. He has witnessed the evolving changes in this technology over this time period.

"When I was a young professor at age 27 in 1982, modeling watershed quality in Minnesota was just becoming possible," recalls Dr. Burley. "I used to have students in my landscape planning classes model land-use development within watersheds for selected lakes and prepare land-use plans to prevent a change in the perceived water quality of that lake. The Minnesota Department of Natural Resources provided publically available information concerning land cover and water quality in the state. Combined with information about rainfall and evapotranspiration, it was possible to estimate potential water quality changes due to changes in the landscape," describes Dr. Burley. "In 1990, when I went to the University of Michigan to work on my PhD this type of watershed modeling was considered as a possible topic for my dissertation. But in discussions with faculty at the School of Natural Resources, such a topic seemed far too ecologically complex to produce a meaningful dissertation. They seemed to be weary of such a long series of equations to predict water quality. I certainly understand their caution. Their concern was that if a series of 10 equations each could only explain 80% of the variance, then by the 10th equation (80% of 80% repeated ten times), the results maybe explaining less than 10% of the variance," expressed Dr. Burley. "So I pursued another topic, addressing surface mine reclamation and in the meantime, I attempted to communicate the findings of various studies undertaken by my students and myself over the past decade. The one that was published is a little known article published by the American Society of Landscape Architects, describing a study concerning how much development could occur within the Lake Itasca watershed (the headwaters of the Mississippi River) to protect the lake's water quality," added Dr. Burley [19]. The article presents a graph of water quality based upon various land cover environments: 100% forested watershed, a catastrophic fire event, unchecked development and, limited development. The development was limited to keep the lake a meso-trophic lake. "Students could develop land-use plans containing various levels of forested land, agricultural land, low density rural housing, and urban land. I thought this was an excellent landscape planning exercise for students to relate development, land management, and natural resource protection," reflected Dr. Burley.

"During this time, I submitted a research article to a journal about the Sauk River watershed in Minnesota, where I had conducted a 1983 study statistically validating the prediction modeling process (originally using hand drawn overlays and a planimeter, but updated in the mid-1980s and replicated with a micro-computer geographical information systems (GIS) application), illustrating the concordance of actual water quality measures with predicted estimates by dividing the watershed into sub-watersheds and comparing measured and predicted scores. And while the results were publishable (no reviewer disputed the findings) and no investigator had ever done this before, by the time I had submitted the manuscript, the GIS technology had radically changed (from mostly hand drawn overlay maps measured with a planimeter, to crude computer maps with over-printing of alpha- numeric characters, to digital colored GIS maps—and the GIS world had changed). The manuscript was rejected for its dated technology, with the reviewers stating that there were better and newer landscape planning tools. They were correct, but I had not interest in updating the research with newer technology that did not change the fundamental results," recalled Dr. Burley. "Reviewers can find numerous reasons to reject an abundance of submitted articles. Journals have reputations to maintain and that may mean not presenting dated technology. Still the fundamentals of that study remain true," confided Dr. Burley. This study of the Sauk River Watershed illustrates the methodology for this investigation. "I am pleased that the Sauk River watershed study conducted in 1983 still has some value and portions of the study can be employed in some useful fashion," added Dr. Burley. "I also want people to recognize the great value and contributions that people like Carlson, Garn, and Parrott made; otherwise it would have been impossible to model the watershed and make any prediction," suggested Dr. Burley [17, 18].

2. Methodology

2.1. Sauk River Watershed, Minnesota

An overview of fundamental methodology is illustrated by a study on the Sauk River, Minnesota, conducted in the 1980s. The Sauk River is in central Minnesota on the border between the Western prairie lands and the Eastern woodlands. Before emptying into the Mississippi River, the water passes through Cedar Island Lake. This lake has experienced algae blooms and diminished water quality. Recreational housing along the lake was blamed for poor water quality. Employing the GIS technology of the time, maps of the watershed's land cover types and soils were generated (**Figures 2** and **3**) [20]. Predicted phosphorus concentrations for each sub-watershed in the study area were statistically compared to measure levels, employing Kendall's Coefficient of Concordance, a nonparametric statistical method to search for significant agreement (note that most statistical tests examine significant difference) [21]. It was discovered that the scores significantly agreed. Therefore, the modeling approach had some degree of validation. This can be surprising, as one might expect the error and increasing amount of variance to be passed from one equation to the next. Yet, at the end of the computations, the process approximates reality. Being able to estimate real conditions, the next step in the process was to examine pre-settlement conditions of the lake with current conditions. The results of the study indicated that the predicted water quality from pre-settlement times should be no different than existing land uses with agriculture and housing in the watershed. Yet, the measured phosphorus levels were higher in the lake than the model predicted. Upstream along the main course of the river, a point source of phosphorus was discovered. The water was not toxic and the effluent met discharge requirements, but the level of suspended phosphorus in the discharged water was enough to influence the water quality downstream. Removing the point-source discharge of phosphorus improved the water quality of the lake and matched predictions. "I believe the interesting part of this study was that farmers (many dairy farms) and recreational home owners were blamed for the problems. But they were not at fault and were unfairly blamed," states Dr. Burley. "A very simple fix at the food processing plant upstream would solve much of the problem. I thought this study was a very practical example concerning how ecological modeling and landscape planning could work together. But sometimes investigators believe their discoveries are much more important than the rest of the academic and professional communities believe. I am afraid I too was susceptible to that disease of over estimating the importance of my study," recalled Dr. Burley.

Back in 1983, BASIC programming was a mathematical tool to make calculations, especially repeated calculations. "In 1974, I had taken a Fortran IV programming class and found Basic programming quite easy [22]. In an afternoon, I could put together the foundation of a computer program to calculate phosphorus concentration for a lake. In a 90 minute lecture in 1985, I gave a detailed explanation of the program to a graduate level landscape architecture class. I thought I was going to be brilliant; instead I was preposterously boring. Unless one wanted to use the model, what I had to say was of little interest," reflected Dr. Burley. "My lecture did not match the interests of the audience."

Today, all this calculation work can be accomplished on a spreadsheet. "Spreadsheets were just being invented and applied back then in the early 1980s," recalled Dr. Burley. "In addition, land cover data, graphic presentation, and area tabulation are much more convenient in this era. Back then it took hundreds of hours just to code and typing to make one map," observed Dr. Burley. The handbooks and machines that could engage the old BASIC computer programming have been discarded for 25 years. "Towards the end of these studies with my students, I was using EPPL 7, a GIS program that could be facilitated with map digitization and color map, making the process must faster to obtain results," notes Dr. Burley [23].

"At Michigan State University, in the early 1990s I imagined that I might pursue such water quality studies for nearby lakes or even for lakes such as Lake Erie. ARC-GIS and other

Figure 2. A map of the Sauk River, Minnesota pre-settlement vegetation. Back in the mid-1980s, black and white alpha-numeric GIS maps were often the standard. The technology had been transferred down from mainframe computers to the recently developed micro-computers (copyright 1983 Jon Bryan Burley, all rights reserved, used by Permission).

more-friendly GIS software was available to be incorporated into the study. But when I inquired about conducting such studies, the university's hydrology research institute was interested in other kinds of investigations and I met the same kind of skepticism at my school as I had encountered at the University of Michigan. So I did not pursue the topic. Yet I am pleased that somebody persevered, and have applied such activities to pursue what I had also envisioned," exclaimed Dr. Burley.

Figure 3. A map of the Sauk River, Minnesota study area comprised of land-uses with soil type. Based upon this information the amount of water runoff and phosphorus contribution can be estimated (copyright 1983 Jon Bryan Burley, all rights reserved, used by Permission).

http://dx.doi.org/10.5772/intechopen.79323

"It takes a special kind of landscape student to be interested in landscape planning studies. At Michigan State University, I found most landscape students were interested in site design," observed Dr. Burley. "Most students expect to make their professional careers from site development. In the landscape planning classes, students who are interested in landscape planning are interested in the shapes and patterns of land cover types (traditional landscape ecology applications) for greenways, plant preservation, wildlife habitat, and to find land that is optimal for development. No one seemed interested in lake and hydrological modeling," proposed Dr. Burley. "But in the past decade something changed. Nations such as P.R. of China, have a renewed interest in managing water, greenways, and water quality. This is a nation with a long history concerning the management of water. International landscape graduate students envision opportunities to model the environment and prepare regional land use plans. They come to study with me at Michigan State University and to learn about how to model site hydrology and water quality. However, I will retire soon. This article is an opportunity to 'pass-along' the present state of knowledge concerning predicting non-toxic water quality to an interested international readership from around the world," replied Dr. Burley.

2.2. Rainbow Lake, Michigan

Rainbow Lake is a small private artificial lake (a lake created by placing a dam across a creek) for housing development in Gratiot County, Michigan (**Figure 4**). The valley that was formed by Pine Creek was produced when the nearby Maple River was the outlet for Glacial Lake Saginaw, sending massive amounts of glacial melt-water down the Maple River, into the Grand River, into post-glacial Lake Michigan (Lake Chicago), eventually reaching the Mississippi River. The valley that was cut by glacial melt-waters also meant that the nearby Pine Creek would also cut a small valley into the landscape to meet the Maple River. This valley was suitable for the creation of a dam and a lake. The surrounding landscape is mostly agricultural as Gratiot County has the largest percentage of agricultural land for any of Michigan's county

Figure 4. An image of Rainbow Lake in Gratiot Country, Michigan looking from east to west. The lake is a long winding water body through a post-glacial valley (copyright 2017 Jon Bryan Burley, all rights reserved, used by Permission).

growing corn, soybeans, winter wheat, dairy herds, beef cattle, horse farms, apple orchards, and hardwood lumber. Surrounding the lake is a residential development, enclosed by the rural environmental agricultural matrix. The lake is used for recreational boating, summer fishing, and winter ice fishing.

Ultra-oligotrophic	<5
Oligo-mesotrophic	5–10
Meso-eutrophic	10–30
Eu-polytrophic	30–100
Polytrophic	>100

Table 1. Trophic state lake classification and the associated Carlson Index Score.

C1 = terrestrial and other water body phosphorus supply in kg/ yr.
C2 = Carlson Index score
E1 = evaporation in meters
F = flushing rate
L2 = maximum lake depth in meters
L4 = nutrient supply
M2 = mean lake depth in meters
P1 = prescription in meters
P2 = present phosphorus level in mg/l
P3 = predicted phosphorus level in mg/l
O1 = areal water load in meters/year
R = retention coefficient
R1 = volume of watershed runoff to lake in cubic meters
S1 = surface area of lake in square meters
T1 = point source phosphorus pollution input
V1 = lake volume in cubic meters
W6, W7, W8 = woodland in square meters
P6, P7, P8 = grassland area in square meters
U6, U7, U8 = urban savanna in square meters
I6, I7, I8 = industrial in square meters
B6, B7, B8 = agriculture in square meters
6 = clayey soils, 7 = loamy soils, 8 = sandy soils.

Table 2. The list of variables employed to calculate non-toxic water quality.

To predict the water quality of the lake, one must know the soil types for the watershed, the land cover types, the amount of yearly rainfall, the evapotranspiration rate, the volume of water in the lake, and the existing phosphorus concentration in the lake. The yearly expected phosphorus concentration can be calculated by the series of equations presented in the results section of this chapter. **Table 1** illustrates the relationship of the final calculated Carlson Index Score with lake quality. The Carlson Index Score can be calculated from either phosphorus concentrations, nitrogen concentrations, or Secchi disk readings. Phosphorus estimations can be used for landscape modeling applications. **Table 2** presents the series of variables employed in calculating the Carlson Index Number.

The trophic state of the lake is associated with the life forms and levels of nutrients found in the water body. Ultra-oligotrophic lakes are very nutrient poor and contain relatively little aquatic life in them. Oligo-mesotrophic lakes are suitable for cold-water fish species such as trout and salmon (*Salmonidae G. Cuvier*). Meso-eutrophic lake contains fish species like walleye (*Sander vitreus*; Michill, 1818). Eu-polytrophic lakes often are suitable for large-mouth bass (*Micropterus salmoides* Lacepede) and bluegill (*Lepomis macrochirus* Rafinesque). Polytrophic lakes often contain no fish.

3. Results

Figure 5 illustrates the land cover types within the study area; while **Figure 6** presents the soil types for the study area. A series of 11 equations (Eq. (1)–(11)) resulted in the final calculated Carlson Index Score, which was 92.52, placing the lake at the high end of a eu-polytrophic lake. The predicted results match the classification results for the lake.
Lake volume in meter cube

$$V1 = ((1.047) * S1/3.14) * (L2) \tag{1}$$

$$V1 = 1760827.261\ \text{m}^3$$

where S1 = 1,320,200 m^2 (derived from CAD and public maps); L2 = 4 m.

Mean lake depth in meters

$$M2 = V1/S1 \tag{2}$$

$$M2 = 1.33\ \text{m (calculate)}.$$

where V1 = 1760827.261 m^3; S1 = 1320200.00 m^2.

Runoff calculations

$$\begin{aligned} R1 = {} & ((U6^* 0.8) + (U7^* 0.6) + (U8^* 0.7) + (B6^* 0.5) + (B7^* 0.1) + (B8^* 0.3) + (P \\ & 6^* 0.4) + (P7^* 0.2) + (P8^* 0.225) + (W6^* 0.2) + (W7^* 0.3) + (W8^* 0.125) \\ & + (I6^* 0.9) + (I7^* 0.85) + (I8^* 0.8) + O1)) * P1 \end{aligned} \tag{3}$$

$$R1 = 38229613.000.$$

where Yearly P1 = 33.22 inches = 0.844 m and **Table 3**.

Phosphorus contribution in calculations

$$C1 = (((U6 + U7 + U8) * 4) + ((B6 + B7 + B8) * 0.35) + ((P6 + P7 + P8) * 0.25) + ((W6 + W7 + W8) * 0.1) + ((I6 + I7 + I8) * 4) + (O1 * 0.46))/10^4 \tag{4}$$

$$C1 = 10753.8.$$

Further calculations

$$O2 = R1 + (S\,1 * (P1 - E1)) \tag{5}$$

$$O2 = 6144742.000 + (1{,}320{,}200 * (0.844 – 0.635)).$$

$$O2 = 6{,}420{,}664.$$

where R1 = 6144742.000; S1 = 1,320,200 m^2 (CAD and map); P1 = 0.844 m/year; E1 = 25 inches = 0.635 m/year.

$$Q1 = O2/S1 \tag{6}$$

$$Q1 = 6,420,664/1320200.$$

$$Q1 = 4.86.$$

$$R = 13/(13 + Q1) \tag{7}$$

$$R = 13/(13 + 4.86).$$

$$R = 0.73.$$

$$F = O2/V1 \tag{8}$$

$$F = 6,420,664/1760827.261.$$

$$F = 3.65.$$

Total present nutrient supply

$$L4 = (((S1 * 0.46 * (10^{-4})) + C1) * 10^{6}) + (P2 * V1 * (10^{-3}))/S1 \quad (9)$$

$$L4 = 8244.93\ mg/m^2.year.$$

where $S1 = 1{,}320{,}200\ m^2$ (CAD and map); $C1 = 10753.8$; $P2 = 40\ mg/m^3$; $V1 = 1760827.261\ m^3$.

Predicted phosphorus level

$$P3 = L4 * (1 - R)/(M2 * F) \quad (10)$$

$$P3 = L4 * 0.6/1.33 * 14.54.$$

$$P3 = 458.57\ mg/m^3.$$

where $L4 = 8244.93$; $R = 0.73$; $M2 = 1.33\ m$; $F = 3.65$.

$$\text{Carlson index} = 14.42 * \text{Ln}(P3) + 4.15 = 92.52 \quad (11)$$

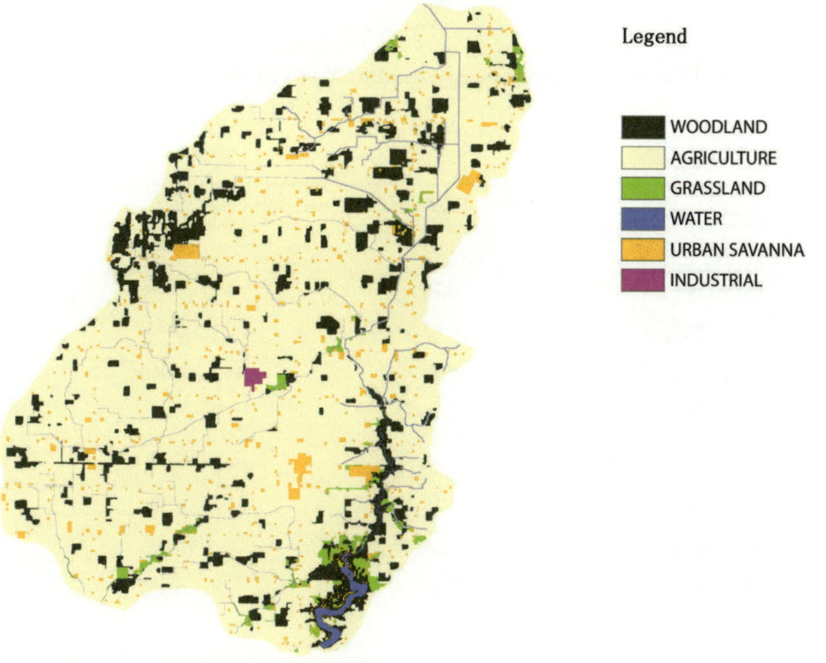

Figure 5. This map presents the land cover types in the study area, with Rainbow Lake in the lower right. Beneath the lake is the Maple River and the Maple River Game Management area. This convergence was the general location of outlet of Glacial Lake Saginaw with flowage from right to left (copyright 2017 Zhen Wu, all rights reserved, used by Permission).

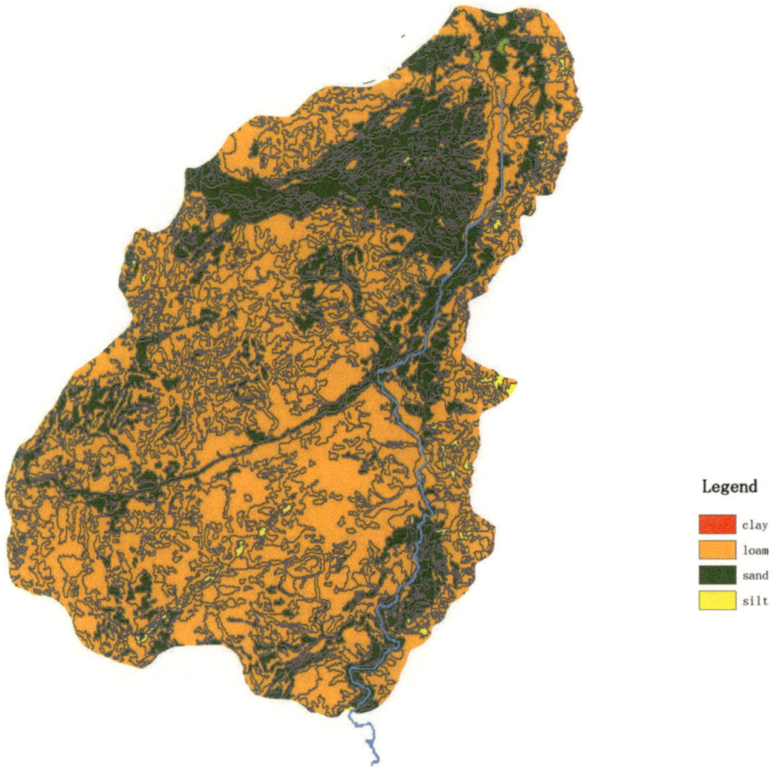

Figure 6. A map of the soil types in the study area (copyright 2017 Zhen Wu, all rights reserved, used by Permission).

Classification	Soil	Area (km²)	Percentage in each land use	Percentage in total research area
Woodland		29.48	1.00	0.12
	Sand (W8)	11.20	0.38	0.046
	Loam (W7)	18.28	0.62	0.074
	Clay (W6)	0.00	0.00	0.000
Grassland		8.52	1.00	0.03
	Sand (P8)	1.11	0.13	0.004
	Loam (P7)	7.50	0.88	0.026
	Clay (P6)	0.00	0.00	0.000
Urban savanna		4	1.00	0.02
	Sand (U8)	1.64	0.41	0.008
	Loam (U7)	2.36	0.59	0.012
	Clay (U6)	0	0.00	0.000

http://dx.doi.org/10.5772/intechopen.79323

Classification	Soil	Area (km^2)	Percentage in each land use	Percentage in total research area
Industrial		3.52	1.00	0.01
	Sand (I8)	1.48	0.42	0.004
	Loam (I7)	2.04	0.58	0.006
	Clay (I6)	0.00	0.00	0.000
Agriculture		203.12	1.00	0.81
	Sand (B8)	40.62	0.20	0.162
	Loam (B7)	162.50	0.80	0.648
	Clay (B6)	0.00	0.00	0
Water	(O1)	2.80	1.00	0.01
	Sand	1.37	0.49	0.005
	Loam	1.43	0.51	0.005
	Clay	0.00	0.00	0.000

Table 3. The area of land in square meters for various land-cover types by soil types.

4. Discussion and conclusion

4.1. Rainbow Lake, Michigan

"The student who conducted the study of Rainbow Lake was surprised how available information was to make the prediction and conduct the modeling," stated Dr. Burley. "He stated that such information is not as freely available in other parts of the world," commented Dr. Burley. "It is true that information can equate to power. But in the United States and Canada, such information is supported by the public and the public has the right to access such information. Farmers, citizens, and researchers are free to access the information to make calculations to refute or support the findings of others. It is expected if not demanded," said Dr. Burley. "In many respects it is comforting that there are checks and balances in the use and application of information. If there are disputes, they can be openly addressed in public forums," mentioned Dr. Burley. "Especially at the township, county and state level, public employees responsible for natural resource management work together with citizens and comparatively, there is a fair amount of respect and trust amongst everyone.," reports Dr. Burley.

"The largest disputes that I have witnessed have been amongst hydrological experts who may debate methods of sampling or the accuracy of equations to predict hydrological variables. These academics often have varying opinions. But any equation is simply an approximation and estimation of physical phenomena. In the engineering field, if the approximation generally works, then it is accepted with no theoretical explanation or search for a better equation. This was true of the Manning formula of confined water flow in swales and pipes. This equation is over 140 years old and is unquestioned by many, but it is simply a mathematical approximation of water flow," observes Dr. Burley. "I am sure that today investigators could develop improved equations and even supply a set of theories to explain the improved equation. For example, the equation assumes laminar flow, however, it has been observed the flow may be at times helical/ spiral with undercurrents. Nevertheless, researchers are often not interested in discovering a new finding to something that by society's measure is considered 'not broken,'" adds Dr. Burley.

"The greatest difficulty by the student was actually determining the watershed boundary by topographical maps," noted Dr. Burley. "If one has had surveying in college or physical geography, reading topographical maps and determining ridge lines can be easily accomplished. I know I could do this by the time I was 12 or thirteen years of age and I am not necessarily the smartest person, even in a very small group," assessed Dr. Burley. "But I understand this is not always easy for some to accomplish and learning how to read topographic maps is an important skill in landscape planning and hydrological studies" added Dr. Burley. "The PhD student was very capable in creating map overlays with the appropriate data to estimate cover types on various soil types. This was very reassuring. I suspect that landscape students around the world could conduct such a study providing the information is available," confirmed Dr. Burley. "It took him about 2 months working full time (40 hours a week) to complete the study. There is much to read, intellectually digest, and apply," Dr. Burley reported.

"Rainbow Lake is near the upper boundary for being a eu-polytrophic lake," advises Dr. Burley. "That means additional inputs of phosphorus into the lake could temperately change the perceived water quality. During large rare storm events, of which I have only witnessed a couple in a 25 year period, massive amounts of suspended soil particles enter the lake. You can see the lake change to temporarily a polytrophic lake. A Secchi disk disappears immediately when placed in the water. The lake may even rise six of more feet above the normal lake level. Before I lived in Michigan, the dam was even 'washed-out' and had to be re-built. For many years the water quality in the lake is quite stable. But in these large storm events, enormous soil erosion takes place at key places along Pine Creek. After the storm, these key spots are radically altered with new gullies and small valleys where many cubic meters of soil had completely disappeared and been removed. Natural resource managers and soil erosion specialists are focusing upon preparing for these large rainfall events by working with land owners to create detention areas to control the outfall during these substantial but rare storms," adds Dr. Burley.

"In addition, over the last two decades, one has witnessed the development of rain gardens and more thoughtful management of shorelines," describes Dr. Burley. "Land owners actually think globally and act locally. There is great pride in many land owners for managing their land independently and thoughtfully, by eliminating the need for mowing their landscape to the water's edge, developing naturalized areas, being respectful for the needs of amphibians, turtles, shorebirds, and wildflowers. Rain gardens convey water through naturalized swales and ravines to the lake. There is a very different attitude by many towards the treatment of the landscape when compared to behavior 40 years ago," observed Dr. Burley.

4.2. Guangzhou wetland parks

In the P.R. of China, it is recognized that wetland open-space/greenway systems are essential to properly manage water. For a time, spaces to hold and retain water were occasionally developed for housing, reducing the ability of the land to hold water and avoid flooding conditions. Initial open-space land-use plans were revised to convert green space to development. Eventually the green space disappeared. However, it was recognized that such conversion could not always continue. In Guangzhou, the P.R. of China, wetland parks were created to store water and improve water quality (**Figures 7–9**). During low water periods the green space can be used as a stroll park. The park is part of a connected system to accommodate high water, being linked to Shiliugang He (Pomegranate Hillock River) and Da Wei Yong (Big Surrounded Water-surge), Da Tang Yong (Big Pond Water-surge), Shang Chong Yong (Out to Rush Water-surge), Yang Wan Yong (Poplar Cove Water-surge), and Xi Lu Yong (West Busy Water-surge). The water in the lake can also supply fresh-water requirements for the city. The Haizhu (sea bead) district contains an exterior ring drainage system and an embankment system circling an internal lake for water storage that is controlled by flood gates. There is an island in the center of the lake, designated as a bird habitat. By the lake, a structure with a night-light beam (like a lighthouse) provides identity and gives meaning to the term "Haizhu." Such symbolism and affiliated meaning can be important in Chinese culture. Numerous floating-leaved aquatic plants and emergent plants are employed in the design to aid in improving water quality.

The design is very park like There are many trails, bridges, leisure boating, buildings, open spaces for group gatherings, cultural exhibits, and concessions. The number of people who can visit the park is limited each day to 3000 people. There are parking facilities, police, and

Figure 7. An image of a wetland park in the Haizhu District of Guangzhou. The lake is 53 hectares in size and the embankment land in the wetland is 41.8 hectares, with additional 55 parkland hectares. Notice the large towers in the distance and the proximity of urban development to the park. (copyright 2015 Jon Bryan Burley, all rights reserved, used by Permission).

Figure 8. A view of the same park in Guangzhou from one of the large towers. Guangzhou experiences torrential rains and typhoons. The proportion of land dedicated with hydrological management network, while substantial, may still be minimal for catastrophic storms. Other parks and improvements are planned and may be implemented (copyright 2015 Jon Bryan Burley, all rights reserved, used by Permission).

Figure 9. A view of the large gate in a Guangzhou Haizhu wetland park. "Other than the many lighthouses in numerous Michigan state parks, I am not familiar with any such substantial features in Michigan natural areas. Nevertheless, the Chinese gates provide a strong sense of entry and identity." stated Dr. Burley. People learn about the value of wetlands, exercise in the park, and socialize (copyright 2015 Jon Bryan Burley, all rights reserved, used by Permission).

medical assistance. The main entrance is connected to the city highway system. Portions of the facility educate visitors about recycling water and reuse.

"An interesting feature of these parks is how much intervention (paths, ornamental plantings, seating, and small structures) occur in the parks," observed Dr. Burley. "The Chinese enjoy water-side stroll parks [3]. Guangzhou has a population greater than the entirety of the state of Michigan. So it is not surprising that these parks would be more developed than many Michigan naturalized settings," commented Dr. Burley. "In Michigan it would be considered a waste of money to expend funds for walkways and ornamental plantings in areas such as the Maple River Game Management area which is adjacent to Rainbow Lake in Michigan. But in China with such a large population, it makes good sense to allow the public access to enjoy these wetland areas," advised Dr. Burley. "I greatly appreciate the approaches of both cultures in the management of use of their wetlands. In Michigan, the goal is to management hydrological resources, provide recreation, facilitate wildlife, and provide green space. In China the role is also to manage hydrological resources, provide stroll gardens, and educate people about the importance of wetlands. I could imagine in just one wetland park, the expenditures for planting is greater than the yearly budget for the state of Michigan in planting vegetation," compared Dr. Burley.

4.3. Sponge city concept

While environments such as Michigan may have a remaining substantial space to accommodate hydrological functions, large urban expanses such as in the P.R. of China may not have the luxury of designating the required amount of space to accommodate hydrological needs. The Chinese have embarked on the "sponge city" concept, where hydrological water storage and water cleansing are accomplished as much as possible at the site level before ever getting to a greenway or water basin.

Recently the Chinese held a competition to explore ideas to accomplish the sponge-city concept. A team at the Michigan State University entered the competition (**Figures 10–12**). In many respects the design incorporates and blends water storage and water-cleansing functions with the aesthetic functions of a soft-scape pocket garden. The figures illustrate the process the team employed to create the design. "What I like about the documentation of this project, is that it demonstrates how a design evolves and is created," notes Dr. Burley. "Experienced professionals understand this process, but novices sometimes believe that a design is just drawn, rendered, and finished. The figures show how hand drawing is often important in initiating a design and as the design emerges, computing technology may be employed to envision the project. The project is drawn, redrawn, and redrawn. During each version, the design is critiqued and assessed," observes Dr. Burley.

"When I teach design to beginning students, it is difficult to get them to evolve the design," describes Dr. Burley. "For the beginning student, being able to create just one version of a design is a milestone. But in truth, this is just the beginning of the process," suggests Dr. Burley. "It takes time for students to be objective about their design and be open to improvements and refinements. In design, there is not 'the answer' there are only 'answers.' Design is not a math problem with a discrete answer. And even in mathematics there are systems and examples where a discrete answer is not possible to discern, such as with irrational numbers

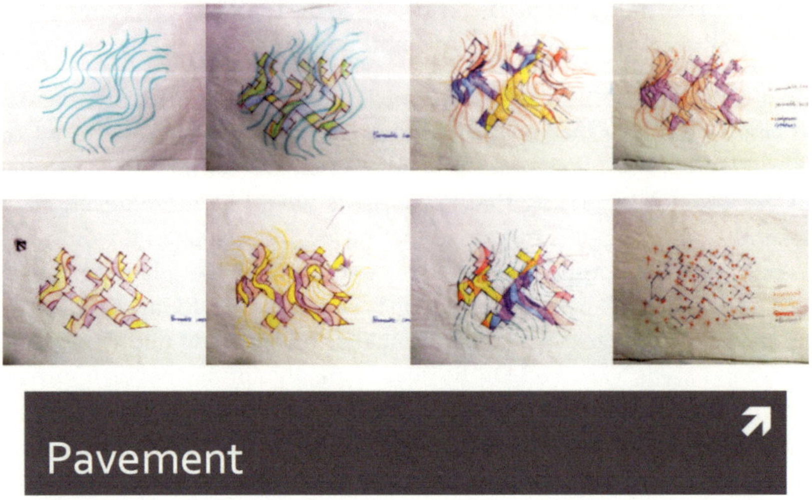

Figure 10. A series of drawings exploring ground plane patterns for the sponge city design. (copyright 2016 Na Li and Yiwen Xu).

Figure 11. A series of 3-D models exploring the three dimensional properties and possibilities in the sponge city design (copyright 2016 Na Li and Yiwen Xu).

or mathematical universes that fade in, fade out, and change in volume across time," advises Dr. Burley. "Creating the design can be a process whereby the dots to be connected may not be apparent and the sequence to connect the dots is not apparent either. Some people are not comfortable with such puzzles and others enjoy the challenge," states Dr. Burley. "The sponge city competition typifies such a challenge. It is creating a new set of relationships and standards previously unexplored and untested," reveals Dr. Burley.

Figure 12. Two perspectives of the sponge city design under different climatological events (copyright 2016 Na Li and Yiwen Xu).

"The sponge city competition illustrates what is occurring and evolving in the use and application of hydrological issues in the management and development of the environment. Needs change, opportunities emerge, and ideas evolve," states Dr. Burley. For example, members of this same sponge city team, name, and date conducted and reported a case study in Grand Rapids, Michigan, where equations and methods were employed to measure and predict water use by vegetation for a developed site [24]. "This is new and evolving knowledge with direct applications to landscape planning and design," concludes Dr. Burley.

"It has been that way for a long time. I remember as a young professor in the 1980s studying shoreline treatments for landscape stabilization for a small lake in eastern North Dakota or developing shoreline treatments along the shores of Lake Superior, or designing for wildlife habitat in reclaimed surface mine lakes near St. Paul, Minnesota or the gallery forest Red River of the North, and advising citizens how to design with plants along the riverine landscape [25–29]. Most projects involve water in some form or manner. There is still much to be learned and discovered," concluded Dr. Burley.

5. Conclusion

Knowledge about water, its management, and use continues to grow and evolve. An analytical approach developed and employed in one part of the world may eventually become important in other parts of the world. This has been the case in the development of non-toxic water-quality estimation methods developed in the USA and being considered elsewhere. In addition, the special needs of selected societies and cultures may necessitate the evolution of new functions and design standards in planning and design. Applications of these standards will continue to evolve and be applied in different manners across the world. In future, some of these relatively newer equations and applications may be included in textbooks and studied by most landscape and civil engineering students.

Author details

Zhen Wu[1], Jon Bryan Burley[2*], Chun-Hua Guo[3], Na Li[4], Yiwen Xu[5] and Zhi Yue[1]

*Address all correspondence to: burleyj@msu.edu

1 Landscape Architecture, Nanjing Forestry University, Nanjing, Jiangsu Province, PR China

2 Landscape Architecture, School of Planning, Design, and Construction, Michigan State University, East Lansing, Michigan, USA

3 Zhongkai University of Agricultural and Engineering, Guangzhou, PR China

4 Design Building, Shenzhen, PR China

5 Design Department 6F, China Shanghai Architectural Design and Research Institute Co., LTD, Shanghai International Purchasing Exhibition Center, Shanghai, PR China

References

[1] Luo P, He B, Takara K, Xiong YE, Nover D, Duan W, Fukushi K. Historical assessment of Chinese and Japanese flood management policies and implications for managing future floods. Environmental Science & Policy. 2015;**48**:265-277. DOI: 10.1016/j.envsci.2014.12.015

[2] Marsh JP. Man and Nature: of Physical Geography as Modified by Human Action. New York, New York: Charles Scribner; 1865. 560 p

[3] Burley JB, Machemer, T. From Eye to Heart: Exterior Spaces Explored and Explained. San Diego, California: Cognella Academic Publishing; 2016. 601 p

[4] Seelye EE. Data Book for Civil Engineers Volume 1 Design. Hoboken, New Jersey: John Wiley & Sons Inc.; 1945

[5] Mono Basin Ecosystem Study Committee. The Mono Basin Ecosystem Effects of Changing Lake Level. Washington, DC: National Academy Press; 1987. 272 p

[6] Fradkin PL. A River No More: The Colordado River and the West. New York, New York: AA Knopf; 1981. 360 p

[7] Allen JE, Burns M, Sargent SC. Cataclysms on the Columbia. Portland. Oregon: Timber Press; 1986 213 p

[8] Munson A. Construction Design for Landscape Architects, New York. New York: McGraw-Hill Book Company; 1971. 212 p

[9] Landphair H, Klatt F. Landscape Architecture Construction. 2nd ed. Amsterdam, Netherlands: Elsevier; 1988. 433 p

[10] Sykes R. Surface water drainage. In: Nelischer M, editor. Handbook of Landscape Architectural Construction: Volume 1. 2nd ed. Washington, DC: Landscape Architecture Foundation; 1985. pp. 133-221

[11] Merritt FS, editor. Standard Handbook for Civil Engineers. 2nd ed. New York, New York: McGraw-Hill Book Company; 1968

[12] Harris CW, Dines NT. Time-Saver Standards for Landscape Architecture: Design and Construction Data. New York, New York: McGraw-Hill Cook Company; 1988

[13] Hopper L, editor. Landscape Architectural Graphic Standards. New York, New York: John Wiley & Sons Inc.; 2006. p. 1096

[14] Strom S, Nathan K, Woland J. Site Engineering for Landscape Architects. 6th ed. Hoboken, New Jersey: John Wiley & Sons, Inc.; 2013. 348 p

[15] Ibanez D, Lyster C, Waldheim C, White M, editors. Third Coast Atlas: Prelude to a Plan. New York, New York: Actar Publishers; 2017. 344 p

[16] Wang M, Hyde RQ, Burley JB, Allen A, Machemer M. Low-Impact Housing: River Rouge, Michigan. Housing and Society. 2015. pp. 1-14. DOI: 10.1080/08882746.2015.1121679

[17] Carlson RE. A trophic state index for lakes. Limnology and Oceanography. 1977; **22**(2):361-369

[18] Garn HS, Parrott HA. Recommended Methods for Classifying Lake Condition, Determining Lake Sensitivity, and Predicting Lake Impacts, Hydrology Paper 2. Washington, DC: US Department of Agriculture, Forest Service, Eastern Region; 1977

[19] Burley JB, Kopperl J, Paliga J, Carter W. Land-use/watershed modeling for Lake Itasca Minnesota: A land-use planning, GIS project at Colorado State University. Landscape and Land Use Planning. Washington, DC: ASLA Open Committee on Landscape and Land Use Planning 1990;**17**:19-25

[20] Tomlin D. An IBM PC-Version of Map Analysis Package. Cambridge, MA: Harvard University; 1985. 59 p

[21] Daniel WW. Applied Nonparametric Statistics. Boston, MA: Houghton Mifflin Company; 1978. 59 p

[22] Sturgul JR, Merchant MJ. Applied Fortran IV Programming. Belmont, California: Wadswworth Publishing Company, Inc.; 1973. 399 p

[23] Minnesota State Planning Agency. EPPL7 User's Guide. St. Paul, Minnesota: Minnesota State Planning Agency; 1987

[24] Burley JB, Li N, Yung J, Tian H, Troost S. Metrics in master planning low impact development for Grand Rapids Michigan. In: Ergen M, editor. Sustainable Urbanization. Rijeka, Croatia: Intech; 2016. pp. 61-86

[25] Burley JB. Bluemont Lakes revegetation shoreline study. In: Erosion Knows No Boundaries. Vancouver, British Columbia: International Erosion Control Association; 1989. pp. 285-295

[26] Burley JB. A creative process to develop revegetation strategies: A North Shore case study. Landscape Architectural Review. 1986;**5**(1):5-7, 9-11

[27] Burley JB. Reclamation landscape generation for wildlife populations on the Upper Grey Cloud Island, utilizing habitat potential models. In: The Landscape: Critical Issues and Resources. Logan, Utah: Council of Educators in Landscape Architecture; 1983. pp. 204-216

[28] Burley JB. Multi-model habitat analysis and design for M.B. Johnson Park in the Red River Valley. Landscape and Urban Planning. 1989:261-280

[29] Burley JB, Burley CJ. Riverscaping. Minnesota Horticulturist. 1989;**117**(6):6-8

Chapter 14

Unlikely Alliances in the Battle for Land and Water Security: Unconventional Gas and the Politics of Risk in NSW, Australia

Meg Sherval

Additional information is available at the end of the chapter

http://dx.doi.org/10.5772/intechopen.78795

Abstract

This chapter, drawing on empirical work in New South Wales, discusses the unlikely alliances forming between environmentalists and farmers against the State which seeks to prioritise extractive development over other alternate futures. In response to a rise in land use conflicts, the State government has recently sought to silence criticism by tightening the laws which for decades have allowed citizens to seek merit and judicial review of government decision-making around development and planning issues. This move, made in conjunction with amendments to the State Environmental Planning Policy (Mining, Petroleum Production and Extractive Industries) 2007, has been met with anger and dismay by farmers, environmentalists and concerned citizens alike. Many places that have traditionally been agricultural strongholds now face an uncertain future as strategic planning moves to increase its focus on enabling energy production. Adopting a qualitative case study approach, this chapter highlights the implications of such decision-making by focusing on one rural region where a vastly different discourse and vision of the future is emerging.

Keywords: strategic planning, energy development, risk, place, water and land security

1. Introduction

Changes to traditional land uses can be contentious, increasingly so when paired with the ongoing challenges of climate change and energy security. In Australia, land use change has a long history as management of landscapes has occupied a central position in our politics ever since re-settlement of the continent over 230 years ago ([1], p. 1). Historically, the agricultural, manufacturing, service and mining sectors have long operated alongside one another to underpin the economic diversity of the nation with rural regions more specifically playing a vital part [2]. Over time, however, as power has devolved from the Federal government to the States, management of the biophysical environment—land, soils and water in particular have taken on greater significance as increased 'scarcity' created by human induced pressures, has become more widely recognised and communities have begun to demand more active responses from government to address these challenges.

In an attempt to 'balance' strong economic growth with protection of agricultural lands, the New South Wales (NSW) State government in 2012 introduced strategic land use plans for the Upper Hunter and New England regions [3, 4]. While these were designed to promote 'co-existence' of diverse land uses, they also set in place long-term strategic goals which focused heavily on increasing the spatial dimensions and needs of energy industries.

For example, in the Upper Hunter and New England regions alone, 11 Petroleum Exploration Licences (PELS) mainly for coal seam gas (CSG, referred to elsewhere 'as coal-bed methane') have been issued, covering a total of 461,000 ha of prime agricultural land [5] (**Figure 1**). This total does not include the minerals titles that also exist in these regions and elsewhere which cover approximately two-thirds of the State.

The increased emphasis on extractive land uses (other than those agricultural in nature) has been met with widespread opposition from farmers, local communities, environmentalists and concerned others, who over the past 5 years in particular have responded with resistance

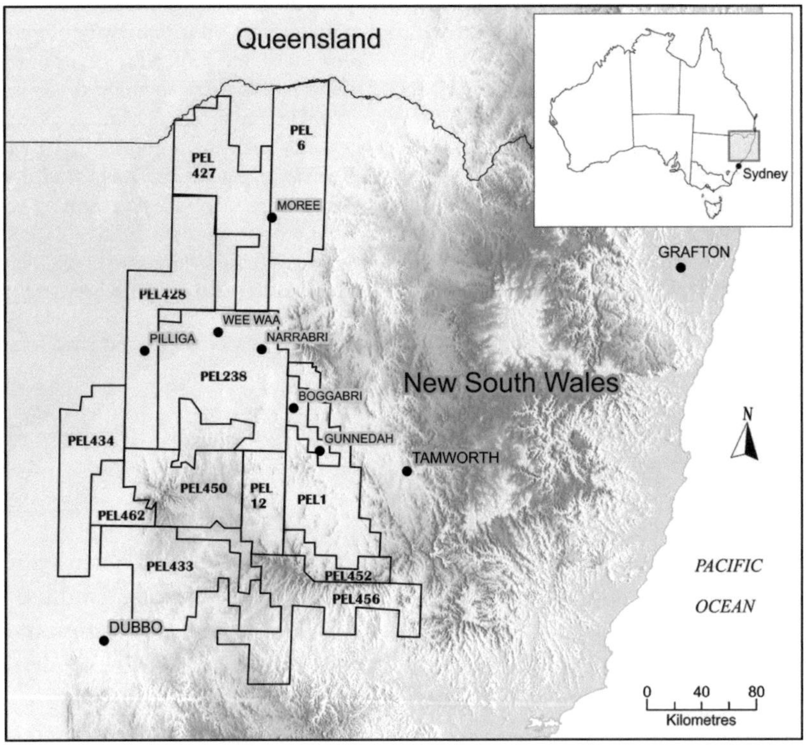

Figure 1. Upper hunter and New England region PEL zones.

in the form of civil disobedience campaigns and blockades. While these campaigns have been mostly against the placement of extractive projects in areas considered to be culturally, ecologically and agriculturally important, they also speak to a larger debate currently being waged throughout Australia [6]. This debate, seeks unproblematically to frame the future of rural places as essential nodes in an ever-expanding fossil fuelled pipeline where economic imperatives drive and surpass all other values that are not seen as being in the 'common' or 'economic good' of the nation [7, 8]. As Peck and Tickell ([9], p. 383) suggest, key to this argument is an all "pervasive metalogic" that draws on appropriate policies to justify its reasoning and effectively silence or marginalise other alternate voices.

This chapter focuses on one rural region in NSW. It reports on the emergence of an unlikely alliance forming between farmers, environmentalists and concerned others in the fight to protect productive land, soil and water from unwanted land use change. Framed by de Rijke ([10], p. 41) as the advent of "agri-gas fields", the chapter considers how the socio-economic and cultural boundaries between place and matter are being contested within the different visions that exist for the region's long-term future. Beginning with an overview of strategic planning in NSW and recent changes to appeal rights, the chapter also contemplates how space is instrumentalised by government for political purposes [11] and how it can also be used by others to challenge and reframe the arena of decision-making.

2. Strategic land use planning and change

When the NSW State government first released its Strategic Regional Land Use Plans, they were heralded as a suite of law reforms which would address what the government itself saw as flaws in the planning process [12]. These plans were to represent the 'government's proposed framework to support growth, protect the environment and respond to competing land uses, while [also] preserving key regional values over the next 20 years " ([3], p. 8). Framed as a way to ensure that land identified as 'Strategic Agricultural Land' – i.e. – highly productive land requiring extra protection mechanisms, the government put in place what it called a 'Gateway Assessment process' [3, 4]. It also appointed a Gateway Assessment Panel of experts (GAP), who would assess the merits of 'State Significant' mining and coal seam gas proposals on strategic agricultural land and then issue a certificate of approval [13].

While the Gateway Assessment process was to mark a significant improvement on previous planning efforts in regional NSW, it did not, however, redress the imbalance between the State's perceived economic 'needs' and the social and environmental needs of specific rural regions [12]. In fact, the Strategic Regional Land Use Plans (which were later to become policy) further entrenched this imbalance by rendering it impossible for the GAP to refuse an application for a Gateway Certificate no matter how questionable the project [12, 14].

Under the Gateway process, the GAP is required to make an upfront assessment of the impacts of a mining/CSG proposal on agricultural activities and water sources. On what basis, however, can this assessment be undertaken if the impacts are cumulative or as of yet unknown [12, 14, 15]? Since CSG mining is still a relatively new industry in Australia, it is possible that the impacts from mining and fracking activities may not only be site specific, but could also potentially affect a greater proportion of a region than initially considered ([16], p. 12).

In an attempt to further assuage these concerns, the NSW government introduced an 'Aquifer Interference Policy' in 2013 [17]. This was seen as 'a key plank' to the Strategic Regional Land Use Policy (SRLUP), but it was essentially left to the Minister for Planning to provide appropriate advice to the GAP on the potential impacts on aquifers from mining, CSG extraction, exploration and other activities based on the minimal impact considerations set out by the Aquifer Interference Policy [17]. There was nothing put in place, however, to either ensure the quality of aquifers and groundwater after CSG activities had occurred, nor penalties or fines for operators if damage was deemed to have been committed. This was seen by many to be a major flaw in the policy suite and did little to quell community disquiet. Other methods employed by government to try to reduce community objections are discussed below in the other sub-sections of this chapter.

2.1. Property law, 'co-existence' and citizen rights

In NSW (like the rest of Australia, and in the UK), all land or 'property' is assumed to be owned by or leased from the crown (or the State) [18]. This control of land tenure also extends to 'free-hold' titles where private ownership is not absolute, with the crown's representatives empowered to withhold certain rights, such as the right to any mineral or petroleum source found on or under the land [19]. It is the prevalence of this ancient paradigm of property rights transcribed onto modern law, that thus allows all land to be viewed as a commodity, as something "fungible, [and] infinitely tradeable" ([8], p. 6). This regime by its very nature is designed to separate people from place and to valorise the material or physical realm as something separate and 'other' to human subjectivity. It is this legacy that provides modern governments with the power not only to licence State intervention in land use practices such as establishing energy infrastructure, but also to determine the path this intervention will take.

As suggested at the start of Section 2, the NSW government through the SRLUP sought to balance land use and promote what it refers to as 'co-existence'. It suggests that: 'agriculture and mining are both vital industries in NSW and share many common beliefs and interests" [20]. It does not attempt to state what these are, only to suggest further that "the successful coexistence of these industries has enormous benefits for the state, particularly in regional areas" [20]. It also makes it very clear, that "although landholders may own the land, most mineral resources in NSW are owned by the state. This means that the royalties and economic benefits from the mining of these resources contribute to the provision of services for the people of NSW" [20]. To ensure the unimpeded continuity of what it calls "the orderly search for minerals"[1], the government has also recently altered how it approves the development of these resources [20].

Under the 2018 amendments to NSW *Environmental Planning and Assessment Act 1979*, all mining and gas operations are to be classified as 'State Significant Development' (SSD). Under this classification, an Independent Planning Commission (IPC) becomes the consent authority instead of the Minister for Planning, especially when there is significant community opposition. Under this guise, once there have been 25 or more public objections to a project

[1]The NSW government includes gas (tight, shale, coal seam, off-shore) and petroleum under its definition of 'minerals' on this site.

application, the IPC can call a public hearing [21]. This is seen by government as a way to build the community's confidence and trust in the Commission's independence by ensuring that assessment and land use planning processes are open and transparent, even though it is the Minister who chooses the make-up of the Commission [21].

While this is considered by some as a way to depoliticise the planning system, in effect it can actually further reduce the rights of the citizen public by closing off avenues to test the legitimacy of decision-making. For example, once a public hearing has been called, this effectively rules out any further questioning of the process or outcome in the Land and Environment Court of NSW where previously, merit and judicial review of planning decisions could be taken for further scrutiny. This essentially means that any matters raised at the public hearing that citizens feel are not dealt with satisfactorily or fairly, can no longer be responded to by the courts; thus removing the only external review mechanism for State government decision-making.

2.2. Exclusion zones, protection and ongoing contestation

While the SRLUP has been heavily critiqued by many, one large alteration made to it in 2013/2014 to try to address community concerns was the introduction of 'exclusion zones'. These zones effectively made areas of the State 'off-limits' to CSG exploration and mining. This was considered to be a much needed addition to the policy to provide certainty for CSG operators and communities alike.

When the list of exclusion zones was announced, however, it was met with mixed emotions. In urban areas of the State where existing residential suburbs were present, as well as in the North West and South West 'Growth Centres' of the State's capital – Sydney, protection from CSG exploration was guaranteed by the government [22]. In the rural regions of the State, however, it was a vastly different story. While the government introduced CSG exclusion zones for seven rural villages across NSW, and the equine and viticulture critical industry clusters of the Upper Hunter region, it did not choose to protect the agricultural areas of the New England region (and other similar regions throughout the state) [22]. This is despite it identifying in its own 'New England North West Regional Plan' that 'the New England North West is one of Australia's most productive agricultural areas" and that the "gross value of agricultural commodities produced in the region for 2016 was worth $2.1 billion" ([23], p. 4).

For many, this lack of protection was seen as continued evidence of an urban/rural (policy) divide where decisions are made at a distance by city-centric government officials who often have little comprehension of their impact locally [2]. In the New England region where the case study (to follow) is located, this and the idea of the urban/rural divide is reinforced through the concept of 'the Sandstone Curtain' [2]. This metaphor is used by many to describe the physical barrier of the Great Dividing (mountain) Range which splits NSW in two with many rural regions located over the range to the west and the urban areas located on the coast and surrounds to the east.

Overall, it is no exaggeration to suggest that changing land uses and increased government support of the extractives sector has been largely responsible for the rising discontent that can be found in many communities throughout rural NSW today. This in conjunction with changes to planning legislation and withdrawal of the right to seek merit and judicial review of most government planning decisions, has essentially been behind the rise in activism over the past few years and the appearance of new mergers between traditionally old foes—farmers and environmentalists. More than this though, the government's framing of its decision-making as in the 'public's economic interest' has led many to question its wisdom and to turn the debate into one about protectionism versus globalised trade. As influential Australian commentator Alan Jones has noted:

> *"This is not just a battle about mining prime farm land or destroying fresh water or covering our land with salt or risking public health. This is about something far more damaging and dangerous: the loss of our rights as Australian citizens, the loss of basic freedoms we have always taken for granted. State and federal governments have conspired to remove our rights over the ownership of our land. They have deliberately conspired to bully, to abuse, and to force Australians into court if they don't comply with the demands of foreign-owned multinational mining companies" [24].*

Jones's comments echo the concerns of many, not just local communities, as they speak to larger issues such as loss of place, the demise of rural landscapes, environments being eroded and overall, rural places beginning to represent what Murton has referred to as 'the countryside under construction" ([25], p. 1). With extractive industries and their associated infrastructure playing a large role in this transformation, it is worth acknowledging that 'there are deeply emotional ramifications [associated with] resource extraction" ([26], p. 1).

As Urry ([27], p. 77) notes further, whether we wish to acknowledge it or not, 'emotions are intimately tied to place "and as such, places are prompts for often intense feelings, springboards for memories and motivators for action. Acknowledging this, is essential if serious attempts are to be made to resolve some of the more contentious issues surrounding changing land uses today. As regional planners such as Godschalk ([28], p. 5) suggest:

> *Twenty-first century land use planning faces both an opportunity and a threat. On the one hand, it is widely counted on and expected to deliver both sustainable development and livable communities. On the other hand, it must cope with serious conflicts in the values related to these two beguiling visions, which represent the big visionary ideas of contemporary… planning. The future of land use planning may well depend on how it resolves these conflicts and creates settlement patterns that are both livable and sustainable. [Italics in original].*

In Narrabri Shire, where the focus of this chapter now shifts (**Figure 1**), government, industry and civil society have all been drawn into a conversation around land use change and the differing visions for the region's long-term future. Why and how this manifests itself on the ground is discussed further in Section 3.

3. Introducing—Narrabri Shire

Narrabri Shire, in the New England North West region is one of many local government areas undergoing land use change as parts of the region re-orientate towards extractive activities [2]. Traditionally, Narrabri has been a 'dryland farming[2]' region but its history has not been a static

[2]Dryland farming systems in Australia combine a rotation of crops, pastures and often livestock. Fallow periods are used to allow soil recovery and account for limited water supply. Australian farmers frequently have to contend with the effects of drought. In Narrabri Shire, wheat, sun flowers, canola and cotton are the dominant crops. In response to ongoing water and soil quality concerns and an increased need for sustainability, much of the cotton industry in the 2000s moved to dryland GM cotton which has greater drought tolerance.

one. Over the past 50 years, it has experienced significant land use change including the introduction of the cotton industry and then later, the advent of genetically modified (GM) cotton. Whilst these developments generated considerable concern in the community at the time, it is the more recent expansion of coal mining and CSG development in the Shire that has engendered the most conflict [2, 6]. Disputes have emerged between those who ultimately see extractive industries facilitating economic growth and diversity, and those who see these activities as a threat to the core agricultural functions upon which Narrabri Shire was founded back in 1848 [2, 6].

The research this chapter discusses, is a sample of the findings from a collaborative study carried out in Narrabri shire in 2015–2016. Narrabri shire consists of 8 towns (including Boggabri, Narrabri, Pilliga and Wee Waa—**Figure 1**), with a total—population of approximately 14,000 residents. The research used qualitative methods that set out to explore the lived experiences of individuals and community in regards to changing land use patterns. It included voluntary, face-to-face interviews with a mix of rural, village and Narrabri town residents [30]. **Figure 2** outlines the methods used for sampling and recruitment of participants.

Ultimately, Narrabri shire was chosen for this study due to its history of significant land use change, the intensification of coal mining and the emergence of CSG interests that are seen to currently challenge the traditional agricultural base of the community ([6], p. 103). Narrabri also presented an opportunity to explore the changing nature of land use contestation and the formation of new allegiances due to the fact that recent civil disobedience campaigns against energy company 'Santos' were situated there ([6], p. 103). These campaigns garnered nationwide attention as environmentalists from across Australia joined local farmers and community groups to protest against what they perceived as an unsustainable and undesirable land use change. Consequently, Narrabri shire presented a unique lens through which to explore the intricacies and complexities that exist around land use change today [2, 6, 31].

This chapter highlights several themes to emerge from this research. These are represented below under Sections 3 and 4. Overall, these articulations represent the values, perceptions,

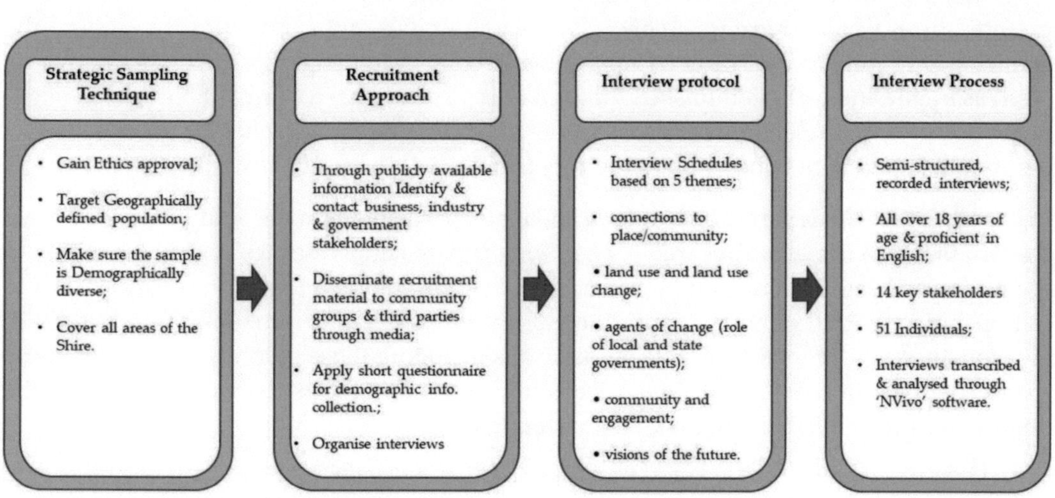

Figure 2. The methodological approach.

aspirations and anxieties that form part of understanding what it is to live in rural places in NSW today.

3.1. Competing for water

In all parts of Australia, access to water is problematic as the continent is one of the most arid on Earth and as such, the availability of its water sources places a fundamental limit on how and where the population can live and work. Likewise, how water is used and thought about is key to our understanding of issues intimately connected with it, such as: access, management and ongoing protection of this limited resource. This is particularly important when considering fresh water sources, 'because the continent has so little of [them], particularly surface flows, and the small amount [Australia does have are] poorly distributed to meet human needs, both spatially and temporally" ([29], p. 422). As such, 'matching supply to demand, geographically and temporally, is a major problem in Australia' ([29], p. 432).

Most participants in this study recognised the value of water in all its forms and the desire to avoid contamination of fresh water sources was thought to be paramount. As a region, Narrabri has had a long history of debilitating droughts, the most recent, having only broken in late 2015 ([6], p. 108). As such, people have long memories and their experiences of continuous drought cycles have clearly influenced their concerns about availability of water and their fear of its loss. As Michael, a local manager in Narrabri shire poignantly noted: 'the water issue is really the touchstone here' ([6], p. 108). This is particularly apparent when participants spoke of the region's relationship with the Great Artesian Basin.

3.2. The great Artesian Basin and continued ecological health

The Great Artesian Basin (GAB) is one of the largest underground water reservoirs in the world. Occupying more than 1.7 million square kilometres beneath the arid and semi-arid parts of Queensland, New South Wales, South Australia and the Northern Territory, it is an essential fresh water source for many inland communities and ecosystems throughout the continent [31]. In Narrabri shire, water is considered a 'life-giving' force but it is also recognised as fragile and easily diminished through over-use, increased competition and the ongoing effects of climate change. It is also considered an integral part of what Narrabri is, and so is strongly linked with issues around identity and place ([6], pp. 107–108); [31].

In Narrabri, the GAB is particularly valued and many participants in this study expressed great concern that governments appeared to be putting this essential resource at risk. According to the National Water Commission [32], planned CSG development in eastern Australia will at full operation, withdraw more than 300 gigalitres of groundwater annually from the GAB, i.e. more than 60% of total allowable withdrawals including those of locals in Narrabri shire. Expressing her concerns about this, Anne Kennedy [2, 6], a sixth-generation farmer in the Shire, sums up the vital role of the GAB by stating that:

> *the Great Artesian Basin is our lifeblood, and if we lose our groundwater, we simply cannot exist here. Not just the farmers, but communities, towns, vast areas of inland Australia will be uninhabitable. It is our only permanent water supply.*

Confirming this further, Dylan, a local farmer also suggested that:

> *no other place has got a big pool of water, millions of gigalitres of water underneath the ground. And... that's been the backbone of the grazing industry for decades. Now if that gets depleted... the worst case scenario is that it will be the death of the grazing industry in Australia.*

The continued health of the GAB was also linked with ideas about continued agricultural and human health more broadly. Many participants in this study, like Dylan above expressed a fear that the potential impacts from new hydro-carbon activities, such as those associated with CSG, might pollute shared water resources and if this were to occur, as local consultant Richard succinctly noted; 'without water, you've got nothing' [2].

3.3. Risk

In Narrabri shire, the perceptions of risk and uncertainty were particularly heightened when these issues were associated with questions about the region's long-term viability. As John from the cotton industry suggested ([6], p. 109):

> *...access to ground water is what drives the economy here. It provides a baseline or a foundation for the agricultural economy. A lot of the water is taken from other systems but the ground water is the backup, it's the foundation. If anything goes wrong...*

The possibility of 'things going wrong' was ever present in the minds of many participants in this study ([6], p. 109). Often referring to the idea of long-term 'sustainability', participants suggested that the 'precautionary principle[3]' should be being applied by governments much more readily than it was. This sense is described by authors such as Beck as the 'materiality of risk', where doubts and concerns are connected with the 'sweeping influence of science and technological change "particularly that associated in new and/or historical industries such as mining, energy production and so on ([33], pp. 4–5). He claims that perceptions of risk associated with these industries, also 'draw attention to the limited controllability "and uncertainty associated with any new or untested techniques or developments utilised ([33], p. 6). In Narrabri shire, the idea of risk and uncertainty may be visualised most clearly through the anxiety expressed about potential fracking operations and toxic by-products that may arise from CSG operations. As Oliver, a local councillor and farm owner noted:

> *So [what happens] if any water ponds ...become toxic? Drilling can sometimes flip it back into our aquifer system and that's very dangerous. I think we haven't got enough scientific evidence to prove that what I'm suggesting won't happen. That's how it is with this industry, the worst case scenario is that it will destroy the agricultural industry and put the grazing industry at risk.*

This lack of certainty around the integrity of infrastructure, would appear to also be strongly connected with the idea of how water (both fresh and discharged) is currently being managed and how it is to be managed to avoid potential contamination into the future and in relation to possible flood or high rainfall events. Both government [34] and academic literature also support this finding with authors such as Mercer, de Rijke and Dressler suggesting that

[3]The 'Precautionary Principle' is a key component of Sustainable Development and environmental legislation throughout Australia. All users of land, are required to assess risk and carefully "evaluate to avoid, wherever practicable, serious or irreversible damage to the environment" [38].

'the impact on water supplies from the mass dewatering of coal seams and the subsequent disposal of saline water are among the strongest concerns people have about unconventional gas extraction' ([7], p. 280).

Echoing this finding, Michael, a local manager suggests:

> *...there's the Leewood or the holding ponds, you know...they're developing out there on [the] flood plain, [but] if there's a major flood during production and the ponds are full, ...then what? You know there's quite a lot of risk. And I'm actually not opposed to gas out in the middle of the desert, or natural gas or mining per se, but this is just something we do not want in this area. There's too much value... it's like this is our irrigation water, our stock water, our drinking water.*

This concern was also linked more generally to scepticism about the burgeoning industry and the feeling of not being told the truth as expressed by Malcolm, a local farmer, who notes:

> *If there was no risk (and we know there's a risk), the companies would guarantee it [but] it's never going to happen. I've asked them; they won't do it. If it was all good, we wouldn't have this dissent. The dissent is enormous; and not just in Narrabri.*

3.4. Food and soil security

At the heart of the debates occurring in Narrabri, is also the recognition of how resource 'scarcities', and water, energy, food and fibre are all interconnected in a web of complex relations [2, 6]. As climate change and other pressures increase, continuing to meet food production and consumption is going to become increasingly challenging. Many farmers and the community in Narrabri acknowledged this and it should not be surprising that the idea of protecting healthy soils not only in this region but Australia-wide was seen as a critical and enduring task.

The ongoing notion of the need to protect soil security in the face of land use change and increasing climate variability has also been expressed federally in the Australian National Soil Research, Development and Extension Strategy [34]. It is noted in this strategy and by Koch et al. [35] that: 'securing soil as a contribution to the current and future competitiveness of Australian agriculture "is essential because 'it is estimated that water erosion is now outstripping soil formation rates across Australia by a factor of several hundred and in some areas, several thousand "and as a result, soil quality is reducing significantly.

This concern is echoed by Murray an inland farmer who commented in regards to mining:

> *'You know [we're] dealing with some of the best soils in the country [here], like these black soil plains [which] don't make up very much of Australia... less than 1% probably. If they could prove to me that it wasn't going to affect anything in the future, I wouldn't worry about it too much. But my biggest worry is you know, once they dig it up, it's gone...'*

Likewise Bryce, another farmer suggested that:

> *The type of country [we're] talking about, it's genuine, high-production, high-value country... There might be something underneath it, but God forbid we ever start digging up that sort of country.*

As Caitlin, another large producer notes:

> *...this is about those things that are priceless. It's about clean air, clean water, and land to grow clean healthy food. I can't impress upon you enough the importance to the State and the nation as a whole of our food producing lands. We have them here, they're clean and green and we produce [a] great product, and to put that at risk...*

For some members of the community, there was little differentiation made between the potential effects of coal mining or CSG as land use types, and while some community members were not against either, the caveat was always that both could proceed as long as they did not have an impact on soil quality or its future security [2, 6]. As Michael, a local manager notes when asked about his perceptions of people's main concerns:

> *Well there are...people [making a] living from that land, but it's such fertile soil and it will be productive forever if it's well looked after. But if that land gets ruined by mines, how can we grow food there?*

As Tania, a local sheep farmer notes:

> *I've never been opposed to coal seam gas...[or] to coal mining, I've always said, and it's on record that 'You can go ahead with your mining, provided it's done with respect. And there's respect to the environment, respect to the laws of the land and respect of the people in the area where you're operating'.*

It is this perceived lack of respect, however, that has some in the community like William, an agricultural supplier, ready to protest:

> *You start digging up the Liverpool plains, I'll be the first bloke standing there with a placard. Because it's just unbelievable that we would even think about doing that... In a hundred years' time, people are still going to need to eat some form of sustenance, and at this point in time I can't see that sustenance coming from anything other than the dirt. And we've got some of the best dirt in the world within 200 kilometres of Narrabri. It's that simple.*

Carl, a retired farmer sums up the general feeling of many by suggesting that: 'it's the farmers; it's always the little people that get hurt. I am not doing [this] for my benefit. This is for the next generations". Likewise, as Bryce, another farmer suggests:

> *...for a lot of people, this is the first time in their life they [have felt] passionate about something, and [want to] stand up... they've never had to do this before. They've never felt threatened or have never gone through this process.*

Overall, what these comments signify is that the nexus between water, soil and food security is widely recognised within the community, as is the connection between stewardship (offered by farmers) and decision-making around land use. The ongoing health of water and soils is stressed as something that is non-negotiable [2, 6]. This is made more apparent when one considers how these land use changes have been met with resistance often from unexpected quarters.

4. Activism and the rise of unlikely alliances

Becoming 'politicised' as Bryce suggests above, is for some an entirely new position in which they find themselves. Many have never protested before, nor felt compelled to stand up and resist the actions of government, until now. In Narrabri shire, unlikely alliances between farmers, environmentalists and concerned others have formed to respond against what people see as undesirable and unsustainable land use practices being imposed upon them. Many perceive that actions taken by industry and inaction by government has allowed social and environmental injustices to be imposed by the State upon communities [6].

Responding to the perceived absence of protection by government, farmers and environmental groups have been drawn together through their experience of what has been framed as 'an antagonistic other', that is, an external force embedded in extractive activities that threatens their individual and collective well-being [2, 6]. On one side of the argument, farmers are concerned that minerals and gas extraction may adversely affect their agricultural pursuits through negative impacts on the environment and, subsequently, impact the lifestyles and well-being of their local communities, local amenity and sense of place ([6], p. 112). On the other side, environmental groups and concerned others are anxious to promote sustainable development, to reduce the human footprint on nature, and to promote alternative and clean energy sources [6]. While traditionally, these values have been quite divergent, in the face of a common antagonist, a sense of collective identity has emerged underpinned by an evolved notion of stewardship, of caring for the land and conserving the environment for future generations [2, 6, 31]. Thus, an interplay between the informal and formal political spaces has emerged. Exemplifying the growing relationship between farmers and those they see as 'like-minded others', Dana a farmer's wife, explains how farming activists, through their association with "greenies" have acquired a sense of support and legitimisation in their opposition to CSG operators seeking access to their land:

> *Right now, all [farmers] perceive is that: "you're going to take away my life, full stop. So, I don't want to hear what you've got to say. You're going to destroy everything I've worked for. I've got every greenie up a tree who's going to back me up, so whatever you've got to say, I don't care".*

Comments like these, help explain how shared common values have become a springboard for collaboration and under this banner, how historical differences have been able to be set aside.

4.1. Temporality and alternate visions of the future

The idea of temporality was referred to many times throughout this research particularly in regard to extractive industries. This issue is important because it is connected with changing social relationships with the material world and as such, it is also reflects ongoing power struggles. Mines and extractive industries are being understood as more than spatial features of the landscape and as such, are intertwined with and against past memories, present experiences and future visions. It is only when these temporal factors are compared, however, against different land use visions such as those we find in Narrabri shire, that we begin to acknowledge how perceptions and understandings of temporal dimensions can become grounds for contestation and dispute.

Recent research conducted by Chen and Randall ([36], p. 17) suggests that while the short-term economic benefits of extractive industries are generally orders of magnitude greater than those of agriculture, in regards to the long-term economic net benefits from agriculture, these tend to exceed those of CSG extraction and/or mixed use (i.e. agriculture and CSG coexistence). This recognition seemed to figure in many conversations with local people who suggested that land use change had been forced upon them and the long-term economic benefits of this was not immediately apparent. For example, as Carl, a retired famer suggests:

> *If the farming industry kind of grows, it evolves slowly over the years... but these industries that we've got coming in now, have been forced upon us. It's not something that the community really has had any input into at all.*

In terms of the influence of CSG on farming land, what many farmers expressed concern about was the footprint of operations. As Malcolm, a large farmer in the Shire notes:

> *As far as the coal seam gas is concerned, there's no doubt regardless of what the mining companies or gas companies say, you've got fields and you've got gas bores in them and they've got roads and pipes and everything. It completely changes the farming programme. Of course, it's a 'no-no' as far as irrigated land [is] concerned.*

In relation to the exporting of coal from the region, Olivier, a local councillor noted that there also seemed to be competition for rail infrastructure occurring and that this was also causing disputes locally:

> *...That's another thing, coal has taken precedence over wheat being moved, and in the middle of a harvest, shifting wheat in a hurry is a big thing. [It's] on a single line track; you can only get x amount of cartage on some trains. So, it's a big thing. The shipping of coal is competing against the shipping of wheat or grain out here.*

In terms of the sustainability of this situation and the changes the community has had to weather, comments from Caitlin, a large producer, re-echo the idea of the urban/rural divide:

> *[I have to ask] - Are we the guinea pigs? Are our families second class citizens? I have heard that the NSW government has bought back CSG licences in the Sydney Basin citing protection of community and protection of water resources as the reason behind this. Why are the residents of Sydney's health protected but we out here in the Narrabri shire neglected? Aren't we deserving of the same protections?*

In terms of visions for the future, almost all of the visions discussed by participants remained firmly situated in agriculture. This was connected with the fact that farmers today, as mentioned above, see themselves very much as stewards caring for the land. In their mind therefore, in terms of temporality, agriculture has proved it is a stalwart that can be maintained long into the future [2, 6]. As Caitlin, notes further:

> *The bottom line is, agriculture will be here in a hundred years' time. Mining won't be, but agriculture will only be here if we do it sustainably. We've got to be very careful; we can't just keep raping and pillaging and plundering the country from an agricultural perspective. We've got to stop using chemical fertilisers, and we've got to genuinely start to look at the biological renewal of our soil. 90% of the growers I deal with agree with that sentiment... We're very fortunate in this region, we're one of the few regions in the country... [where] we're blessed with a beautiful mix of an ideal climate, soils and clean water resources, and we can grow two crops a year here. Very, very few places in the country can do that.*
>
> *The guiding principle that we run our business by, [is] that we are effectively borrowing this land from our children, whether that be our own children or just future generations, and we're very mindful of that. With any borrowing, comes interest, and in terms of the interest in this particular case, whilst we own that land in a legal sense, we're borrowing it, and the interest we have to pay is to return that land, or pass that on to the next generation, in a better way than we found it.*

Echoing this and the future of agriculture, Keith, a local farmer notes:

> *The thing is we've been in agriculture for six generations. The last thing I'm ever going to do is to be turned onto another industry that has no future. Even if it had a 30-year future here, that's not the sort of future that you can then hand over to your grandchildren and their children.*

As researchers Langridge et al. ([37], p. 4) suggest overall though, while there is public support nationally for agriculture, it is science and technology that will assist farmers in changing the face of farming and ensuring its long term sustainability. In their opinion ([37], p. 1), 'agriculture today is a very sophisticated and highly technical industry, and in Australia it has been one of our most innovative and efficient industries "which has allowed us to meet 'our moral commitment to food security in the region". To ensure that this continues to happen, they suggest that a combination of old knowledge and new is needed and that modern farmers, will need for example:

> *the traditional knowledge of cropping systems, fertiliser regimes, field pathology and so on but will also need to know techniques for assessing crop health based on analysis of the light reflected from crops and captured on images generated from drones or satellites. In the future, farmers will also be capturing data from even more diverse sources, linking this to genetic information and predictive climate models and using the result to help them decide when to sow their crops, when to apply fertilisers, how to protect crops from disease and when to harvest ([37], p. 4).*

Preparing for this brave new world, is deemed essential as 'food security is inextricably linked to the political stability of our region "and that of others globally ([37], p. 2).

5. Conclusion

As this chapter has shown, changes to traditional land uses can be contentious, none more so, than when differing visions exist for a region's future. In the NSW shire of Narrabri, land use change has been occurring for some time, although garnering the most attention has been the more recent arrival of coal seam gas development. This, in conjunction with the approval of large coal mines in an area which prides itself on its agricultural heritage, State forests, and National parks, has been met by many with anger and disbelief. These emotions have translated into ongoing acts of resistance including blockades and civil disobedience campaigns by farmers, environmentalists and concerned citizens, who traditionally have been the most unlikely of allies. Much of this conflict could have been avoided, however, if the State had moved to protect vital water sources and productive lands through the creation of exclusion zones in the New England North West as it did elsewhere. Likewise, if its' policy 'reforms' had not effectively disempowered local communities, by disallowing those affected by its strategic planning decisions to seek merit or judicial review, it might not have found itself in the position it does today. Instead, by promoting the economic imperative as the only measure of 'worth', the State has effectively signalled that it considers 'the rural' as merely a geographic location; a ubiquitous space ripe for development of energy's spatial project. Given this, confrontation by those articulating a different vision for the future, is as inevitable as it is predictable for land use change forced upon people, is rarely welcome or sought. Therefore, if the State wishes its citizens to be receptive to change, it needs to find an appropriate way to engage with their concerns and to offer them a valuable stake in the decisions made. Only then, might there be a real chance for 'co-existence'.

Acknowledgements

The author wishes to sincerely thank all of the residents of Narrabri shire who participated in this research project. This research received ethics clearance from the University of Newcastle's Human Research Ethics Committee (Approval No. H-2015-0193) before recruitment of participants took place. All quotations from participants used throughout the chapter are with permission; though the participants have been de-identified.

Many thanks also to Olivier Rey-Lescure, cartographer at the University of Newcastle, who provided mapping assistance for this chapter.

Conflict of interest

No potential conflict of interest is reported by the author.

Author details

Meg Sherval

Address all correspondence to: meg.sherval@newcastle.edu.au

The University of Newcastle, Callaghan, NSW, Australia

References

[1] Lane M, Robinson C, Taylor B. Introduction. In: Lane M, Robinson C, Taylor B, editors. Contested Country—Local and Regional Natural Resources Management in Australia. 1st ed. Collingwood: CSIRO Publishing; 2009. pp. 1-13

[2] Askland HH, Askew M, Hanley J, Sherval M, Farrugia D, Threadgold S, Coffey J. Local Attitudes to Changing Land Use—Narrabri Shire [Report]. Newcastle: The University of Newcastle and the NSW Department of Primary Industries; 2016. 118p

[3] NSW Government Department of Planning and Infrastructure. Strategic regional land use plan—Upper hunter [Internet]. 2012. Available from: http://www.nsw.gov.au/sites/default/files/initiatives/upperhunterslup_sd_v01.pdf [Accessed: 2018-04-12]

[4] NSW Government Department of Planning and Infrastructure. Strategic regional land use plan–New England north west [Internet]. 2012. Available from: http://www.

planning.nsw.gov.au/~/media/Files/DPE/Plans-and-policies/strategic-regional-land use-plan-new-england-north-west-2012-09.ashx [Accessed: 2018-04-12]

[5] NSW Government Department of Planning and Environment. Resources and geoscience. Commonground. [internet]. 2018. Available from: http://commonground.nsw.gov.au//#!/titlemap/Petroleum%20and%20Gas/Exploration%20Licence/Standard/7/-30.56226095049943/149.4964599609375 [Accessed: 2018-04-12]

[6] Sherval M, Askland HH, Askew M, Hanley J, Farrugia D, Threadgold S, Coffey J. Farmers as modern-day stewards and the rise of new rural citizenship in the battle over land use. Local Environment. 2018;**23**:100-116. DOI: 10.1080/13549839.2017.1389868

[7] Mercer A, De Rijke K, Dressler W. Silences in the boom: Coal seam gas, neoliberalizing discourse, and the future of regional Australia. Journal of Political Ecology. 2014;**21**:279-302

[8] Graham N. Lawscape: Property, Environment, Law. 1st ed. Oxford: Routledge; 2011. 236p

[9] Peck J, Tickell A. Neoliberalizing space. Antipode. 2002;**34**:380-404

[10] de Rijke K. The Agri-gas fields of Australia: Black soil, food, and unconventional gas. Culture, Agriculture, Food and Environment. 2013;**35**:41-53

[11] Purcell M. Excavating Lefebvre: The right to the city and its urban politics of the inhabitant. GeoJournal. 2002;**58**:99-108

[12] Sherval M, Graham N. Missing the connection: How the SRLU policy fragments landscapes and communities in NSW. Alternate Law Journal. 2013;**38**:176-180

[13] NSW Government Department of Planning and Environment. Gateway assessment and site verification [Internet]. 2017. Available from: http://www.planning.nsw.gov.au/gateway-assessment-and-site-verification [Accessed: 2018-04-14]

[14] Poisel T. Coal seam gas exploration and production in new South Wales: The case for better strategic planning and more stringent regulation. Environmental and Planning Law Journal. 2012;**29**:129-151

[15] Owens K. Strategic regional land use plans: Presenting the future for coal seam gas projects in NSW. Environmental and Planning Law Journal. 2012;**29**:113-128

[16] Randall A. Coal seam gas—Toward a risk management framework for a novel intervention. Environmental and Planning Law Journal. 2012;**29**:152-162

[17] NSW Government Department of Primary Industries—Office of Water. NSW aquifer interference policy—Fact sheet 1 [Internet]. 2013. Available from: http://www.water.nsw.gov.au/__data/assets/pdf_file/0005/547934/law_key_aquifer_interference_fact-sheet_1_groundwater_and_aquifers.pdf [Accessed: 2018-04-14]

[18] House of Commons. Hansard written answers: Land [Internet]. 2009. Available from: https://www.publications.parliament.uk/ [Accessed: 2017-07-06]

[19] Reed H. Mining. In: Farrier D, Stein P, editors. The Environmental Law Handbook. 5th ed. Pyrmont: Thomson Reuters; 2011. pp. 717-768

[20] NSW Government Department of Planning and Environment. Resources and Energy. Agriculture and mining—A co-existence [Internet]. 2018. Available from: https://www.resourcesandenergy.nsw.gov.au/landholders-and-community/minerals-and-coal/template-for-land-access [Accessed: 2018-04-16]

[21] NSW Government. Independent Planning Commission. About us [Internet]. 2018. Available from: http://ipcn.nsw.gov.au/about-us [Accessed: 2018-04-16]

[22] NSW Government Department of Planning and Environment. Policy and Legislation. CSG exclusion zones [Internet]. 2017. Available from: http://www.planning.nsw.gov.au/Policy-and-Legislation/Mining-and-Resources/Coal-Seam-Gas [Accessed: 2018-04-16]

[23] NSW Government Department of Planning and Environment. New England North West Regional Plan—2036 [Internet]. 2017. Available from: http://www.planning.nsw.gov.au/Plans-for-your-area/Regional-Plans/New-England-North-West/New-England-North-West-Regional-Plan [Accessed: 2018-04-14]

[24] Jones A. Australia's National Interest—Food security and the protection of the Australian regional way of life [Internet]. 2011. Available from: https://www.npc.org.au/archive_status/archived/ [Accessed: 2016-05-21]

[25] Murton J. Creating a Modern Countryside. Vancouver: UBC Pres. 2007. 280 p

[26] Ey M. "Soft, airy fairy stuff"? Re-evaluating 'social impacts' in gendered processes of natural resource extraction. Emotion, Space & Society. 2018;**27**:1-8. https://doi.org/10.1016/j.emospa.2018.02.002

[27] Urry J. The place of emotions within place. In: Davidson J, Bondi L, Smith M, editors. Emotional Geographies. 1st ed. Hampshire: Ashgate Publishing Ltd; 2007. pp. 77-86

[28] Godschalk DR. Land use planning challenges: Coping with conflicts in visions of sustainable development and livable communities. Journal of the American Planning Association. 2004;**70**:5-13

[29] Aplin G. Australians and their Environment. Melbourne: Oxford University Press; 2001. 536 p

[30] Australian Government – Department of Agriculture and Water resources. Great Artesian Basin [Internet]. 2017. Available from: http://www.agriculture.gov.au/water/national/great-artesian-basin [Accessed: 2018-04-18]

[31] Farrugia D, Hanley J, Sherval M, Askland HH, Askew M, Coffey J, Threadgold S. The local politics of rural land use: Place, extraction industries and narratives of contemporary rurality. Journal of Sociology. DOI: 10.1177/1440783318773518

[32] National Water Commission. Water Markets in Australia—A Short History. Canberra: Australian Government. 2011. 138 p

[33] Beck U. World Risk Society. 3rd ed. Cambridge: Polity Press; 2001; 185 p

[34] Australian Government, Department of Agriculture. The National Soil Research, Development and Extension Strategy—Securing Australia's soil for profitable industries

and healthy landscapes [Internet]. 2014. Available from: http://www.agriculture.gov.au/Style%20Library/Images/DAFF/__data/assets/pdffile/0012/2379585/soil.pdf [Accessed: 2018-04-18]

[35] Koch A, Chappell A, Eyres M, Scott E. Monitor soil degradation or triage for soil security? An Australian challenge. Sustainability. 2015;**7**:4870-4892

[36] Chen C, Randall A. The economic contest between coal seam gas mining and agriculture on prime farmland: It may be closer than we thought. Journal of Economic and Social Policy. 2013;**15**:1-30

[37] Langridge P, Cordell D, Occhio M. Agriculture in Australia: Growing more than our farming future [Internet]. 2014. Available from: https://theconversation.com/agriculture-in-australia-growing-more-than-our-farming-future-22843 [Accessed: 2018-04-24]

[38] Farrier D, Stein P. The Environmental Law Handbook. 5th ed. Pyrmont: Thomson Reuters; 2011. 10 p